Gmelin Handbook of Inorganic and Organometallic Chemistry

8th Edition

Gmelin Handbook of Inorganic and Organometallic Chemistry

8th Edition

Gmelin Handbuch der Anorganischen Chemie

Achte, völlig neu bearbeitete Auflage

PREPARED
AND ISSUED BY

Gmelin-Institut für Anorganische Chemie
der Max-Planck-Gesellschaft
zur Förderung der Wissenschaften

Director: Ekkehard Fluck

FOUNDED BY

Leopold Gmelin

8TH EDITION

8th Edition begun under the auspices of the
Deutsche Chemische Gesellschaft by R. J. Meyer

CONTINUED BY

E. H. E. Pietsch and A. Kotowski, and by
Margot Becke-Goehring

Springer-Verlag Berlin Heidelberg GmbH 1993

Organometallic Compounds in the Gmelin Handbook

The following listing indicates in which volumes these compounds are discussed or are referred to:

Ag   Silber B5 (1975)

Au   Organogold Compounds (1980)

Be   Organoberyllium Compounds 1 (1987)

Bi   Bismut-Organische Verbindungen (1977)

Co   Kobalt-Organische Verbindungen 1, 2 (1973), Kobalt Erg.-Bd. A (1961), B 1 (1963), B 2 (1964)

Cr   Chrom-Organische Verbindungen (1971)

Cu   Organocopper Compounds 1 (1985), 2 (1983), 3 (1986), 4 (1987), Index (1987)

Fe   Eisen-Organische Verbindungen A 1 (1974), A 2 (1977), A 3 (1978), A 4 (1980), A 5 (1981), A 6 (1977), A 7 (1980), Organoiron Compounds A 8 (1986), A 9 (1989), A 10 (1991), Eisen-Organische Verbindungen B 1 (partly in English; 1976), Organoiron Compounds B 2 (1978), Eisen-Organische Verbindungen B 3 (partly in English; 1979), B 4, B 5 (1978), Organoiron Compounds B 6, B 7 (1981), B 8, B 9 (1985), B 10 (1986), B 11 (1983), B 12 (1984), B 13 (1988), B 14, B 15 (1989), B 16a, B 16b, B 17 (1990), B 18 (1991), B 19 (1992), Eisen-Organische Verbindungen C 1, C 2 (1979), Organoiron Compounds C 3 (1980), C 4, C 5 (1981), C 6a (1991), C 6b (1992), C 7 (1985), and Eisen B (1929–1932)

Ga   Organogallium Compounds 1 (1986)

Ge   Organogermanium Compounds 1 (1988), 2 (1989), 3 (1990), 5 (1993)

Hf   Organohafnium Compounds (1973)

In   Organoindium Compounds 1 (1991)

Mo   Organomolybdenum Compounds 5 (1992), 6 (1990), 7 (1991), 8 (1992), 9 (1993)

Nb   Niob B 4 (1973)

Ni   Nickel-Organische Verbindungen 1 (1975), 2 (1974), Register (1975), Nickel B 3 (1966), and C 1 (1968), C 2 (1969)

Np, Pu Transurane C (partly in English; 1972)

Os   Organoosmium Compounds A 1 (1992), B 6 (1993) **present volume**

Pb   Organolead Compounds 1 (1987), 2 (1990), 3 (1992)

Po   Polonium Main Volume (1941)

Pt   Platin C (1939) and D (1957)

Re   Organorhenium 1, 2 (1989), 3 (1992)

Ru   Ruthenium Erg.-Bd. (1970)

Sb   Organoantimony Compounds 1, 2 (1981), 3 (1982), 4 (1986), 5 (1990)

Sc, Y,   Rare Earth Elements D 6 (1983)
La to Lu

Sn   Zinn-Organische Verbindungen 1, 2 (1975), 3, 4 (1976), 5 (1978), 6 (1979), Organotin Compounds 7 (1980), 8 (1981), 9 (1982), 10 (1983), 11 (1984), 12 (1985), 13 (1986), 14 (1987), 15, 16 (1988), 17 (1989), 18 (1990), 19 (1991), 20 (1993)

Ta   Tantal B 2 (1971)

Ti   Titan-Organische Verbindungen 1 (1977), 2 (1980), Organotitanium Compounds 3 (1984), 4 and Register (1984), 5 (1990)

U   Uranium Suppl. Vol. E 2 (1980)

V   Vanadium-Organische Verbindungen (1971), Vanadium B (1967)

Zr   Organozirconium Compounds (1973)

Gmelin Handbook of Inorganic and Organometallic Chemistry

8th Edition

Os
Organoosmium
Compounds

Part B 6

With 95 illustrations

AUTHOR   Kerstin Behrends

EDITOR   Cornelia Weber

CHIEF EDITOR   Johannes Füssel

Springer-Verlag Berlin Heidelberg GmbH 1993

LITERATURE CLOSING DATE: END OF 1992
IN MANY CASES MORE RECENT DATA HAVE BEEN CONSIDERED

Library of Congress Catalog Card Number: Agr 25-1383

ISBN 978-3-662-07538-8   ISBN 978-3-662-07536-4 (eBook)
DOI 10.1007/978-3-662-07536-4

© by Springer-Verlag Berlin Heidelberg 1993
Originally published by Springer-Verlag Berlin Heidelberg New York in 1993.
Softcover reprint of the hardcover 8th edition 1993

Preface

The present volume, "Organoosmium Compounds" B 6, systematically covers the literature through 1992, including many later references.

This volume is the first published of Series B. This series is devoted to compounds containing two or more osmium atoms.

The volume forms a unit with "Organoosmium Compounds" B 5 (in preparation). Both volumes deal with trinuclear compounds with ligands other than CO which are bonded to Os by one carbon atom ("^{1}L ligands"), regardless of whether the ligand is additionally coordinated to Os by heteroatoms. Generally CO groups are additional ligands. As is usual in the organometallic Gmelin series, the term "trinuclear" means three osmium atoms in the molecule without regard to any additional metals that may be present.

The content and the subdivision of both volumes are described on p. 1. Volume B 5 will deal with homometallic compounds in which the bonding C atom of the leading ^{1}L ligand is bonded to Os by one non-bridging Os–C bond. The first part of the present volume, B 6, is devoted to homometallic compounds in which the bonding C atom of the ^{1}L ligand bridges two or three Os atoms. A second part deals with all heterometallic compounds with ^{1}L ligands other than CO.

An Empirical Formula Index and a Ligand Formula Index for both volumes B 5 and B 6 will be included in volume B 5.

For abbreviations and dimensions used throughout this volume, see p. X.

Frankfurt am Main, October 1993   Johannes Füssel

Remarks on Abbreviations and Units

Most compounds in this volume are presented in tables. For the sake of conciseness, some abbreviations are used and some dimensions are omitted in the tables. This necessitates the following clarification.

Formulas. For conciseness or for better comparison of associated compounds, labeling can deviate from IUPAC numbering. **Geometric isomers** are designated according to the IUPAC rules. Structural labels are missing when authors fail to report structural details.

Abbreviations used with **temperatures** are m.p. for melting point, b.p. for boiling point, subl. for sublimation temperature, dec. for decomposition.

Solvents or the physical state are given in parentheses immediately after the spectral symbol if reported.

Preparations and reactions were generally carried out under an inert-gas atmosphere, whereas workup was mostly done in air. The products were usually purified by preparative thin-layer or column chromatography on silical gel plates or silica gel filled columns, if not otherwise stated.

Nuclear magnetic resonance is abbreviated as NMR. Noise decoupling (if mentioned by the authors) is indicated by braces { }. Chemical shifts are given as δ values in ppm with the positive sign for downfield shifts; differing signs given by authors are corrected if possible from the given information. Reference substances are $Si(CH_3)_4$ for ^1H and ^{13}C NMR, $BF_3 \cdot O(C_2H_5)_2$ for ^{11}B NMR, $CFCl_3$ for ^{19}F NMR, and H_3PO_4 for ^{31}P NMR, if not stated otherwise.

Multiplicities of the signals are abbreviated as s, d, t, q (singlet to quartet), quint, sext, sept (quintet to septet), and m (multiplet); terms like dd (double doublet), s's (singlets), and pt (pseudo triplet) are also used. Assignments referring to labeled structures are given in the forms such as H-3,5 or C-4. Coupling constants ^nJ in Hz are given as $J(A,B)$ or as $J(1,3)$ referring to labeled structural formulas; n is the number of bonds between the coupled nuclei.

IR (infrared) and **Raman** bands are given in cm^{-1}; in the band assignments the symbols ν, δ, ϱ, τ, χ, and γ (for stretching, deformation, rocking, twisting, wagging, and out-of-plane bending vibrations) are omitted if not necessary; FT means Fourier Transformations, asym and sym mean asymmetric and symmetric.

Mass spectral data are given as the most important ions or the m/e values followed by the relative intensities in parentheses. EI means electron impact, FAB means fast atom bombardment, and FD means field desorption. The molecular ion is abbreviated with $[M]^+$.

Further abbreviations:

conc.	concentrated	c-	cylco-
D_c	calculated density	c-C_6H_{11}	cyclohexyl
D_m	measured density	i-C_3H_7	$CH(CH_3)_2$
THF	tetrahydrofuran	t-C_4H_9	$C(CH_3)_3$
TLC	preparative thin-layer chromatography		

Table of Contents

Organoosmium Compounds B 6

3 Trinuclear Compounds (continued)

3.1 Compounds with Ligands Bonded by One Carbon Atom (continued)

3.1.6 Compounds with Other Ligands Bonded by One Carbon Atom (continued)

General Remarks. Trinuclear osmium compounds containing one or more ^1L ligands other than CO are described in "Organoosmium Compounds" B 5 (in preparation) and B 6 (present volume). In the linearized formulas the leading ligand is always given immediately after Os.

The approximately 700 compounds are classified into homo- and heterometallic compounds and with respect to the bonding mode of the C atom of the leading ^1L ligand to Os. Compounds with two types of ^1L ligands are described under that ligand which is in the last position in the order given above. Only compounds with terminal isocyanide or carbene ligands are described in the corresponding chapters regardless of the presence of other ^1L ligands.

"Organoosmium Compounds" B 5 will deal with homometallic compounds in which the leading ligand is bonded by one non-bridging Os-C bond (including double bonds). It comprises compounds with terminal ligands such as alkyls or other σ-bonded groups, CS, isocyanides, and carbenes, see structures I to IV (carbonyl groups and ligands bonded only by heteroatoms are omitted for clarity in all structures). The main part of the volume will be devoted to compounds with polydentate ligands in which heteroatoms such as O, S, N, or P are coordinated to Os in addition to one non-bridging C atom; these compounds are represented by structure V, but compounds containing ligands with bridging heteroatoms or with more than one coordinated heteroatom are also included.

Organoosmium Compounds B5

In "Organoosmium Compounds" B 6, the present volume, most compounds have a triangular Os_3 core, and all contain additional CO ligands. The first sections deal with homometallic compounds with the bonding C atom of the leading ^1L ligand bridging two or three Os atoms, regardless of whether the ligand is additionally coordinated to Os by heteroatoms. Sections 3.1.6.2.1 to 3.1.6.2.5 are devoted to homometallic compounds with ^1L ligands bonded by μ_2-bridging C atoms (stuctures VI to X), Section 3.1.6.3 is devoted to homometallic compounds with ^1L ligands bonded by μ_3-bridging C atoms (structure XI). Finally, Section 3.1.6.4 deals with heterometallic triosmium compounds (i.e., compounds with Sn or transition metals directly bonded to Os), in which ^1L ligands other than CO are coordinated to Os, regardless of their bonding type.

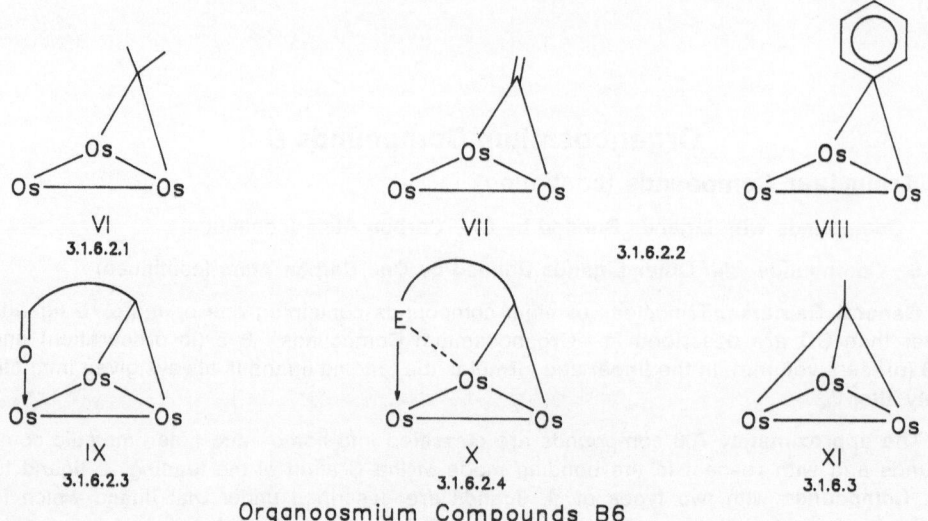

Organoosmium Compounds B6

3.1.6.2 Compounds with ¹Ligands Bonded by μ_2-Bridging C Atoms

3.1.6.2.1 Compounds with μ_2,η^1-Bridging ¹L Ligands of the CRR' Type

Most compounds dealt with in this section have skeletons of the type $(\mu\text{-Y})_m Os_3(\mu_2,\eta^1\text{-}CRR')(CO)_{10-n}D_n$ with μ-alkylidene or μ-carbene ligands CRR' (Formulas I to VI; Y = H, anions X (such as halogens, NO_2, NCO) or CO; m = 1, 2; n = 0 or 1; D = $NCCH_3$, a phosphane, or a phosphite) or bridging cyclic groups. Only No. 46 deviates completely from these formu-

IV   V   VI

las. It contains three μ-H ligands besides a formal μ_3,η^2-HCR bridge with an "agostic" bonded hydrogen, resulting in a cationic complex.

The structures of the compounds generally consist of an approximate isosceles triangle of Os atoms, having two Os-Os bonds of ca. 2.863 Å, and one relatively short doubly bridged Os-Os bond of ca. 2.785 Å [55]. A structure which is assigned to a given compound, but which is not confirmed by X-ray analysis, can only be considered as a proposal based on spectroscopic and analytical data.

Nos. 14 to 33 are zwitterionic compounds that best describe the bridging carbene as a three-electron 1,3-dipolar iminium or phosphonium ligand with two μ-C-Os σ bonds and with the negative charge located at the Os_3 moiety (see Formulas IIIa and IV, and related compounds) [2, 5, 11, 55]. In contrast to the phosphonium-containing complexes, the relatively short C-C bonds in the μ_2,η^1-$CR^3CR^4NR^1R^2$ compounds imply some double-bond characters (evaluated on X-ray analysis) consistent with a fixed trans orientation of R^3 and R^4 (see Formula IIIb); the μ_2,η^1-$CR^3CR^4NR^1R^2$ ligand acting as a three-electron donor thus requires one Os-C σ bond and one Os-C carbenoid bond with the metal-bound carbon bearing a negative charge [55].

Nos. 3 to 13 (Formula II) and Nos. 34 to 38 (Formula V) are anionic complexes. The complexes of Formula V are 50-electron clusters and thus require only two Os-Os bonds. The X-ray analysis of Nos. 36 and 38 show all Os-Os distances to be within or close to bonding values with the dibridged Os-Os distance of ca. 3.1 Å being the longest. The bonding uncertainty between the dibridged Os atoms is represented by a dashed line in Formula V [32].

Most compounds listed in Table 1 were prepared by the following methods:

Method I: Starting from $(\mu$-H$)Os_3(\mu_2,\eta^2$-CH=CHR$)(CO)_{10}$ (R = H, n-C_4H_9, C_6H_5).

a. By treatment with an excess of $(C_2H_5)_2$NH/NaOCH$_3$, phosphanes or P(OCH$_3$)$_3$, leading to compounds of the type $(\mu$-H$)Os_3(\mu_2,\eta^1$-CHCH= N(C$_2$H$_5$)$_2$)(CO)$_{10}$ (No. 15) and $(\mu$-H$)Os_3(\mu_2,\eta^1$-CHCHRPR$^1_2$R^2)(CO)$_{10}$ (R^1 = R^2 = n-C_4H_9, C_6H_5, $C_6H_4CH_3$-4, $C_6H_4OCH_3$-4, OCH$_3$; R^1 = CH$_3$, R^2 = C_6H_5; Nos. 24 to 31), respectively (see Formulas III and IV). The preparations were performed in a nonpolar solvent [2, 3, 26], in CH$_2$Cl$_2$, CH$_3$NO$_2$ [26, 50] at ca. 25 °C, in CH$_3$OH at reflux temperature, or in acetone at −70 °C (No. 31) [26]. The products could be separated from unreacted starting material and by-products such as 1,1- and 1,2-$Os_3(CO)_{10}$(P(CH$_3$)$_2C_6H_5$)$_2$ by TLC with petroleum ether/CH$_2$Cl$_2$ as eluant [26].

b. By treatment with an excess of the corresponding nucleophilic reagent to give $[(\mu$-H$)Os_3(\mu_2,\eta^1$-CHCHRR^1)(CO)$_{10}]^-$ (Nos. 4 to 13; see Formula II); R^1 was derived from LiBH(C$_2$H$_5$)$_3$ [36], KCN, KOH?, NaOCH$_3$, HSC$_6$H$_4$CH$_3$-4,

$HN(C_2H_5)_2$ [26], or $NH_2C_6H_{11}$-c [50]. The reactions were performed in CH_3OH, C_2H_5OH [26], THF [36], or CH_3NO_2 [50] at room temperature for 5 to 30 min, in some cases followed by isolation with $[N(P(C_6H_5)_3)_2]Cl$ or $[NR_4]Br$ (R = CH_3, C_2H_5) in aqueous THF; most products were not isolated [26, 36, 50].

Method II: Starting from $Os_3(CO)_{12-n}(NCCH_3)_n$, n = 1, 2.

a. $Os_3(\mu_2,\eta^1$-CHR)(μ-CO)(CO)$_{10}$ (R = H, CH_3, $Si(CH_3)_3$; Nos. 43 to 45, see Formula VI) were prepared starting from $Os_3(CO)_{11}NCCH_3$ and an excess of an ethereal solution of $RCHN_2$ in hot cyclohexane [15, 19].

b. $(\mu$-H)$Os_3(\mu_2,\eta^1$-CHCH=NR^1R^2)(CO)$_{10}$ (Nos. 14 to 17; Formula III; R^3, R^4 = H) were prepared from $Os_3(CO)_{10}(NCCH_3)_2$ and HNR^1R^2 in CH_2Cl_2 (or $C_6H_5CH_3$ [56]) at ca. 22 °C for 5 min to 40 h [8, 51, 55] in the dark [55], or in C_6H_6 at 40 to 50 °C for 6 to 8 h [51, 55], or at 80 °C for 30 min [8]. In the case of HNR^1R^2, the reactions proceeded by transalkylation/C–H bond activation on the alkyl groups on the amine. On the contrary, in the case of $HN(C_3H_7$-i)$_2$ only C–H bond activation leading to $(\mu$-H)$Os_3(\mu_2,\eta^1$-CHC(CH_3)=NHC_3H_7-i)(CO)$_{10}$ (No. 18) was observed [51]. Reactions with N-methylpyrrolidine, N-methylindole, or N-methylpyrrole giving Nos. 20 to 22 were performed in refluxing C_6H_6 or cyclohexane for 30 min and 2 h [52, 53].

The products, always obtained as a mixture with various by-products, were separated by TLC with hexane/CH_2Cl_2, pentane/CH_2Cl_2 [51 to 53, 55, 56], or petroleum ether [8, 52]. $(\mu$-H)$_2Os_3(CO)_{10}$ and $(\mu$-H)$Os_3(CO)_{10}(\mu$-OH) were obtained as the usually observed by-products in ca. 5 to 15% yield; the formation of $(\mu$-H)$Os_3(CO)_{10}(\mu$-OH) probably resulted from traces of water in the precursor amine [8, 51, 53, 55, 56].

Method III: Starting from $Os_3(\mu_2,\eta^1$-CH_2)(μ-CO)(CO)$_9$D, D = CO or $NCCH_3$.

a. $[Os_3(\mu_2,\eta^1$-CH_2)(CO)$_{10}(\mu$-X)][N(P(C_6H_5)_3)_2]$ (Nos. 34 to 37; see Formula V) were prepared by dropping $[N(P(C_6H_5)_3)_2]X$ (X = Cl, Br, I, NO_2; dissolved in CH_2Cl_2) or $[N(P(C_6H_5)_3)_2]N_3$ (to give No. 38 with μ-X = NCO; dissolved in acetone) into a solution of $Os_3(\mu_2,\eta^1$-CH_2)(μ-CO)(CO)$_{10}$ (No. 43) in CH_2Cl_2 or THF (room temperature, 1 to 10 min). Solids were obtained by treating the produced crude oils with pentane to remove unreacted starting material [31 to 33].

b. $Os_3(\mu_2,\eta^1$-CH_2)(μ-CO)(CO)$_9$D (Nos. 39 and 42) were prepared from $Os_3(\mu_2,\eta^1$-CH_2)(μ-CO)(CO)$_{10}$ (No. 43) by treatment with $(CH_3)_3NO$ (dissolved in CD_3CN [15]) or with $P(C_6H_5)_3$ (dissolved in THF [29]) in CD_2Cl_2 at -78 °C [15], or THF at ca. 22 °C [29]. No. 41 was obtained from $Os_3(\mu_2,\eta^1$-CH_2)(μ-CO)(CO)$_9NCCH_3$ (No. 39) and $P(CH_3)_2C_6H_5$ in CD_2Cl_2 at 0 °C [15]. By-products were the unreacted starting complex (No. 43), $Os_3(CO)_{12}$, $Os_3(CO)_{11}P(C_6H_5)_3$, and $Os_3(CO)_{10}(P(C_6H_5)_3)_2$, which were separated by TLC with CH_2Cl_2/hexane as eluant [29].

Table 1
Compounds with μ_2,η^1-Bridging ^{1}L Ligands of the CRR' Type.
An asterisk preceding the compound number indicates further information at the end of the table, pp. 22/36.
Explanations, abbreviations, and units on p. X.

No. compound	method of preparation (yield in %) properties and remarks

compounds of the type $(\mu\text{-H})_2Os_3(\mu_2,\eta^1\text{-CHR})(CO)_{10}$ (Formula I)

*1 $(\mu\text{-H})_2Os_3(\mu_2,\eta^1\text{-CH}_2)(CO)_{10}$

from $(\mu\text{-H})_2Os_3(CO)_{10}$ in CH_2Cl_2 and CH_2N_2 in ether at ca. 25 °C for 10 min; work-up by TLC with hexane (77%, together with the tautomer $(\mu\text{-H})Os_3(\eta^1\text{-HCH}_2)(CO)_{10}$, "Organoosmium Compounds" B 5, Section 3.1.2.2, in preparation) [4]; see also [6, 7]

by hydrogenation of $Os_3(\mu_2,\eta^1\text{-CH}_2)(\mu\text{-CO})$-$(CO)_9NCCH_3$ (No. 39), together with the tautomer $(\mu\text{-H})Os_3(\mu_2,\eta^1\text{-CH}_3)(CO)_{10}$ and small amounts of $(\mu\text{-H})_2Os_3(CO)_{10}$ [15]

yellow, air-stable crystals from the reaction mixture at -2 °C [4, 6, 11]

^{1}H NMR $(CDCl_3)$: -20.71 (m, $\mu\text{-H}$; $J(\mu\text{-H},\mu\text{-H}) = 0.8$), -15.38 (m, $\mu\text{-H}$), 4.32 (m, 1 H, CH_2; $J(\mu\text{-H},H) = 0.7$ and 2.4, $^2J(H,H) = 5.9$), 5.12 (m, 1 H, CH_2; $J(\mu\text{-H, H}) = 2.1$ and 3.0) [4]

^{13}C NMR (CD_2Cl_2): 25.8 (dd, CH_2; $J(\mu\text{-H,C}) = 3$, $J(H,C) = 140$ and 143) [4, 10]; ^{13}C NMR relaxation time $T_1 = 1.10$ s and nuclear Overhauser enhancement value $\eta_{CH} = 1.99$ were evaluated for CH_2 in CD_2Cl_2 at 32 °C [10]; see also p. 23

IR (CsBr or CsI, 27 °C): 660 ($\nu_s(OsC)$, a_1), 811 ($\rho(CH_2)$, b_2), 869 ($\tau(CH_2)$, a_2), 961 ($\chi(CH_2)$, b_1), 1428 ($\delta(CH_2)$, a_1), 2935 ($\nu_s(CH)$, a_1), 2984 ($\nu_{as}(CH)$, b_2), assignments according to a C_{2v} symmetry [20]

mass spectrum: $[M]^+$ [4]

reaction with $P(CH_3)_2C_6H_5$ at 25 °C yielded $Os_3(CO)_{10}(P(CH_3)_2C_6H_5)_2$ [4]

thermolysis under N_2 in xylene at 110 °C yielded $(\mu\text{-H})_3Os_3(\mu_3,\eta^1\text{-CH})(CO)_9$ (Section 3.1.6.3) [4]

reacted with $(\eta\text{-C}_5H_5)W(\equiv CC_6H_4CH_3\text{-4})(CO)_2$ in THF at 60 °C to give the $Os(\mu_3,\eta^2\text{-C}_2(C_6H_4CH_3\text{-4})_2)(CO)_7W_2(\eta\text{-C}_5H_5)_2$ isomers VIIa and VIIb (p. 21) [13, 22]

reacted with $(C_2H_4)_2PtP(C_6H_{11}\text{-c})_3$ at ca. 25 °C in toluene or THF to yield an isomeric mix-

References on pp. 34/6

Table 1 (continued)

No. compound	method of preparation (yield in %) properties and remarks
*1 (continued)	ture of $(\mu\text{-H})_2Os_3(\mu_2,\eta^1\text{-CH}_2)(CO)_9Pt(CO)\text{-}P(C_6H_{11}\text{-c})_3$ (Section 3.1.6.4.2) [14, 30] reacted with $3,3\text{-}(CH_3)_2C_3H_2\text{-c}$ in refluxing hexane resulting in $(\mu\text{-H})Os_3(\mu_2,\eta^1\text{-}CCHC(CH_3)_2)(CO)_{10}$ (Section 3.1.6.2.2) and CH_4 [27, 46]
2　$(\mu\text{-H})_2Os_3(\mu_2,\eta^1\text{-CHCH}_3)(CO)_{10}$	by warming of $(\mu\text{-H})Os_3(\eta^1\text{-HCHCH}_3)(CO)_{10}$ (compound VIII, p. 21; "Organoosmium Compounds" B 5, Section 3.1.2.2, in preparation; formed in situ by protonation of No. 4) in CD_2Cl_2 above $-20\,°C$; the equilibrium constant No. 2/compound VIII amounts to 7.3 ± 0.7 and is temperature independent between $-16\,°C$ and $+16\,°C$; raising of the temperature above $+19\,°C$ led to decomposition of the tautomers, giving $(\mu\text{-H})_2Os_3(CO)_{10}$ and C_2H_4 [36] by treatment of $(\mu\text{-H})_2Os_3(CO)_{10}$ with excess C_2H_4 (5 atm) in CD_2Cl_2 for ca. 3 h (ca. 10% as a mixture with compound VIII; not isolated) [36] ^1H NMR (CD_2Cl_2): -20.43 (s, μ-H), -14.26 (s, μ-H), 2.49 (d, CH_3; J = 7.0), 6.76 (q, CH) [36]

compounds of the type $[(\mu\text{-H})Os_3(\mu_2,\eta^1\text{-CHR})(CO)_{10}]^-$ (Formula II)

3　$[(\mu\text{-H})Os_3(\mu_2,\eta^1\text{-CH}_2)(CO)_{10}][N(C_2H_5)_4]$	from $(\mu\text{-H})Os_3(\mu_3,\eta^1\text{-CH})(CO)_{10}$ (Section 3.1.6.3) and $LiBH(C_2H_5)_3$ in CH_2Cl_2 at $-60\,°C$; isolated with $[N(C_2H_5)_4]Br$ [25] ^1H NMR (CD_2Cl_2): -17.00 (d, μ-H), 4.26 (d, 1 H, CH_2), 5.23 (dd, 1H, CH_2); $^2J(H,H) = 5$, $J(\mu\text{-H},H) = 3$ [25] protonation at $-40\,°C$ yielded $(\mu\text{-H})Os_3(\eta^1\text{-HCH}_2)(CO)_{10}$ ("Organoosmium Compounds" B 5, Section 3.1.2.2, in preparation) [25]
4　$[(\mu\text{-H})Os_3(\mu_2,\eta^1\text{-CHCH}_3)(CO)_{10}][NR_4]$ 　　$R = CH_3, C_2H_5$	Ib (using $LiBH(C_2H_5)_3$, ca. 1:2, in THF at room temperature, 90%) [36] orange–brown air-sensitive solid [36] ^1H NMR (CD_2Cl_2, $[N(C_2H_5)_4]^+$ salt): -15.95 (d, μ-H; J(H,H) = 2.8), 2.41 (d, CH_3; J(H,H) = 7.5), 7.00 (qd, CH); resonances of the cation: 1.35 (t, CH_3; J = 7), 3.21 (q, CH_2) [36]

References on pp. 34/6

Table 1 (continued)

No. compound	method of preparation (yield in %) properties and remarks

| | IR (KBr, $[N(CH_3)_4]^+$ salt): 1931, 1980, 2009, 2048, 2069, 2090 (all CO) [36] protonation with $HBF_4 \cdot O(C_2H_5)_2$ in CD_2Cl_2 at $-70\,°C$ yielded $(\mu\text{-}H)Os_3(\eta^1\text{-}HCHCH_3)(CO)_{10}$ ("Organoosmium Compounds" B 5, Section 3.1.2.2, in preparation) [36] |

5 $[(\mu\text{-}H^1)Os_3(\mu_2,\eta^1\text{-}CH^2CH_2^3CN)(CO)_{10}][N(P(C_6H_5)_3)_2]$

Ib (84%) [26]
orange crystals from ether [26]
^{1}H NMR $(CDCl_3)$: -16.46 (d, $\mu\text{-}H^1$; $J(H^1,H^2) = 3.3$), 2.95 (d, CH_2^3; $^1J(H^2,H^3) = 8.4$), 6.26 (dt, $\mu_2,\eta^1\text{-}CH^2$) [26]
IR (ethanol): 1942, 1953, 1982, 1991, 2021, 2029, 2078 (all CO), 2220 (νCN) [26]
reaction with HCl gas in $CDCl_3$ gave quantitatively C_2H_5CN and $HOs_3(CO)_{10}Cl$ [26]
reaction with CR_3CO_2H (R = H, F) gave $(\mu\text{-}H)Os_3(CO)_{10}(O_2CCR_3)$ [26]

6 $[(\mu\text{-}H)Os_3(\mu_2,\eta^1\text{-}CHCH_2OH)(CO)_{10}]^-$

Ib (not isolated) [26]
IR (C_2H_5OH): 1947, 1975, 1985, 1994, 2017, 2024, 2075 (all CO) [26]

7 $[(\mu\text{-}H^1)Os_3(\mu_2,\eta^1\text{-}CH^2CH_2^3OCH_3)(CO)_{10}][N(P(C_6H_5)_3)_2]$

Ib [26]
yellow oil (impure) [26]
^{1}H NMR $(CDCl_3)$: -16.57 (d, $\mu\text{-}H^1$; $J(H^1,H^2) = 3.3$), 3.25 (s, CH_3), 3.71 (d, CH_2^3; $J(H^2,H^3) = 8.4$), 6.47 (dt, $\mu_2,\eta^1\text{-}CH^2$) [26]
IR (ether): 1946, 1964, 1983, 2016, 2023, 2074 (all CO) [26]
reaction with CF_3CO_2H yielded $(\mu\text{-}H)Os_3(\mu_2,\eta^2\text{-}CH=CH_2)(CO)_{10}$ and CH_3OH [26]

8 $[(\mu\text{-}H)Os_3(\mu_2,\eta^1\text{-}CHCH_2SC_6H_4CH_3\text{-}4)(CO)_{10}]^-$

Ib (not isolated) [26]
IR (C_2H_5OH): 1938, 1949, 1975, 1986, 2017, 2024, 2075 (all CO) [26]

9 $[(\mu\text{-}H)Os_3(\mu_2,\eta^1\text{-}CHCH_2N(C_2H_5)_2)(CO)_{10}]^-$

Ib (not isolated); with an excess of $HN(C_2H_5)_2$ probably $[(\mu\text{-}H)Os_3(\mu_2,\eta^1\text{-}CHCH_2N(C_2H_5)_2)\text{-}(CO)_{10}][NH_2(C_2H_5)_2]$ is formed [26]
IR $(C_6H_5N(CH_3)_2)$: 1943, 1982, 1994, 2022, 2029, 2079 (all CO) [26]
presumably initially formed in the reaction of $(\mu\text{-}H)Os_3(\mu_2,\eta^2\text{-}CH=CH_2)(CO)_{10}$ with

References on pp. 34/6

Table 1 (continued)

No. compound	method of preparation (yield in %) properties and remarks

9 (continued)

$HN(C_2H_5)_2/NaOCH_3$ in CH_3OH to give No. 15; see Preparation Method Ia [26]

10 $[(\mu-H)Os_3(\mu_2,\eta^1-CHCH_2NHC_6H_{11}-c)(CO)_{10}]^-$

Ib (not isolated) [50]

^1H NMR (CD_3NO_2): 3.30 (m, CH_2), 5.70 (m, CH) [50]

IR (CH_3CN): 1940, 1985, 2020, 2030, 2075 (all CO) [50]

proposed structure based on IR and ^1H NMR [50]

11 $[(\mu-H^1)Os_3(\mu_2,\eta^1-CH^2CH^3(C_6H_5)CN)(CO)_{10}][N(P(C_6H_5)_3)_2]$

Ib (impure) [26]

orange solid [26]

^1H NMR $(CDCl_3)$: -16.46 (d, $\mu-H^1$; $J(H^1,H^2) = 3.2$), 3.50 (d, CH^3; $J(H^2,H^3) = 12.2$), 6.82 (dd, μ_2, η^1-CH^2) [26]

IR (CH_2Cl_2): 1948, 1986, 2022, 2028, 2077 (all CO), 2220 (vCN) [26]

12 $[(\mu-H^1)Os_3(\mu_2,\eta^1-CH^2CH^3(n-C_4H_9)CN)(CO)_{10}][N(P(C_6H_5)_3)_2]$

Ib (impure) [26]

orange solid [26]

^1H NMR $(CDCl_3)$: -16.48 (d, $\mu-H^1$; $J(H^1,H^2) = 3.1$), 0.83 (t, CH_3 of $n-C_4H_9$), 1.70 (m, CH_2 of $n-C_4H_9$), 2.27 (ddd, CH^3; $J(H^2,H^3) = 11.8$), 6.32 (dd, μ_2,η^1-CH^2) [26]

IR (C_2H_5OH): 1942, 1952, 1980, 1990, 2020, 2029, 2079 (all CO), 2220 (vCN) [26]

13 $[(\mu-H)Os_3(\mu_2,\eta^1-CHOCH_3)(CO)_{10}][NR_4]$
    $R = CH_3, C_2H_5$

from $(\mu-H)Os_3(\mu_2,\eta^1-COCH_3)(CO)_{10}$ (Section 3.1.6.2.2) in CH_2Cl_2 $(CD_2Cl_2?)$ and $LiBH(C_2H_5)_3$ in THF (THF-d_8?), ca. 1:2, at or below room temperature; isolated by addition of ca. 2 equivalents of $[NR_4]Br$ in aqueous THF (ca. 70%) [25]

orange, slightly air-sensitive solids [25]

^1H NMR $(CD_2Cl_2/THF-d_8)$: -15.56 (d, $\mu-H$; $J(H,H) = 2.4$), 8.85 (d, CH); directly measured from the reaction mixture [25]; $(CD_2Cl_2?$, $[N(C_2H_5)_4]^+$ salt): 1.35 (t, CH_3; $J(H,H) = 7$), 3.21 (q, CH_2), 3.26 (s, OCH_3) [25]

^{13}C NMR $(CD_2Cl_2, -20\,°C; [N(C_2H_5)_4]^+$ salt): 112.3 (CH; $^1J(H,C) = 154$), 174.5 (s, 2 CO-c), 179.2 (s, 2 CO-d), 181.5 (d, 2 CO-b; $^2J(\mu-H,C) = 10$), 182.0 (s, 1 CO-a; d, 1 CO-a;

Table 1 (continued)

No. compound	method of preparation (yield in %) properties and remarks

^{2}J(C,C) = 13), 184.6 (s, 0.5 CO-f; d, 0.5 CO-f; ^{2}J(C,C) = 34), 187.9 (s, 0.5 CO-e; d, 0.5 CO-e; ^{2}J(C,C) = 34); resonances e and d were broad at 25 °C due to a 3-fold exchange of the CO groups, measured from a 50% ^{13}C enriched product, to assign the signals; see Formula II [25]

IR (KBr, [N(CH$_3$)$_4$]$^+$ salt): 1931, 1942, 1981, 2014, 2062, 2072, 2112 (all CO) [25]

protonation with CF$_3$CO$_2$H in CD$_2$Cl$_2$ at −60 °C yielded (μ-H)Os$_3$(μ$_3$,η^1-CH)(CO)$_{10}$ (Section 3.1.6.3) [25]

compounds of the type (μ-H)Os$_3$(μ$_2$,η^1-CR^{3}CR^4=NR^{1}R^2)(CO)$_{10}$ (Formula III) and similar compounds with bridging heterocycles

*14 (μ-H^1)Os$_3$(μ$_2$,η^1-CH^{2}CH^3=N(CH$_3$)C$_2$H$_5$)(CO)$_{10}$

equilibrium of two anti isomers (Isomers 1 and 3 in Scheme 1, p. 26) and two syn isomers (Isomers 2 and 4) in solution; ^{1}H NMR data indicated an isomeric 1:3:2:4 ratio of 64:32:3:1 in CD$_2$Cl$_2$ at room temperature but syn-anti averaging in the NMR time scale in toluene-d$_8$ at 60 °C [51, 55]

IIb (15 to 20%) yielded an inseparable mixture with small amounts of (μ-H)Os$_3$(μ$_2$,η^2-C(CH$_3$)=NCH$_3$)(CO)$_{10}$, (μ-H)Os$_3$(μ$_2$,η^2-CH$_2$NHC$_2$H$_5$)(CO)$_{10}$, (μ-H)Os$_3$(μ$_2$,η^2-CH$_2$CH=NCH$_3$)(CO)$_{10}$, (all "Organoosmium Compounds" B 5, Sections 3.1.6.1.2.5 and 3.1.6.1.2.7, in preparation), and (μ-H)Os$_3$(μ-NHCH$_3$)(CO)$_{10}$ [51, 55]

from (μ-H)Os$_3$(μ$_2$,η^2-CH$_2$NHC$_2$H$_5$)(CO)$_{10}$ ("Organoosmium Compounds" B 5, Section 3.1.6.1.2.7, in preparation) and an excess of HN(CH$_3$)C$_2$H$_5$ in refluxing cyclohexane for 1 h; purified by TLC as under Preparation Method IIb (67%) [55]

orange crystals from CH$_2$Cl$_2$/hexane −20 °C [55]

^{1}H NMR (CD$_2$Cl$_2$):
Isomer 1 (or less probable, Isomer 3):
−16.81 (d, μ-H; J(H^1,H^2) = 2.4), 1.28 (t, CCH$_3$; J(H,H) = 7.3), 2.95 (s, NCH$_3$), 3.35 (q, CH$_2$; J(H,H) = 7.3), 4.74 (dd, μ$_2$,η^1-CH; J(H^1,H^2) = 2.4, J(H^2,H^3) = 13.7), 7.14 (d, CH^3; J(H^2,H^3) = 13.7);

References on pp. 34/6

Table 1 (continued)

No. compound	method of preparation (yield in %) properties and remarks

*14 (continued)

Isomer 3 (or less probable, Isomer 1):
− 16.72 (d, μ-H; $J(H^1,H^2) = 2.4$), 1.29 (t, CCH_3; $J(H,H) = 7.3$), 3.03 (s, NCH_3), 3.35 (q, CH_2; $J(H,H) = 7.3$), 4.81 (dd, μ_2,η^1-CH; $J(H^1,H^2) = 2.4$, $J(H^2,H^3) = 13.7$), 7.08 (d, CH^3; $J(H^2,H^3) = 13.7$);
Isomer 2 (or less probable, Isomer 4):
− 16.00 (d, μ-H; $J(H^1,H^2) = 1.9$), 5.03 (dd, μ_2,η^1-CH; $J(H^1,H^2) = 1.9$, $J(H^2,H^3) = 14.4$), 8.67 (d, CH^3; $J(H^2,H^3) = 14.4$);
Isomer 4 (or less probable, Isomer 2):
− 16.07 (d, μ-H; $J(H^1,H^2) = 1.5$), 5.01 (dd, μ_2,η^1-CH; $J(H^1,H^2) = 1.5$, $J(H^2,H^3) = 14.3$), 8.61 (d, CH^3; $J(H^2,H^3) = 14.3$);
methyl and ethyl signals of Isomers 2 and 4 are not distinguishable from those of the Isomers 1 and 3 [55]; see also [51]
IR (cyclohexane): 1963, 1977, 1986, 1992, 2009, 2035, 2045, 2057, 2090 (all CO) [51, 55]

*15 $(\mu\text{-}H^1)Os_3(\mu_2,\eta^1\text{-}CH^2CH^3=N(C_2H_5)_2)(CO)_{10}$

anti isomer (Isomer 1/3 in Scheme 1, p. 26) in the solid state; ^1H NMR data indicated an equilibrium with the minor syn isomer No. 16 at room temperature in CD_2Cl_2 (9:1) and toluene-d_8 (8:1) but in toluene-d_8 at 60 °C reversible averaging of the isomers in the NMR time scale ($\delta = -16.53$ ppm, d, μ-H; $J(H^1,H^2) = 2.3$ Hz) [55]
Ia (5%, along with 22% of $(\mu\text{-}H)Os_3(CO)_{10}(\mu\text{-}OCH_3)$ and 15% $(\mu\text{-}H)Os_3(CO)_{10}(\mu\text{-}OH))$ [26]
IIb (36%) [8], IIb (ca. 23%, along with less than 5% of a minor product) [51], IIb (20%, together with the minor isomer No. 16 and 4% $(\mu\text{-}H)Os_3(\mu_2,\eta^2\text{-}CH_2CH=NC_2H_5)(CO)_{10})$ ("Organoosmium Compounds" B 5, Section 3.1.6.1.2.7, in preparation) [56]
from No. 16 in refluxing hexane for 15 min (80%) [56]
yellow crystals [8, 9]
^1H NMR ($CDCl_3$, 30 °C): − 16.75 (d, μ-H; $J(H^1,H^2) = 2.5$), 1.29, 1.3 (2 t's, both CH_3; $J(H,H) = 7.2$), 3.2 to 3.6 (m, CH_2; $J(H,H) = 7.2$), 4.8 (dd, μ_2,η^1-CH; $J(H^2,H^3) = 13.9$), 7.1 (d, CH^3) [8, 55]

References on pp. 34/6

Table 1 (continued)

No. compound	method of preparation (yield in %) properties and remarks

IR (cyclohexane): 1966, 1979, 1987, 1995, 2011, 2039, 2048, 2093 (all CO) [8]

mass spectrum: $[M]^+$ [8]

UV irradiation in hexane yielded a mixture of the two isomers $(\mu\text{-H})_2Os_3(\mu_3,\eta^1\text{-}CCH=N(C_2H_5)_2)(CO)_9$ (Section 3.1.6.3) and $(\mu\text{-H})_2Os_3(\mu_3,\eta^2\text{-}CHCN(C_2H_5)_2)(CO)_9$ [44, 56]

*16 $(\mu\text{-H}^1)Os_3(\mu_2,\eta^1\text{-}CH^2CH^3=N(C_2H_5)_2)(CO)_{10}$

syn isomer (Isomer 2/4) according to Scheme 1, p. 26

IIb (7%); see also No. 15 [56]

orange crystals from CH_2Cl_2/hexane at $-10\,°C$; thermally unstable, isomerized slowly into the anti isomer No. 15 [56]

^1H NMR (CDCl$_3$): -15.98 (d, μ-H; $J(H^1,H^2)=1.8$), 1.29 (t, CH$_3$; $J(H,H)=7.4$), 3.37 (q, CH$_2$; $J(H,H)=7.4$), 5.01 (dd, μ_2,η^1-CH; $J(H^2,H^3)=12.5$), 8.62 (d, CH^3; $J(H,H)=15.2$) [56], similar in toluene-d_8 or CD$_2$Cl$_2$ [55]

IR (hexane): 1967, 1979, 1989, 1997, 2012, 2041, 2049, 2094 (all CO) [56]

*17 $(\mu\text{-H}^1)Os_3(\mu_2,\eta^1\text{-}CH^2CH^3=N(C_2H_5)C_3H_7\text{-}n)(CO)_{10}$

equilibrium of two pairs of anti/syn isomers (Isomers 1 to 4, Scheme 1, p. 26) in solution [51, 55]; ^1H NMR spectroscopy revealed an isomeric 1:2:3:4 ratio of 40:40:12:8 in CD$_2$Cl$_2$ at room temperature, but in toluene-d_8 at 60 °C all isomers average in the NMR time scale to $\delta = -16.51$ ppm (d, μ-H; $J(H^1,H^2)=2.1$ Hz) [51, 55]

IIb (15%, along with 13% $(\mu\text{-H})Os_3(\mu\text{-NHC}_3H_7\text{-}n)(CO)_{10}$) [55]

red crystals from CH_2Cl_2/hexane at $-20\,°C$ according to the anti Isomer 1 in Scheme 1, p. 26 [55]

^1H NMR (CD$_2$Cl$_2$):

Isomer 1 or 2: -16.70 (s, μ-H), 0.98 (m, CH$_3$), 1.29 (m, CH$_3$), 1.69, 1.77, 3.22, 3.27, 3.34, 3.45 (6 m's, each 1 H of CH$_2$), 4.82 (d, μ_2,η^1-CH; $J(H^2,H^3)=14.3$), 7.14 (d, CH^3; $J(H^2,H^3)=14.3$)

Isomer 1 or 2: -16.69 (d, μ-H; $J(H^1,H^2)=3.3$), 0.98 (m, CH$_3$), 1.29 (m, CH$_3$), 1.69, 1.77, 3.22, 3.27, 3.34, 3.45 (6 m's, each 1 H of CH$_2$), 4.80 (dd, μ_2,η^1-CH;

References on pp. 34/6

Table 1 (continued)

No. compound	method of preparation (yield in %) properties and remarks

*17 (continued)

$J(H^1,H^2) = 3.3$, $J(H^2,H^3) = 14.2$), 7.11 (d, CH^3; $J(H^2,H^3) = 14.2$)

Isomer 3 or 4: -15.98 (d, μ-H; $J(H^1,H^2) = 2.1$), 5.07 (dd, μ_2,η^1-CH; $J(H^1,H^2) = 2.1$, $J(H^2,H^3) = 14.6$), 8.62 (d, CH^3; $J(H^2,H^3) = 14.6$)

Isomer 3 or 4: -15.99 (s, μ-H), 5.08 (d, μ_2,η^1-CH; $J(H^2,H^3) = 14.5$), 8.69 (d, CH^3; $J(H^2,H^3) = 14.5$);

methyl and ethyl signals of Isomers 3 and 4 not distinguishable from those of the Isomers 1 and 2 [55]

IR (hexane): 1964, 1978, 1993, 2010, 2036, 2046, 2090 (all CO) [55]

*18 $(\mu$-H)Os$_3(\mu_2,\eta^1$-CHC(CH$_3$)=NHC$_3$H$_7$-i)(CO)$_{10}$

^{1}H NMR spectroscopy revealed an inseparable mixture of two anti isomers (6:4 ratio) in CDCl$_3$ solution (compare Isomers 1/3 in Scheme 1, p. 26), which probably differ in the relative orientation around the C=N bond rather than around the μ-CH-C bond, the isomers do not interconvert up to 70 °C [51]

IIb (ca. 30%) [51]

crystals from hexane/CH$_2$Cl$_2$ at -20 °C according to the anti Isomer 3 in Scheme 1, p. 26 [51]

^{1}H NMR (CDCl$_3$):

Isomer 1: -16.19 (d, μ-H; $J(H,H) = 3.0$), 1.27, 1.31 (2 d's, both CH$_3$ of i-C$_3$H$_7$; $J(H,H) = 7.2$), 2.04 (s, CCH$_3$), 3.85 (sept, CH of i-C$_3$H$_7$; $J(H,H) = 7.2$), 4.86 (d, μ_2,η^1-CH; $J(H,H) = 3.0$), 6.62 (NH)

Isomer 3: -16.32 (d, μ-H; $J(H,H) = 3.0$), 1.27, 1.3(?) (2d's, both CH$_3$; $J(H,H) = 7.2$), 2.00 (s, CCH$_3$), 3.85 (sept, CH; $J(H,H) = 7.2$), 5.12 (d, μ_2,η^1-CH; $J(H,H) = 2.4$), 6.22 (NH) [51]

IR (hexane): 1943, 1955, 1963, 1978, 1993, 1999, 2010, 2034, 2045, 2091 (all CO) [51]

*19 $(\mu$-H)Os$_3(\mu_2,\eta^1$-C(CH$_3$)CH=N(CH$_3$)$_2$)(CO)$_{10}$

from $(\mu$-H)$_2$Os$_3$(CO)$_{10}$ and an excess of CH$_3$C≡CN(CH$_3$)$_2$ in hexane at 25 °C for 1.5 h; purified by column chromatography on Florisil with CH$_2$Cl$_2$/hexane as the eluant (68%) [47]

Table 1 (continued)

No. compound	method of preparation (yield in %) properties and remarks
	orange crystals from CH_2Cl_2/hexane at 0 °C [47]

	orange crystals from CH_2Cl_2/hexane at 0 °C [47] ^1H NMR ($CDCl_3$): -15.01 (s, μ-H), 2.83 (s, CCH_3), 3.10 (s, NCH_3), 8.21 (s, CH) [47] IR (hexane): 1968, 1979, 1990, 2014, 2041, 2049, 2094 (all CO) [47] thermolysis in refluxing hexane yielded a mixture of (μ-H)Os_3(μ_2,η^2-CH(CH_3)C=N(CH_3)$_2$)-(CO)$_{10}$, (μ-H)Os_3(μ_2,η^2-C(C_2H_5)N(CH_3)CH_2)-(CO)$_{10}$, and (μ-H)$_2Os_3$(μ_3,η^2-C(CCH_3)-N(CH_3)$_2$)(CO)$_9$ [47]

20

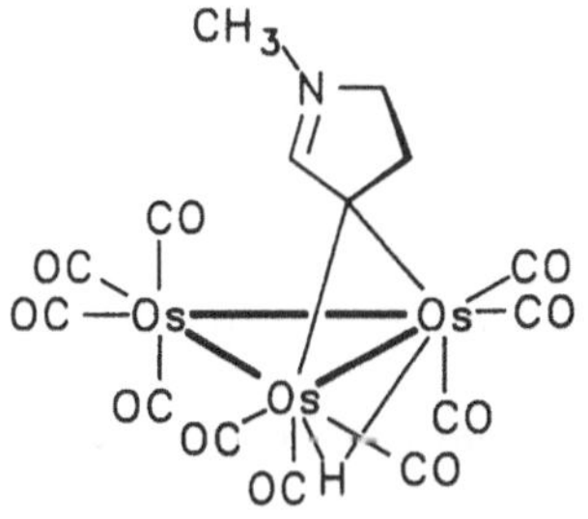

1

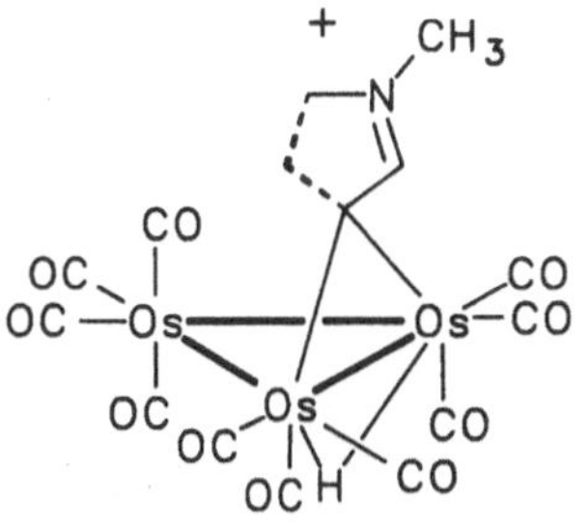

+

2

^1H NMR spectroscopy (solvent not given) revealed the existence of two isomers (ratio 5:1), probably differing in the orientation of the N-methylpyrrolidine ligand to the Os_3 core, which did not interconvert up to 80 °C [53]

IIb (16%, along with 6% of (μ-H)Os_3(μ,η^2-$C_4H_5NCH_3$)(CO)$_{10}$) [53]
red solid [53]
^1H NMR (no solvent given):
Isomer 1: -15.83 (s, μ-H), 2.96 (s, CH_3), 2.1, 3.17 (2 t's, both CH_2; J(H,H)=8.3), 8.33 (s, CH)
Isomer 2: -16.23 (s, μ-H), 7.06 (s, CH) [53]
IR (cyclohexane): 1963, 1975, 1985, 1922, 2011, 2038, 2047, 2057, 2091 (all CO) [53]

*21

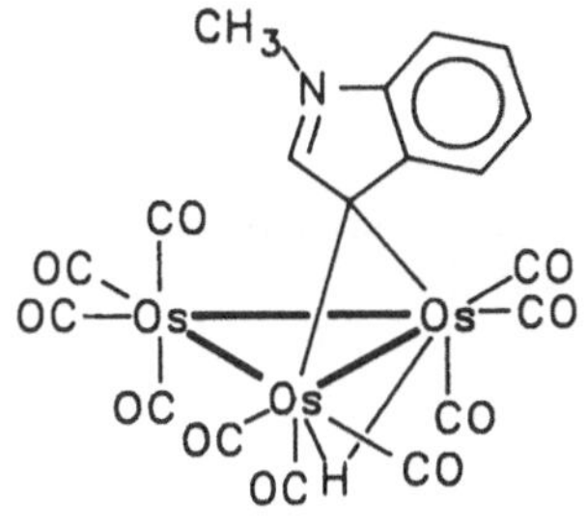

IIb (56%, along with 18% of (μ-H)$_2Os_3$(μ_3-$C_8H_4NCH_3$(CO)$_9$) [52]
orange crystals from hot cyclohexane [52]
^1H NMR (CD_2Cl_2): -14.73 (s, μ-H), 3.75 (s, CH_3), 7.35 (m, C_6H_4), 8.80 (s, CH) [52]
IR (cyclohexane): 1974, 1985, 1955, 2014, 2044, 2055, 2096 (all CO) [52]

Table 1 (continued)

No. compound	method of preparation (yield in %) properties and remarks

*22

IIb (35%) [52]
orange-red crystals from cyclohexane [52]
^{1}H NMR (CD$_2$Cl$_2$): -15.20 (s, μ-H), 6.43 (dd, H^4; J(H^4,H^2) = 4.2), 8.21 (m, H^3; J(H^3,H^4) = 2.3), 8.60 (dd, H^2; J(H^2,H^3) = 1.2) [52]
IR (cyclohexane): 1978, 1982, 2005, 2017, 2046, 2058, 2100 (all CO) [52]

23

from $(\mu$-H)Os$_3(\mu_3,\eta^1$-CH)(CO)$_{10}$ (Section 3.1.6.3) and 4-methylpyridine in CH$_2$Cl$_2$ at $-60\,°C$ (78%) [25]
solid, stable at 25 °C [25]
^{1}H NMR (CD$_2$Cl$_2$): -16.00 (d, μ-H; J(H,H) = 3), 2.27 (s, CH$_3$), 7.28, 8.48 (2 d's, each 2 H of NC$_5$H$_4$; J(H,H) = 5.9), 7.84 (d, CH; J(H,H) = 3) [25]
IR (CH$_2$Cl$_2$): 1957, 2002, 2022, 2030, 2040, 2066, 2088, 2110 (all CO) [25]
protonation reformed the starting complex $(\mu$-H)Os$_3(\mu_3,\eta^1$-CH)(CO)$_{10}$, indicating that H$^+$ added at the nitrogen atom [25]

compounds of the type $(\mu$-H)Os$_3(\mu_2,\eta^1$-CHCHR^{3}PR$_2^1$R^2)(CO)$_{10}$ (Formula IV) and similar compounds with bridging cyclic groups

24   $(\mu$-H)Os$_3(\mu_2,\eta^1$-CHCH$_2$P(OCH$_3$)$_3$)(CO)$_{10}$

Ia [2]
thermolysis in refluxing CHCl$_3$ gave P(OCH$_3$)$_3$ and $(\mu$-H)Os$_3(\mu_2,\eta^2$-CH=CH$_2$)(CO)$_{10}$ [2]

25   $(\mu$-H)Os$_3(\mu_2,\eta^1$-CHCH$_2$P(C$_4$H$_9$-n)$_3$)(CO)$_{10}$

Ia [2]

*26 $(\mu$-H^1)Os$_3(\mu_2,\eta^1$-CH^{2}CH$_2^3$P(CH$_3$)$_2$C$_6$H$_5$)(CO)$_{10}$

Ia (quantitative) [2, 3]
yellow crystals [2]
^{1}H NMR (CDCl$_3$): -16.31 (dd, μ-H^1; J(H^1,H^2) = 3.1, J(H^1,P) = 1.5), 2.37 (d, CH$_3$; J(H,P) = 13.8), 3.38 (dd, CH$_2^3$; J(H^3,P) = 12.3), 5.77 (ddt, μ_2,η^1-CH^2; J(H^2,H^3) = 7.7, J(H^2,P) = 17.2) [26]; similar in CD$_2$Cl$_2$ [2]; (toluene-d$_8$, 27 °C): -15.97 (μ-H^1; J(H^1,H^2) = 3.3, J(H^1,P) = 1.7), 1.06 (CH$_3$; J(H,P) = 13.0), 2.82 (CH$_2^3$; J(H^3,P) = 11.0), 5.39 (μ_2,η^1-CH^2; J(H^2,H^3) = 7.3, J(H^2,P) = 16.7), 7.0 (C$_6$H$_5$) [3]
^{31}P NMR: 20.86 [2]

References on pp. 34/6

Table 1 (continued)

No. compound	method of preparation (yield in %) properties and remarks

IR (cyclohexane): 1927, 1946, 1954, 1972, 1981, 1985, 2000, 2002, 2022, 2030, 2083 (all CO) [3]; similar in cyclohexane/acetone [26] and CH_2Cl_2 [2]

mass spectrum: only ions arising from $(\mu\text{-H})Os_3(\mu_2,\eta^2\text{-CH=CH}_2)(CO)_{10}$ and $P(CH_3)_2C_6H_5$ [2]

dissociated in refluxing $CHCl_3$ after 48 h [2]

thermolysis in refluxing hexane for 2.5 h gave $Os_3(CO)_{10}(P(CH_3)_2C_6H_5)_2$ and $(\mu\text{-H})Os_3(\mu_2,\eta^2\text{-CH=CH}_2)(CO)_{10}$ [3]

treatment with CH_3I in $CHCl_3$ reformed $(\mu\text{-H})Os_3(\mu_2,\eta^2\text{-CH=CH}_2)(CO)_{10}$ [3]

reaction with HCl gas in refluxing CH_3OH [2] or $CHCl_3$ [5] gave $[C_6H_5P(CH_3)_2C_2H_5]Cl$ [2, 26] and small amounts of $(\mu\text{-H})Os_3(CO)_{10}Cl$ [26]

27 $(\mu\text{-H}^1)Os_3(\mu_2,\eta^1\text{-CH}^2CH_2^3P(C_6H_5)_3)(CO)_{10}$

Ia (not isolated) [50]

^{1}H NMR (CD_3NO_2): -16.5 (dd, $\mu\text{-H}^1$; $J(H^1,H^2)=3.4$), 4.04 (dd, CH_2^3; $J(H^2,H^3)=6.7$, $J(H^3,P)=12.4$), 5.96 (ddt, $\mu_2,\eta^1\text{-CH}^2$; $J(H^2,H^3)=6.8$, $J(H^2,P)=18.5$), 7.7 to 7.9 (m, C_6H_5) [50]

IR (CH_3NO_2): 1944, 1965, 1995, 2020, 2028, 2080 (all CO) [50]

28 $(\mu\text{-H})Os_3(\mu_2,\eta^1\text{-CHCH}_2P(C_6H_4CH_3\text{-4})_3)(CO)_{10}$

Ia (not isolated) [50]

29 $(\mu\text{-H}^1)Os_3(\mu_2,\eta^1\text{-CH}^2CH_2^3P(C_6H_4OCH_3\text{-4})_3)(CO)_{10}$

Ia (not isolated) [50]

^{1}H NMR (CD_3NO_2): -16.3 (dd, $\mu\text{-H}^1$; $J(H^1,H^2)=3.3$), 3.9 (m, CH_2^3), 3.94 (s, CH_3), 5.98 (ddt, $\mu_2,\eta^1\text{-CH}^2$; $J(H,P)=18.8$), 7.8 (m, C_6H_4) [50]

IR (hexane): 1945, 1965, 1989, 2025, 2080 (all CO) [50]

30 $(\mu\text{-H}^1)Os_3(\mu_2,\eta^1\text{-CH}^2CH^3(C_6H_5)P(CH_3)_2C_6H_5)(CO)_{10}$

Ia (66% in CH_2Cl_2, failed in cyclohexane) [26]

orange crystals from acetone in the presence of $P(CH_3)_2C_6H_5$ [26]

^{1}H NMR (acetone-d_6, in the presence of $P(CH_3)_2C_6H_5$): -16.41 (dd, $\mu\text{-H}^1$; $J(H^1,H^2)=2.9$, $J(H^1,P)=1.8$), 2.26, 2.37 (2 d's, CH_3; $J(H,P)=13.7$), 4.04 (dd, CH^3; $J(H^3,P)=10.2$), 6.36 (ddd, $\mu_2,\eta^1\text{-CH}^2$;

References on pp. 34/6

Table 1 (continued)

No. compound	method of preparation (yield in %) properties and remarks

30 (continued)

$^1J(H^2,H^3) = 12.5$, $J(H^2,P) = 15.9$); diastereotopic CH_3 groups due to the asymmetric CH^3 [26]

IR (acetone, in the presence of $P(CH_3)_2C_6H_5$): 1948, 1962, 1983, 1992, 2023, 2031, 2080 (all CO) [26]

decomposed into the starting materials upon redissolving in $CHCl_3$ [26]

31   $(\mu-H^1)Os_3(\mu_2,\eta^1-CH^2CH^3(C_4H_9-n)P(CH_3)_2C_6H_5)(CO)_{10}$

Ia (not isolated); after warming to 25 °C decomposed into the starting material [26]

^1H NMR (acetone-d_6, in the presence of $P(CH_3)_2C_6H_5$, -70 °C): -16.41 (dd, $\mu-H^1$; $J(H^1,H^2) = 3.1$, $J(H^1,P) = 1.9$), 2.27, 2.37 (2 d's, both CH_3; $J(H,P) \approx 13$), 2.82 (m, CH^3), 6.28 (ddd, μ_2,η^1-CH^2; $^1J(H^2,H^3) = 8.5$, $J(H^2,P) = 14.0$); diastereotopic CH_3 groups due to the asymmetric CH^3 [26]

IR (acetone, in the presence of $P(CH_3)_2C_6H_5$, -70 °C): 1946, 1963, 1993, 2026, 2081 (all CO) [26]

*32

from $Os_3(CO)_{10}(NCCH_3)_2$ and c-$C_5H_4P(C_6H_5)_3$ under the conditions of Preparation Method IIb (61%) [54]

ruby red crystals from CH_2Cl_2/heptane [54]

^1H NMR ($CDCl_3$, 27 °C): -15.34 (s, $\mu-H$), 6.26 (dt, H^3; $J(H^1,H^3) = 1.7$, $J(H^2,H^3) = 4.3$, $J(H^3,P) = 1.6$), 6.95 (dt, H^2; $J(H^1,H^2) = 1.7$, $J(H^2,P) = 4.3$), 7.55 to 7.75 (m, C_6H_5), 8.17 (dt, H^1; $J(H,P) = 6.8$) [54]

IR (CH_2Cl_2): 1965, 1994, 2031, 2039, 2067, 2083 (all CO) [54]

isomerized in CH_2Cl_2/heptane (1:1) in $CDCl_3$ at room temperature giving No. 33, separated by TLC; equilibrium with ca. 40% No. 33 reached after several days; Nos. 32 and 33 probably differ by the position of the bridging carbene ligand relative to the Os_3 core [54]

thermolysis of a mixture of Nos. 32 and 33 in refluxing toluene yielded $(\mu-H)Os_3(\mu_3,\eta^3-C_5H_3P(C_6H_5)_3)(CO)_9$ [54]

References on pp. 34/6

Table 1 (continued)

No. compound	method of preparation (yield in %) properties and remarks

33

formed by isomerization of No. 32 in
   CH_2Cl_2/heptane or $CDCl_3$ (40%) [54]; see
   also No. 32
pink in petroleum ether/ether solution [54]
^{1}H NMR ($CDCl_3$, 27 °C): −15.59 (s, μ-H), 6.39
   (dt, H^3), 7.02 (dt, H^1), 7.55 to 7.75 (m, C_6H_5),
   7.95 (dt, H^2); coupling constants as that of
   No. 32 [54]
IR (CH_2Cl_2) indistinguishable from that of No.
   32 [54]

compounds of the type $[Os_3(\mu_2,\eta^1\text{-}CH_2)(CO)_{10}(\mu\text{-}X)]^-$ (Formula V)

*34 $[Os_3(\mu_2,\eta^1\text{-}CH_2)(CO)_{10}(\mu\text{-}Cl)][N(P(C_6H_5)_3)_2]$

IIIa (89%) [31, 33]
orange powder [33]
^{1}H NMR ($CDCl_3$): 2.92, 3.70 (2 d's, each 1 H of
   CH_2; J(H,H) = 8.60) [31, 33]
IR (CH_2Cl_2): 1935, 1946, 1977, 2016, 2030, 2072
   (all CO) [31, 33]

*35 $[Os_3(\mu_2,\eta^1\text{-}CH_2)(CO)_{10}(\mu\text{-}Br)][N(P(C_6H_5)_3)_2]$

IIIa (92%) [31, 33]
orange solid [33]
^{1}H NMR ($CDCl_3$): 3.19, 3.81 (2 d's, each 1 H of
   CH_2; J(H,H) = 8.55) [31, 33]
IR (CH_2Cl_2): 1933, 1950, 1979, 2018, 2029, 2070
   (all CO) [31, 33]

*36 $[Os_3(\mu_2,\eta^1\text{-}CH_2)(CO)_{10}(\mu\text{-}I)][N(P(C_6H_5)_3)_2]$

IIIa (97%) [31, 33]
red-orange powder [33]
^{1}H NMR ($CDCl_3$): 3.58, 3.83 (2 d's, each 1 H of
   CH_2; J(H,H) = 8.01) [31, 33]
IR (CH_2Cl_2): 1933, 1950, 1979, 2018, 2027, 2072
   (all CO) [31, 33]
reaction with SO_2 yielded $[Os_3(\mu_2,\eta^2\text{-}CH_2SO_2)$-
   $(CO)_{10}(\mu\text{-}I)][N(P(C_6H_5)_3)_2]$ [39]

*37 $[Os_3(\mu_2,\eta^1\text{-}CH_2)(CO)_{10}(\mu\text{-}O\text{-}N=O)][N(P(C_6H_5)_3)_2]$

IIIa (92%) [33]
yellow-orange powder [33]
^{1}H NMR ($CDCl_3$): 2.66, 2.84 (2 d's, each 1 H of
   CH_2; J(H,H) = 10.05) [33]
IR (CH_2Cl_2): 847 (δONO), 1024, 1175, 1445
   (νNO),1956, 1986, 2020, 2037, 2076 (all νCO);
   assignments of the ONO bands uncertain,
   structure only suggested [33]

References on pp. 34/6

Table 1 (continued)

No. compound	method of preparation (yield in %) properties and remarks

*38 $[Os_3(\mu_2,\eta^1\text{-}CH_2)(CO)_{10}(\mu\text{-}NCO)][N(P(C_6H_5)_3)_2]$

IIIa (92%) [31, 33]

orange, microcrystalline solid [33]

^1H NMR $(CDCl_3)$: 2.79, 3.83 (2 d's, each 1 H of CH_2; J(H,H) = 8.06) [31, 33]

IR (CH_2Cl_2): 1930, 1948, 1977, 2016, 2031, 2070 (all CO), 2207 (NCO); the sharpness of the νNCO band is typical for μ_2,η^1-NCO groups in contrast to terminal ones [31, 33]

compounds of the type $Os_3(\mu_2,\eta^1\text{-}CHR)(\mu\text{-}CO)(CO)_9D$ (Formula VI)

39 $Os_3(\mu_2,\eta^1\text{-}CH_2)(\mu\text{-}CO)(CO)_9NCCH_3$

IIIb (ca. 90% at −78 °C, not isolated due to instability) [15]

^1H NMR $(CD_2Cl_2, 0\,°C)$: 8.42 (d, 1 H, CH_2; J(H,H) = 6.0), 8.83 (d, 1 H, CH_2) [15]

reacted with diazoalkanes $RCHN_2$ (R = H, CH_3, $Si(CH_3)_3$) at 0 to 25 °C resulting in the alkenyl compounds $(\mu\text{-}H)Os_3(\mu_2,\eta^2\text{-}CH=CHR)(CO)_{10}$ [15]

reaction with $P(CH_3)_2C_6H_5$ in CD_2Cl_2 yielded No. 41 [15]

hydrogenation yielded a tautomeric mixture of $(\mu\text{-}H)Os_3(\eta^1\text{-}HCH_2)(CO)_{10}$ ("Organoosmium Compounds" B 5, Section 3.1.2.2, in preparation) and $(\mu\text{-}H)_2Os_3(\mu_2,\eta^1\text{-}CH_2)(CO)_{10}$ (No. 1), and some $(\mu\text{-}H)_2Os_3(CO)_{10}$ [15]

40 $Os_3(\mu_2,\eta^1\text{-}CH_2)(\mu\text{-}CO)(CO)_9P(OC_6H_5)_3$

by protonation of $[Os_3(\eta^1\text{-}CHO)(CO)_{10}\text{-}P(OC_6H_5)_3]K$ (pregenerated in situ from $Os_3(CO)_{11}P(OC_6H_5)_3$ and $K[BH(OC_3H_7)_3\text{-}i]$ in THF at 0 °C) by dropwise addition of an excess of 20% H_3PO_4, followed by extraction with CH_2Cl_2 and column chromatography on SiO_2 with CH_2Cl_2/hexane; inseparable mixture of two isomers containing some starting material [29]

red solid [29]

^1H NMR (acetone-d_6): Isomer 1: 6.39 (d, 1H, CH_2; J(H,H) = 6.7), 7.3 to 7.8 (m, C_6H_5); the second methylene proton is probably obscured by C_6H_5; Isomer 2: 6.47 (d, 1 H, CH_2; J(H,H) = 7.3), 6.55 (d, 1 H, CH_2; J(H,H) = 7.3, J(H,P) = 16.5), 7.3 to 7.8 (m, C_6H_5) [29]

Table 1 (continued)

No. compound	method of preparation (yield in %) properties and remarks
	mass spectrum: $[M-CO]^+$, $[M-x\ CO-y\ H]^+$ (x = 1 to 8; y = 1, 2) [29] proposed structure based on ^1H NMR and mass spectra [29]
41 $Os_3(\mu_2,\eta^1\text{-}CH_2)(\mu\text{-}CO)(CO)_9P(CH_3)_2C_6H_5$	IIIb [15] red solid [15] ^1H NMR (CDCl$_3$, 0 °C): 2.04, 2.06 (2 d's, each CH$_3$; J(H,P) = 9.5), 5.25 (dd, 1 H, CH$_2$; J(H,P) = 15.3), 7.00 (d, 1 H, CH$_2$; J(H,H) = 7.0), 7.39 to 7.53 (m, C$_6$H$_5$); the upfield shift of one of the methylene resonances and its strong coupling to phosphorus indicates that the substitution site appears to be on one of the Os atoms bridged by the methylene group [15] IR (Nujol): 1839 (μ-CO), 1932, 1958, 1998, 2021, 2031, 2092 (all CO); (cyclohexane): 1969, 1973, 1986, 1998, 2011, 2022, 2039, 2053, 2067, 2093 (all CO) [15] mass spectrum: $[M]^+$ [15]
42 $Os_3(\mu_2,\eta^1\text{-}CH_2)(\mu\text{-}CO)(CO)_9P(C_6H_5)_3$	IIIb (inseparable mixture with Os$_3$(CO)$_{12}$ and Os$_3$(CO)$_{11}$P(C$_6$H$_5$)$_3$) [29] orange solid [29] ^1H NMR: 3.58 (s, CH$_2$) [29] mass spectrum: $[M-CO]^+$ and fragment ions due to successive loss of all CO [29] structure tentatively suggested from ^1H NMR and mass spectra [29]
*43 $Os_3(\mu_2,\eta^1\text{-}CH_2)(\mu\text{-}CO)(CO)_{10}$	IIa (50%) [15, 19] from Os$_3$(CO)$_{10}$(NCCH$_3$)$_2$ and CH$_2$=C=O under the conditions of Preparation Method IIb (49%) [17] by protonation of [Os$_3$(η^1-CHO)(CO)$_{11}$]K (pre-generated in situ from Os$_3$(CO)$_{12}$ and K[BH(OC$_3$H$_7$-i)$_3$] in THF at 0 °C) by dropwise addition of an excess of 20% H$_3$PO$_4$, fol-lowed by extraction with CH$_2$Cl$_2$ and column chromatography on SiO$_2$ with hexane as eluant (37%) [16, 29]; by protonation of [Os$_3$(η^1-CHO)(CO)$_{11}$]K with CF$_3$CO$_2$H or [(CH$_3$)$_3$O]BF$_4$ (ca. 20%) [16, 29]

Table 1 (continued)

No. compound	method of preparation (yield in %) properties and remarks

*43 (continued)

by warming of $Os_3(\mu_2,\eta^2\text{-}CH_2C\text{=}O)(CO)_{12}$ in $CDCl_3$ in a sealed evacuated NMR tube to 60 to 64 °C for 30 min (8% to 10%) [24, 28]

deep red or orange-red, air-stable crystals [15 to 18, 29]; soluble in all common solvents [29]

^1H NMR (acetone-d_6, 35 °C): 6.68, 7.83 (2 d's, each 1 H of CH_2; $^2J(H,H) = 6.9$) [15]; ($CDCl_3$): 6.47, 7.75 (2 d's, each 1 H of CH_2; $J(H,H) = 7.2$); spectrum depicted [16, 29]; ^1H NMR spin-magnetization-transfer experiments revealed a slow exchange of the non-equivalent methylene protons with a rate constant k of 0.8 ± 0.4 s^{-1}(at room temperature?) compared to $k = 4.0$ s^{-1} at 17 °C for $(\mu\text{-}CO)Ru_3(\mu\text{-}CH_2)(CO)_{10}$ [41]

^{13}C NMR (CD_2Cl_2): 62.5 (dd, CH_2; $J(H,C) = 144$, $J(H,C) = 147$), 171.9, 172.6, 173.9, 174.3 (each 2 CO), 180.9, 184.0 (each 1 CO), 193.4 (μ-CO); in the fully coupled spectrum CH_2 appeared as a pseudo triplet; spectrum depicted [16, 29]; ^{13}C NMR magnetization-transfer experiments show some CO goup exchange [41]

IR (CsBr or CsI, 27 °C): 796 ($\rho(CH_2)$, b_2), 850 ($\tau(CH_2)$, a_2), 974 ($\chi(CH_2)$, b_1), 1426 ($\delta(CH_2)$, a_1), 2949 ($\nu_s(C\text{-}H)$, a_1), 2990 ($\nu_{as}(C\text{-}H)$, b_2), assignment according to C_{2v} symmetry [20]; (cyclohexane): 2004, 2014, 2027, 2035, 2060, 2068, 2113 [15]; (Nujol): 1867, 1993, 2026, 2058, 2113 [15]; (hexane): 1869 (μ-CO), 1920, 1995, 2010, 2031, 2063, 2116 (all CO); the stretching vibration of the bridging CO group was only observed in concentrated solution or using a FT/IR spectrometer [16, 29]

mass spectrum: $[M]^+$, $[M - x\ CO - y\ H]^+$ ($x = 1$ to 11, $y = 1, 2$) [15, 16, 29]

44 $Os_3(\mu_2,\eta^1\text{-}CHCH_3)(\mu\text{-}CO)(CO)_{10}$

IIa (< 10%) [15]
red, air-stable [15]
^1H NMR ($CDCl_3$, 20 °C): 2.55 (d, CH_3), 9.82 (q, CH; $J(H,H) = 7.8$) [15]
IR (cyclohexane): 2003, 2014, 2026, 2033, 2057, 2068, 2099, 2110 (all CO) [15]
mass spectrum: $[M]^+$ [15]

Table 1 (continued)

No. compound	method of preparation (yield in %) properties and remarks
*45 Os$_3$(μ_2,η^1-CHSi(CH$_3$)$_3$)(μ-CO)(CO)$_{10}$	IIa (> 50%) [15] red, air-stable plate-like crystals [15, 23] ^{1}H NMR (acetone-d$_6$, 35 °C): 0.06 (s, CH$_3$), 8.72 (s, CH) [15] IR (Nujol): 1862 (μ-CO), 1979, 1992, 2004, 2027, 2032, 2040, 2055, 2111(all CO); similar in cyclohexane [15] mass spectrum: [M]$^+$ [15]

other compound

46
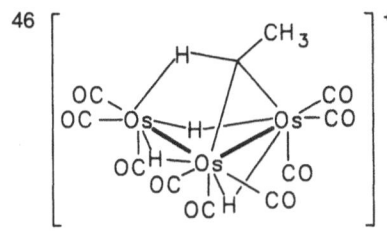

earlier formulated as [(μ-H)$_4$Os$_3$(μ_3,η^1-CCH$_3$)-(CO)$_9$]$^+$ [1] based on ^{1}H NMR data which are probably due to decomposition products [1, 35, 37, 48]

from (μ-H)$_3$Os$_3$(μ_3,η^1-CCH$_3$)(CO)$_9$ (Section 3.1.6.3) in HSO$_3$F or CF$_3$SO$_3$H at 22 °C (not isolable; work-up of the acid solution recovered the starting complex) [35, 48]; see also [1]

^{1}H NMR (HSO$_3$F): − 19.25 (s, 3 μ-H), − 10.64 (s, Os − H − C), 3.39 (s, CH$_3$); similar values in CF$_3$SO$_3$H; spectrum unchanged at −65 °C, indicating fast migration of the "agostic" bonded hydrogen among all three Os-C bonds, thus creating apparent C$_{3v}$ symmetry [35, 48]

decomposition overnight in HSO$_3$CF$_3$ yielded probably [(μ-H)$_2$Os$_3$(CO)$_9$(μ_3,η^3-O$_3$SCF$_3$)]$^+$ or (μ-H)Os$_3$(CO)$_9$(μ_3,η^3-O$_3$SCF$_3$) [35, 48]

due to the agostic bonded hydrogens, the three tautomers of No. 46 are thought to be involved in catalytic reactions of hydrocarbons with polymetallic systems [48]

VII

a   b

VIII

References on pp. 34/6

*Further information:

$(\mu-H)_2Os_3(\mu_2,\eta^1-CH_2)(CO)_{10}$ (Table 1, No. 1) crystallizes in the orthorhombic space group $Pn2_1a$ $(Pna2_1)-C_{2v}^9$(No. 33) with a = 18.502(9), b = 10.096(5), c = 8.763(4) Å; Z = 4, D_c = 3.517 g/cm³, R = 0.077. The molecular structure, evaluated by neutron diffraction analyses, is shown in **Fig. 1**. The structure consists of an unsymmetrical Os_3 triangular array with the Os(1)–Os(3) distance being the longest due to the bond lengthening effect of the bridging hydride, whereas for Os(1)–Os(2) this effect is counterbalanced by the bridging carbene exhibiting the shortest Os–Os distance. The $(\mu-H)Os_2(\mu_2,\eta^1-C)$ moiety is nonplanar with a dihedral angle of 154.0(5)° between the Os(1)–μ_2,η^1–C–H(4) and Os(2)–μ_2,η^1–C–H(4) planes. Comparison of the structure with that of No. 26 (p. 29) having a dipolar bridging carbene ligand, revealed an obvious similarity of the geometries of the μ_2,η^1–C groups with the spatial arrangements of the atoms around the coordinated methylene carbons best characterized as distorted tetrahedral. In the 48–electron system of No. 1 the $(\mu-H)Os_2(\mu_2,\eta^1-C)$ moiety produces a six–electron, four–center bond with the combination of four sp^3d^2 orbitals from Os(1) and Os(2), the 1s orbital from H(4) and the (μ_2,η^1-C) sp^2 orbital. The average bond distances of Os–C and C≡O are 1.930 and 1.138 Å, respectively. For each of the three Os atoms, the CO ligands trans to the hydrides have the shortest Os–C distances [11]; see also [6].

Indirect location of the hydride positions was performed by potential–energy evaluations giving a mean value of 1.82(27) Å for the Os–H distance. The deviation between the calculated

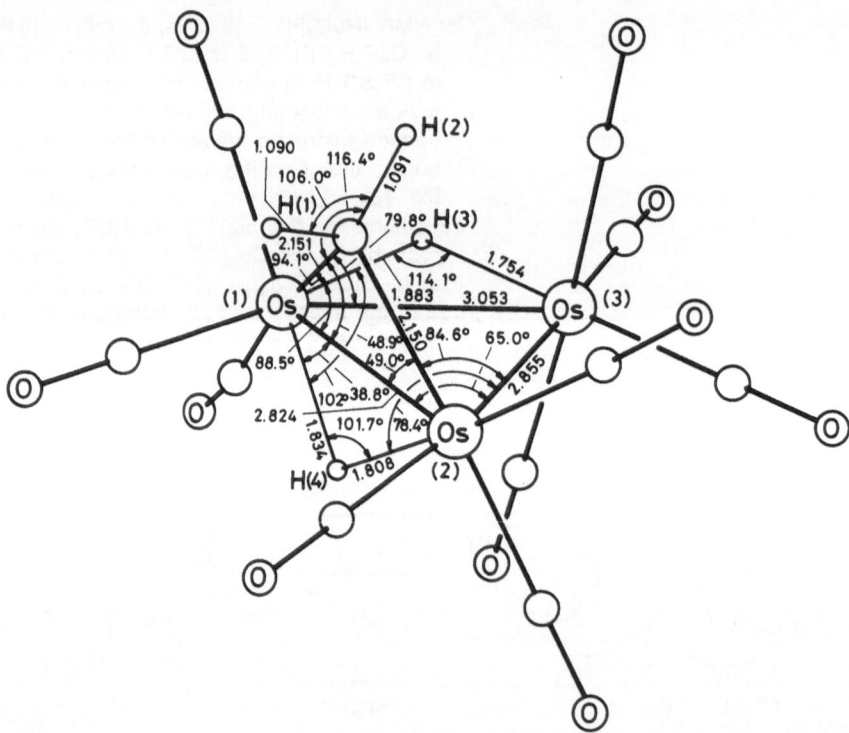

Fig. 1. Molecular structure of $(\mu-H)_2Os_3(\mu_2,\eta^1-CH_2)(CO)_{10}$ (No. 1) with selected bond distances (in Å) and angles [6, 11].

References on pp. 34/6

and the measured bond distances is due to the asymmetry of the Os–H–Os bond, which was not taken into account in the calculations [12].

For the title compound ^{1}H and ^{13}C NMR spectroscopy in CH_2Cl_2 solution at room temperature indicated an equilibrium with its hydridomethyl tautomer $(\mu\text{-H})Os_3(\eta^1\text{-HCH}_2)(CO)_{10}$ ("Organoosmium Compounds" B 5, Section 3.1.2.2, in preparation) [4, 6, 7]. For the hydridomethyl tautomer an Os^1-$(\eta^1\text{-CH}_2)$ moiety was evaluated showing significant C–H–Os^2 interactions [7] (contrary to earlier results assuming a symmetrical $\mu\text{-CH}_3$ moiety [4]), whereas the title compound exhibits only Os^1-$(\mu\text{-H})(\mu\text{-CH}_2)$-$Os^2$ bonds (compare Formula I, p. 2; R = H). Below $-20\,°C$ only the ^{1}H and ^{13}C NMR signals of the hydridomethyl tautomer were observed. The rate constant $k = 1 \times 10^{-3}$ s^{-1} at $14\,°C$ and the activation energy $\Delta G^+ \approx$ 20 kcal/mol were calculated for its rearrangement into No. 1. The equilibrium constant [No. 1]/[$\eta^1\text{-HCH}_2$ tautomer] is 3.5 ± 0.1 in CD_2Cl_2 at $32\,°C$ [4] and 2.65 ± 0.16 in $CDCl_3$ at $35\,°C$ [6]; it decreases in the order acetone > dichloromethane > benzene but changes little with temperature between 30 and $115\,°C$ when evaluated in toluene [4].

The two possible coordination sites for the hydride in the tautomers, Os–H–Os and Os–H–C, were found to be similar, evaluated in terms of the differences in electronegativities of the Os, H, and C elements. The overall enthalpy change for the conversion of the tautomers as a function of the electronegativities was calculated as $\Delta H = 18.8$ kcal/mol using Mullay's electronegativities [49].

^{1}H NMR spectroscopic site population determinations of a partially deuterated sample, prepared from $(\mu\text{-D})_2Os_3(CO)_{10}$ and CH_2N_2, indicated an equilibrium isotope effect that favors the deuterium being located at the methylene or the methyl group in both tautomers, and H over D in the metal hydride sites with respect to the different vibrational zero point energies. All possible configurations of the partially deuterated tautomers were considered and their relative concentrations expressed in terms of the equilibrium constants K_1, K_2, K_{eq} (relating the hydridomethyl and methylene tautomers (in $CDCl_3$ at $35\,°C$?)), and d_1 and d_2 (mono- and bideuterated species). K_2 ($K_2 = 1.58(21)$) is defined as the equilibrium constant for a pairwise H/D interchange in the methylene tautomer exhibiting in total six possible configurations. For the hydridomethyl tautomer at each level of deuteration there are only two configurations which are related by the equilibrium constant K_1 ($K_1 = 1.74(23)$); K_{eq} between the two tautomers amounted to 2.45, and $d_1/d_2 = 0.20$ [6].

Neutron diffraction analysis of the partially deuterated No. 1 showed a K_2 value of 2.30(10), and $d_1/d_2 = 0.17$ at $-2\,°C$, and, adjusting to $35\,°C$ this equilibrium constant is reduced to $K_2 = 2.08$. The percentage of ^{1}H (versus D) in each of the hydrogen sites was evaluated as follows (compare Fig. 1): 42% for H(1), 41% for H(2), 70% for H(3), and 64% for H(4). [6].

$(\mu\text{-H})Os_3(\mu_2,\eta^1\text{-CHCH=NR}^1R^2)(CO)_{10}$ ($NR^1R^2 = N(CH_3)C_2H_5$, $N(C_2H_5)_2$, $N(C_2H_5)_2$, $N(C_2H_5)C_3H_7$-n; Table 1, Nos. **14, 15, 16, 17**). X-ray analysis of Nos. 15, 16, and 17 revealed quite a similarity of the structures, all consisting of an isosceles triangle with one relatively short doubly bridged Os–Os bond and two longer, nearly identical bonds to the $Os(CO)_4$ moiety. The three-electron $\mu_2,\eta^1\text{-CHCH=NR}^1R^2$ unit is anti oriented in Nos. 15 and 17, and syn in No. 16. The relatively short bond lengths C–CN imply some double bond character with a fixed trans orientation of the hydrogens at these C atoms [9, 55, 56]; see also Nos. 18, 19, pp. 26/28, and p. 3.

The anti isomer $(\mu\text{-H})Os_3(\mu_2,\eta^1\text{-CHCH=N}(C_2H_5)_2)(CO)_{10}$ (No. 15) crystallizes in the monoclinic space group $P2_1/c - C_{2h}^5$ (No. 14) with $a = 7.676(2)$, $b = 18.392(5)$, $c = 16.026 (4)$ Å, $\beta = 97.57(2)°$; $Z = 4$, $D_c = 2.813$ g/cm^3, $R = 0.0503$. The $(\mu\text{-H})Os_3(CO)_{10}$ portion of the molecule (see **Fig. 2**) has approximate C_s symmetry, which is mainly violated by the anti-bridging

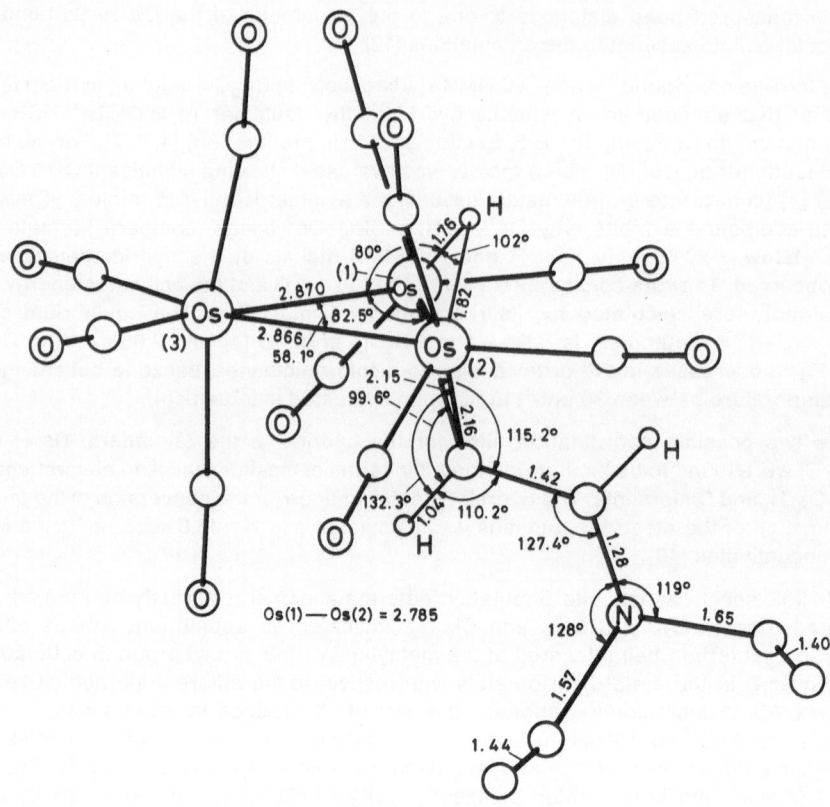

Fig. 2. Molecular structure of $(\mu-H)Os_3(\mu_2,\eta^1-CHCH=N(C_2H_5)_2)(CO)_{10}$ (No. 15) with selected
bond lengths (in Å) and angles [8, 9].

$CHCH=N(C_2H_5)_2$ ligand. The Os(1)–Os(2) bond is reduced due to the shortening effect of
the iminium ion, counteracting the lenghtening effect of the bridging hydride. The geometry
around the nitrogen is planar; the N–CHCH bond distance is consistent with a C=N linkage,
whereas the other N–C bond lengths reveal single bonds on a substituted iminium cation.
Only the bridging hydride and the hydrogen bonded to μ_2,η^1–C were observed crystallographically; the other one, CH=N, is given in an idealized position in Fig. 2. The average
bond distances of Os–CO and C≡O are 1.90 and 1.13 Å [8, 9].

The trans geometry around the μ_2,η^1–CH–CHN bond was also evidenced by the large
three-bond coupling $^3J(H,H)=13.9$ Hz, based on ^1H NMR data. The rotation around this
bond is slow on the NMR time scale up to 70 °C, suggesting considerable double bond
character (see also Formula IIIa, p. 2) [51].

The syn isomer $(\mu-H)Os_3(\mu_2,\eta^1-CHCH=N(C_2H_5)_2)(CO)_{10}$ (No. 16) crystallizes in the monoclinic space group $P2_1/m-C_{2h}^2$ (No. 11) with a = 7.644(1), b = 12.706(2), c = 11.912(2) Å, β =
108.02(1)°; Z = 2, D_c = 2.87 g/cm³, R = 0.032. The structure is quite similar to that of No. 15
with the exception that the bridging ligand is syn in No. 16. A crystallographically imposed
mirror plane passes through μ–H, **CHCH=N**, the methylene groups of C_2H_5 and the two
axial coordinated CO groups at the nonbridged Os atom resulting in a 50:50 disorder for

the methyl groups occupying two sites equally displaced from the plane. Statistically, one methyl group lies on each side of the plane but it is also possible, that both methyl groups lie on the same side of the plane as in No. 15 (compare Fig. 2). Selected interatomic distances and bond angles are (for numbering, see Fig. 2) [56]:

atoms	distance (Å)	atoms	angle (°)
Os(1)–Os(2)	2.789(1)	Os(1)–Os(2)–Os(3)	61.01
Os(1)–Os(3)	2.8780(9)	Os(1)–Os(3)–Os(2)	57.97
Os(2)–Os(3)	2.8780(9)		
Os(1)–μ^2,η^1-C	2.22(1)	Os(1)–Os(2)–μ^2,η^1-C	51.0
Os(3)–μ^2,η^1-C	2.22(2)	Os(3)–Os(1)–μ^2,η^1-C	91.1
μ^2,η^1-C–C	1.33(2)	Os(1)–μ^2,η^1-C–Os(2)	77.9
C–N	1.32(2)	Os(1)–μ^2,η^1-C–C	125
Os–C	1.90	μ^2,η^1-C–C–N	126

(μ-H)Os$_3$(μ_2,η^1-CHCH=N(C$_2$H$_5$)C$_3$H$_7$-n)(CO)$_{10}$ (No. 17) crystallizes in the monoclinic space group P2$_1$/c – C$_{2h}^5$ (No. 14) with a = 7.679(2), b = 19.250(5), c = 17.091(5) Å, β = 112.63(3)°; Z = 4, D$_c$ = 2.75 g/cm^3, R = 0.061. The molecular structure, shown in **Fig. 3**, reveals anti bridging of the μ_2,η^1-CHCH=N(C$_2$H$_5$)C$_3$H$_7$-n for the solid complex according to Isomer 1 in Scheme 1, p. 26. The asymmetry in the Os–C–Os bond distances is probably due to the steric crowding indicated by the n-C$_3$H$_7$ group. The bridging hydride is given in a calculated position. Average bond distances of Os–CO and C≡O are 1.90 and 1.15 Å [55].

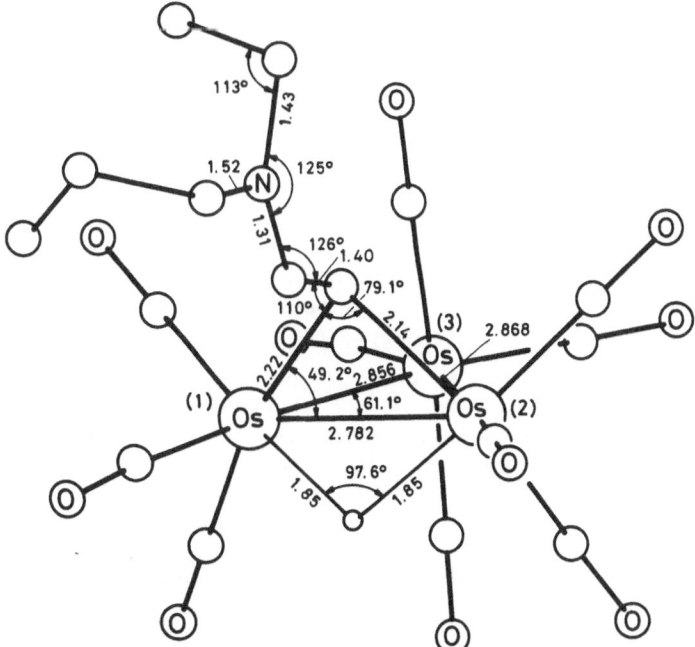

Fig. 3. Molecular structure of (μ-H)Os$_3$(μ_2,η^1-CHCH=N(C$_2$H$_5$)C$_3$H$_7$-n)(CO)$_{10}$ (No. 17) with selected bond distances (in Å) and angles [55].

$(\mu-H)Os_3(\mu_2,\eta^1-CHCH=N(CH_3)C_2H_5)(CO)_{10}$ (No. 14) and $(\mu-H)Os_3(\mu_2,\eta^1-CHCH=N(C_2H_5)-C_3H_7-n)(CO)_{10}$ (No. 17) were both obtained as an equilibrium of two pairs of anti and syn isomers in solution (see Scheme 1). Detailed ^1H NMR investigations (see also pp. 9/12) showed that for No. 14 the anti conformation (Isomers 1 and 3) is favored over the syn conformation (Isomers 2 and 4), but the major isomers in each pair were assigned to the isomers with the ethyl group pointing away from the cluster, indicating that at higher temperature a faster interchange between Isomer 1 and 2 or Isomer 3 and 4 than between Isomer 1 and 3 or Isomer 2 and 4. For No. 17 the major isomer pair was assigned to an anti–syn pair (Isomers 1 and 2) with the n–propyl group pointing away from the cluster [55].

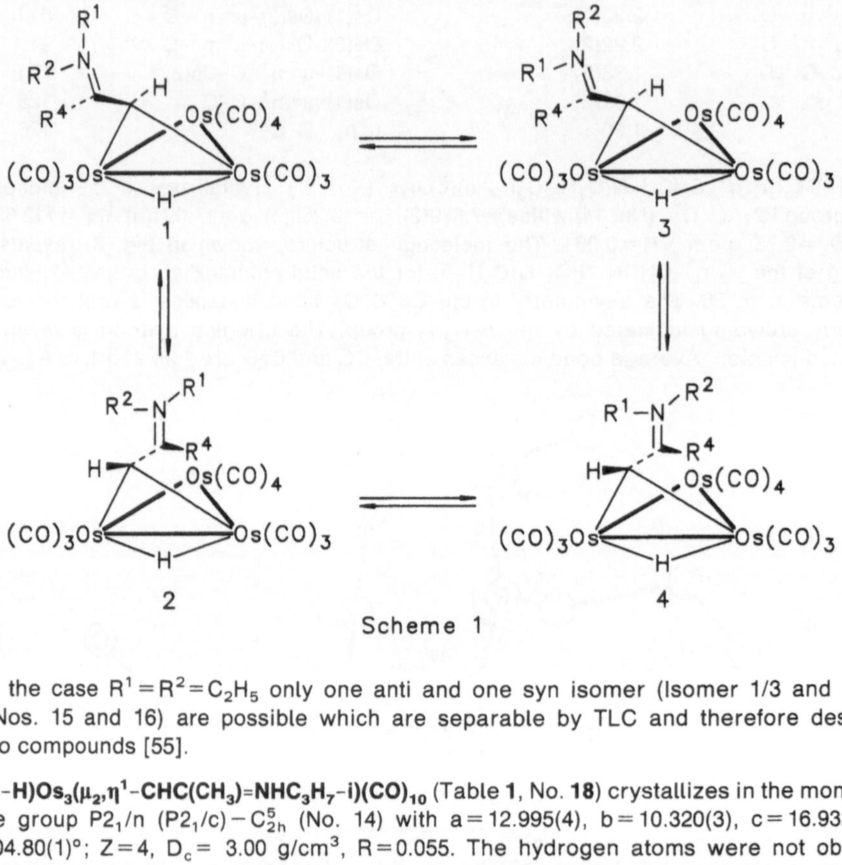

Scheme 1

In the case $R^1 = R^2 = C_2H_5$ only one anti and one syn isomer (Isomer 1/3 and Isomer 2/4, Nos. 15 and 16) are possible which are separable by TLC and therefore described as two compounds [55].

$(\mu-H)Os_3(\mu_2,\eta^1-CHC(CH_3)=NHC_3H_7-i)(CO)_{10}$ (Table 1, No. 18) crystallizes in the monoclinic space group $P2_1/n$ $(P2_1/c) - C_{2h}^5$ (No. 14) with $a = 12.995(4)$, $b = 10.320(3)$, $c = 16.932(3)$ Å, $\beta = 104.80(1)°$; $Z = 4$, $D_c = 3.00$ g/cm^3, $R = 0.055$. The hydrogen atoms were not observed crystallographically but their positions were calculated with the program "Hydro" for NH and μ_2,η^1-CH, and with the program "Hydex" for the bridging hydrides. The structure is quite similar to that of No. 15; selected atomic distances and bond angles are (for numbering, see Fig. 2, p. 24) [51]:

atoms	distance (Å)	atoms	angle (°)
Os(1)–Os(2)	2.870(1)	Os(1)–Os(2)–Os(3)	60.90
Os(1)–Os(3)	2.782(1)		
Os(2)–Os(3)	2.868(1)		

References on pp. 34/6

Table (continued)

atoms	distance (Å)	atoms	angle (°)
Os(1)–μ_2,η^1-C	2.20(2)	Os(1)–Os(3)–μ_2,η^1-C	50.8
Os(3)–μ_2,η^1-C	2.19(2)	Os(3)–Os(1)–μ_2,η^1-C	50.5
μ_2,η^1-C–C	1.45(2)	Os(1)–μ_2,η^1-C–Os(3)	78.7
μ_2,η^1-C–H	1.0	Os(1)–Os(3)–μ-H	41
C–N	1.30(2)	Os(3)–Os(1)–μ-H	42.6
Os–C	1.91	Os(1)–μ-H–Os(3)	97
C–O	1.14	μ_2,η^1-C–Os(1)–μ-H	83

The structural similarity between the anti complexes No. 18 and No. 15 (compare Scheme 1, Isomers 1 and 3; $R^4 = CH_3$ or H; see p. 26) around the μ_2,η^1-C–CR^4 bond was confirmed by the quite similar μ_2,η^1-C–CR^4 and CR^4–N bond lengths, and by the same size of the three-bond coupling between the hydride and the hydrogen at μ_2,η^1-C observed by ^{1}H NMR investigations [51].

(μ–H)Os$_3$(μ_2,η^1-C(CH$_3$)CH=N(CH$_3$)$_2$)(CO)$_{10}$ (Table 1, No. **19**) crystallizes in the monoclinic space group P2$_1$/c – C$^5_{2h}$ (No. 14) with a = 8.842(2), b = 17.275(8), c = 14.211(2) Å, β = 101.04(1)°; Z = 4, D$_c$ = 2.92 g/cm^3, R = 0.051. The structure, shown in **Fig. 4**, is generally similar to that of No. 16 with the μ_2,η^1-C(CH$_3$)CH=N(CH$_3$)$_2$ ligand occupying syn conformation with respect

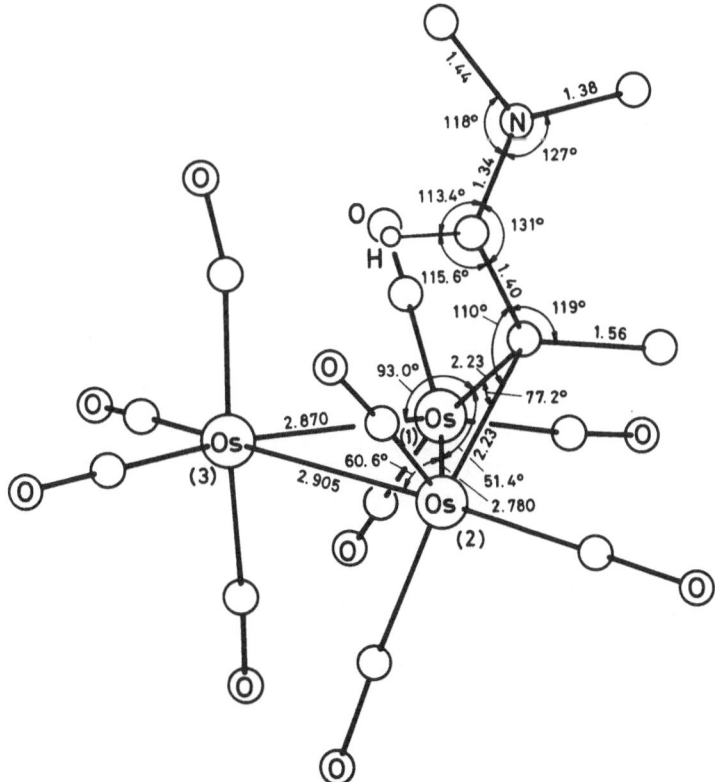

Fig. 4. Molecular structure of (μ–H)Os$_3$(μ_2,η^1-C(CH$_3$)CH=N(CH$_3$)$_2$)(CO)$_{10}$ (No. 19) with selected bond distances (in Å) and angles [47].

to the cluster, instead of the anti orientation of Nos. 15, 17, and 18. Average bond distances of Os–CO and C≡O are 1.91 and 1.14 Å. The hydride atom was not observed crystallographically, but it is thought to bridge the Os(1)–Os(2) bond [47].

$(\mu$–H)Os$_3(\mu_2,\eta^1$-C$_8$H$_5$NCH$_3$)(CO)$_{10}$ (Table 1, No. **21**) crystallizes in the monoclinic space group P2$_1$/c − C$_{2h}^5$ (No. 14) with a = 8.718(4), b = 11.270(6), c = 23.20(1) Å, β = 95.15(4)°; Z = 4, D$_c$ = 2.87 g/cm^3, R = 0.0771. The molecular structure is shown in **Fig. 5**. The dihedral angle between the heterocycles and the Os$_3$ plane is 86.0° [52].

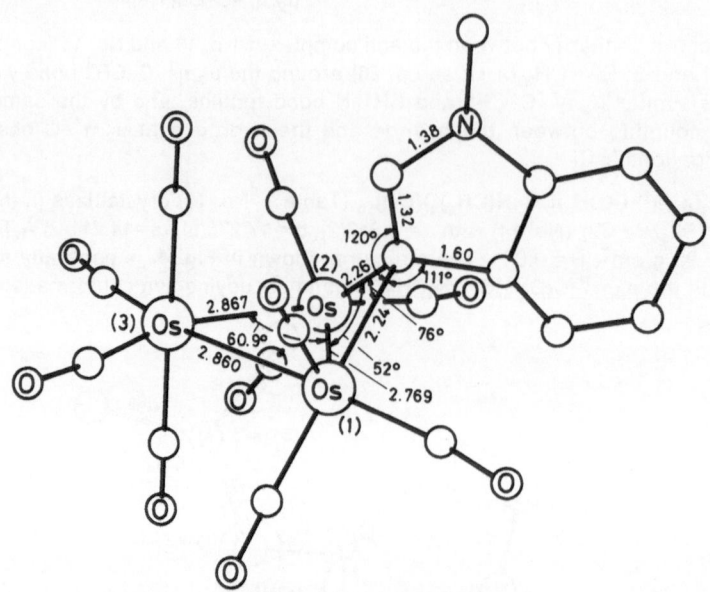

Fig. 5. Molecular structure of $(\mu$–H)Os$_3(\mu_2,\eta^1$-C$_8$H$_5$NCH$_3$)(CO)$_{10}$ (No. 21) with selected bond distances (in Å) and angles [52].

$(\mu$–H)Os$_3(\mu_2,\eta^1$-C$_4$H$_3$NCH$_3$)(CO)$_{10}$(Table **1**, No. **22**) crystallizes in the monoclinic space group P2$_1$/c − C$_{2h}^5$ (No. 14) with a = 12.138(2), b = 12.086(3), c = 13.532(4) Å, β = 96.45(2)°; Z = 4, D$_c$ = 3.14 g/cm^3, R = 0.0665. The molecular structure is shown in **Fig. 6**. The angle between the heterocycles and the Os$_3$ plane is 90.6° [52].

$(\mu$–H)Os$_3(\mu_2,\eta^1$-CHCH$_2$P(CH$_3$)$_2$C$_6$H$_5$)(CO)$_{10}$ (Table 1, No. **26**) crystallizes in the monoclinic space group P2$_1$/n − C$_{2h}^5$ (No. 14) with a = 11.3389(18), b = 16.4265(25), c = 13.8840(20) Å, β = 100.64(1)°; Z = 4, D$_c$ = 2.657 g/cm^3, D$_m$ = 2.66(1) g/cm^3, R = 0.0361. The C$_s$(m) symmetry of the $(\mu$–H)Os$_3$(CO)$_{10}$ portion of the molecule (see **Fig. 7**) is not continued into the bridging μ_2,η^1-CHCH$_2$P(CH$_3$)$_2$C$_6$H$_5$ ligand which exhibits an all-staggered conformation. The shortness of the dibridged Os(1)–Os(2) bond is due to the CHCH$_2$P(CH$_3$)$_2$C$_6$H$_5$ ligand, which counterbalances the normal lengthening effect of a bridging hydride. Average bond distances of Os–CO and C≡O are 1.90 and 1.146 Å; the Os–C distances of the mutually trans CO groups at Os(3), and of those to the bridging ligand trans-standing CO's are found to be slightly longer than the other ones [2, 5].

References on pp. 34/6

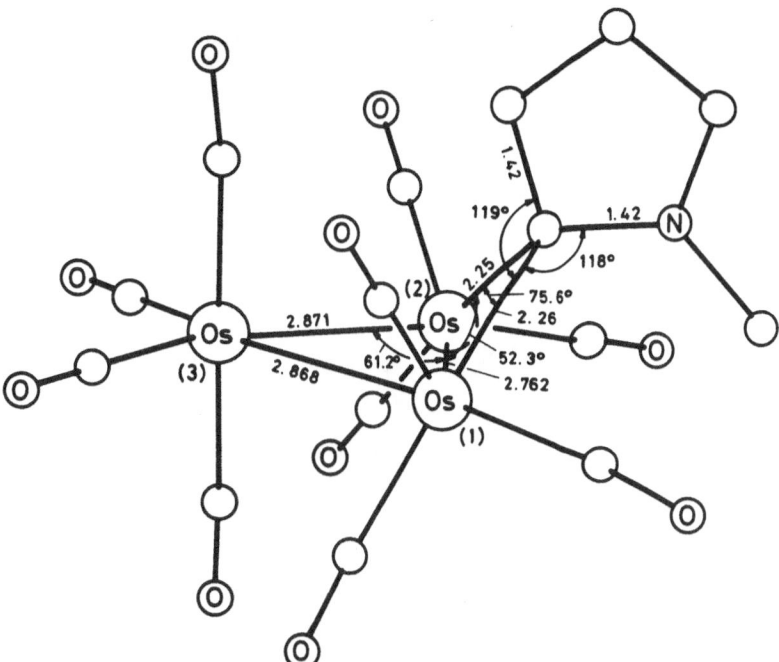

Fig. 6. Molecular structure of $(\mu-H)Os_3(\mu_2,\eta^1-C_4H_3NCH_3)(CO)_{10}$ (No. 22) with selected bond distances (in Å) and angles [52].

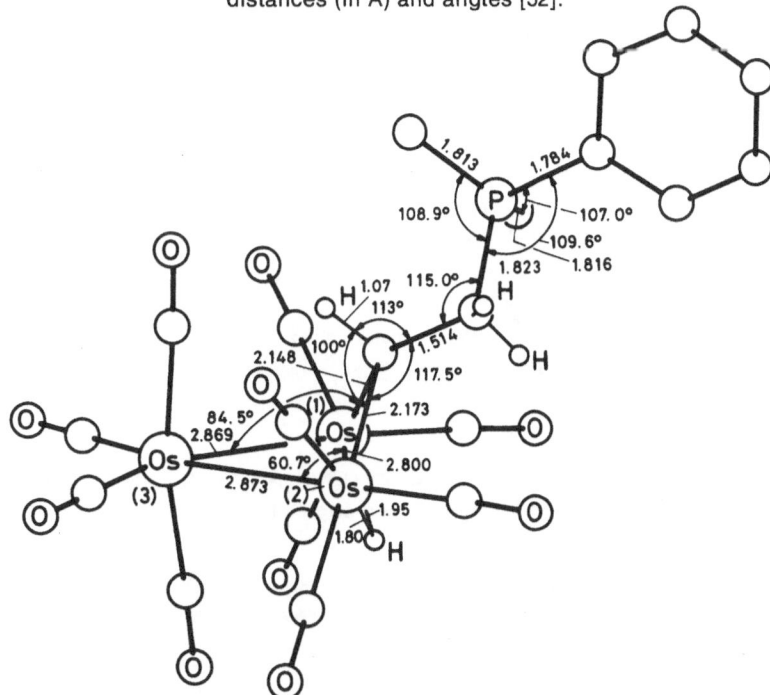

Fig. 7. Molecular structure of $(\mu-H)Os_3(\mu_2,\eta^1-CHCH_2P(CH_3)_2C_6H_5)(CO)_{10}$ (No. 26) with selected bond distances (in Å) and angles [2, 5].

References on pp. 34/6

(μ–H)Os₃(μ₂,η¹-C₅H₃P(C₆H₅)₃)(CO)₁₀ (Table 1, No. **32**) crystallizes in the orthorhombic space group P2₁cn (Pna2₁)−C²v⁹ (No. 33) with a = 9.358(3), b = 15.420(5), c = 23.033(4) Å; Z = 4, D_c = 2.35 g/cm³, R = 0.0403. The molecular structure, see **Fig. 8**, indicates that the C₅H₃P(C₆H₅)₃ ligand is coordinated through one C atom; the other interatomic distances between the C₅H₃ moiety and Os(1) or Os(3) are too long. The C₅H₃ plane is 88.9° to the Os₃ plane and essentially perpendicular. The hydride ligand is not observed crystallographically, but it is assumed to bridge the Os(1)–Os(3) bond [54].

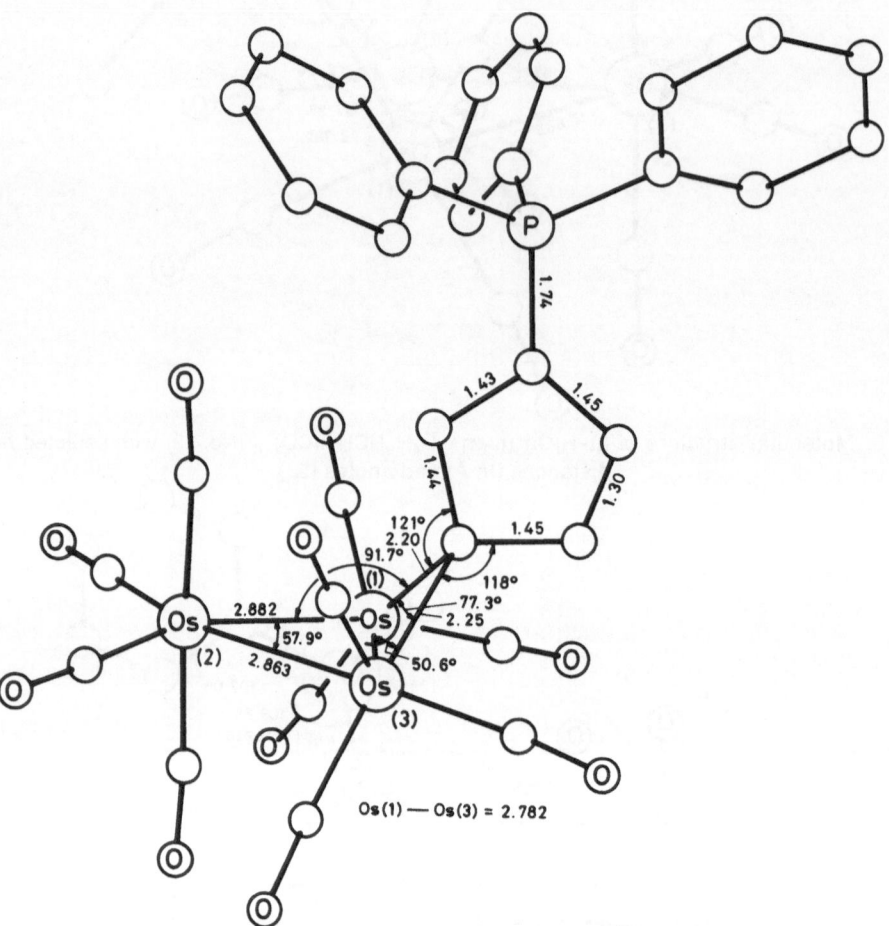

Fig. 8. Molecular structure of (μ–H)Os₃(μ₂,η¹-C₅H₃P(C₆H₅)₃)(CO)₁₀ (No. **32**) with selected bond distances and angles [54].

[Os₃(μ₂,η¹-CH₂)(CO)₁₀(μ-X)][N(P(C₆H₅)₃)₂] (X = Cl, Br, I, ON=O, NCO) (Table **1**, Nos. **34**, **35, 36, 37, 38**). Nos. 36 and 38 were investigated by X-ray analysis.

[Os₃(μ₂,η¹-CH₂)(CO)₁₀(μ-I)][N(P(C₆H₅)₃)₂] (No. 36) crystallizes in the triclinic space group P1̄−C_i¹ (No. 2) with a = 12.978(3), b = 13.131(3), c = 14.074(4) Å, α = 90.21(2)°, β = 96.40(2)°, γ = 90.63(2)°; Z = 2, D_c = 2.13 g/cm³, R = 0.041. The structure is quite similar to that of No. 38 shown in Fig. 9 (p. 31). Differences in the bond distances and angles of

References on pp. 34/6

the dibridged unit are due to the larger atomic radius of iodine compared with nitrogen. Selected atomic distances and angles (for numbering, see Fig. 9) [31, 33]:

atoms	distance (Å)	atoms	angle (°)
Os(1)–Os(2)	2.927(1)	Os(1)–Os(2)–Os(3)	64.1(1)
Os(1)–Os(3)	3.112(1)	Os(1)–Os(3)–Os(2)	57.8(1)
Os(2)–Os(3)	2.934(1)	Os(2)–Os(1)–Os(3)	58.0(1)
Os(1)–I	2.863(1)	Os(2)–Os(1)–I	94.7(1)
Os(3)–I	2.844(1)	Os(2)–Os(3)–I	94.9(1)
Os(1)–μ_2,η^1–C	2.145(14)	Os(2)–Os(1)–μ_2,η^1–C	85.5(4)
Os(3)–μ_2,η^1–C	2.180(13)	Os(2)–Os(3)–μ_2,η^1–C	84.7(4)
		Os(1)–I–Os(3)	66.1(1)
		Os(1)–μ_2,η^1–C–Os(3)	92.0(5)
		μ_2,η^1–C–Os(1)–I	83.2(4)
		μ_2,η^1–C–Os(3)–I	83.1(4)

$[Os_3(\mu_2,\eta^1-CH_2)(CO)_{10}(\mu-NCO)][N(P(C_6H_5)_3)_2]$ (No. 38) crystallizes in the triclinic space group $P\bar{1} - C_i^1$ (No. 2) with a = 12.874(3), b = 13.291(3), c = 13.976(3) Å, α = 89.07(2)°, β = 83.39(2)°, γ = 89.51(2)°; Z = 2, D_c = 2.00 g/cm³, R = 0.05. The molecular structure of the anion is shown in **Fig. 9**. The Os(1)–Os(3) distances (compare also No. 36) are slightly outside the normal Os–Os single bond lengths but the internal angles of the bridging CH_2 ligands (and in particular the bridging iodide in No. 36) clearly indicate some degree of metal–metal interaction [33].

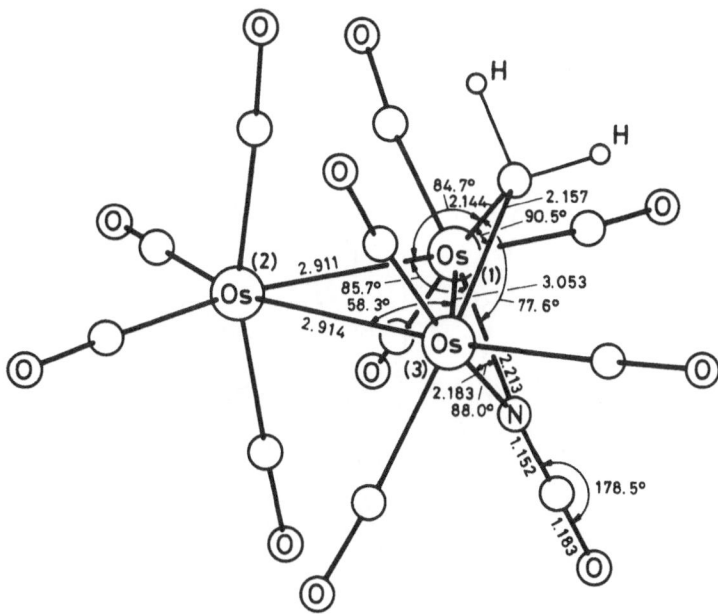

Fig. 9. Molecular structure of the anion of $[Os_3(\mu_2,\eta^1-CH_2)(CO)_{10}(\mu-NCO)][N(P(C_6H_5)_3)_2]$ (No. 38) with selected bond distances (in Å) and angles [33].

Nos. 34 to 38 in CH_2Cl_2 reversibly insert CO yielding the unstable $[Os_3(\mu_2,\eta^2-CH_2C=O)-(CO)_{10}(\mu-X)][N(P(C_6H_5)_3)_2]$ clusters (see Formula IX). The reaction is complete with 1 atm CO at room temperature after 1 min; the rate is apparently limited by the dissolution of CO in CH_2Cl_2. Equilibrium constants $[Os_3(\mu_2,\eta^2-CH_2C=O)(CO)_{10}(\mu-X)]^-/[Os_3(\mu_2,\eta^1-CH_2)-(CO)_{10}(\mu-X)]^-$ [CO] of ca. 3.5 atm^{-1} for No. 36 (between 0.149 and 1.03 atm) and ca. 3.0 atm^{-1} for No. 38 (between 0.153 and 0.589 atm) at 25 °C were evaluated by ^1H NMR and IR spectra and by computer simulation of the IR spectra. Mechanistic studies with ^{13}CO indicated that the ketene carbonyl is derived from the starting cluster demonstrating a coordinatively unsaturated ketene intermediate [31, 32]; see also [43]. Such a coordinatively unsaturated ketene intermediate was also believed to be involved in the formation of Nos. 34 to 38 from No. 43 (see Scheme 2, p. 33) [33]. The $\mu-X$ ligands accelerate the CO insertion step by a factor of at least 10^2 in contrast to $\mu-CO$ in No. 43, see p. 33 [31, 32].

IX

Nos. 34 to 38 reacted in CH_2Cl_2 at room temperature with $HBF_4 \cdot O(C_2H_5)_2$ or CF_3SO_3H in CH_2Cl_2 to yield $Os_3(\eta^1-CH_3)(CO)_{10}(\mu-X)$ ("Organoosmium Compounds" B 5, Section 3.1.2.2, in preparation). The reaction was studied in detail only for No. 36 (X=I) [38]. Reaction of Nos. 36 and 38 with CNC_4H_9-t in CH_2Cl_2 at −78 °C, or in $CDCl_3$ at −40 °C gave $[Os_3(\mu_2,\eta^2-CH_2C=O)(CO)_{10}(\mu-X)CNC_4H_9-t][N(P(C_6H_5)_3)_2]$ (X=I or NCO) [45].

$Os_3(\mu_2,\eta^1-CH_2)(\mu-CO)(CO)_{10}$ (Table 1, No. **43**) crystallizes in the monoclinic space group $P2_1/n - C_{2h}^5$ (No. 14) with a = 8.5620(14), b = 11.2352(22), c = 9.1486(16) Å, β = 96.274(13)°; Z = 2, $D_c = 3.39$ g/cm³, R = 0.079. The structure is 4-fold disordered caused by an inversion disorder coupled with a mirror disorder about the Os_3 triangle, which scrambles μ_2,η^1-CH_2 and $\mu-CO$. **Fig. 10** shows a composite image of two mirror-related molecules with the disordered μ-methylene and μ-carbonyl carbon atoms C′ and C″ [18].

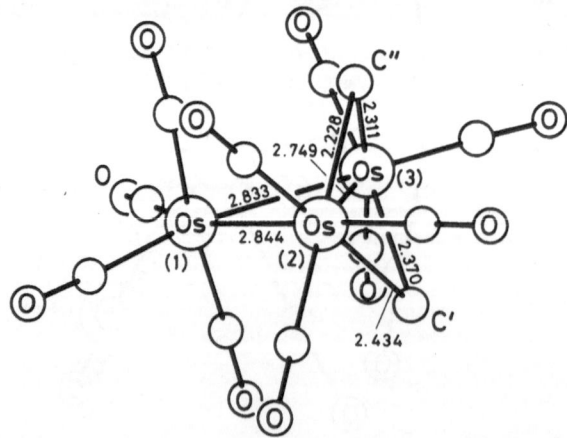

Fig. 10. Structure of $Os_3(\mu_2,\eta^1-CH_2)(\mu-CO)(CO)_{10}$ (No. 43) with selected bond distances (in Å) [18].

References on pp. 34/6

No. 43 reacted easily with various nucleophiles, which is possibly due to an equilibrium between the coordinatively saturated title compound and a highly reactive unsaturated ketene intermediate having an open coordination site at one Os atom. Most reactions of No. 43 are summarized in Scheme 2.

Reaction with CO (1 atm) in CH_2Cl_2 at room temperature for 2 to 8 h yielded $Os_3(\mu_2,\eta^2-CH_2C{=}O)(CO)_{12}$. [13]C labeling experiments ruled out the incorporation of an initial $Os_3(\mu_2,\eta^1-CH_2)(\mu-CO)(CO)_{10}$ carbonyl ligand in the μ_2,η^1-CH_2 bond; a possible mechanism for this process is discussed in detail [24, 28]. Stirring of No. 43 at ca. 22 °C under a CO atmosphere with CH_3OH in $CDCl_3$, with H_2O in THF [24, 28], or with $n-C_4H_9NHR$ (R = H, $n-C_4H_9$) [45] in CH_2Cl_2 for several hours always resulted in $Os_3(CO)_{12}$, in addition to methyl acetate, acetic acid, or $CH_3CONRC_4H_9-n$, respectively [24, 28, 45]. These observations led to the assumption that a ketene complex with an open coordination site must be involved as intermediate in this reaction. Such an insertion of CO into a metal–methylene bond is suggested to play an important role in the high-yield formation of C_2H_5OH from CO/H_2 over $Rh/ZrO_2/SiO_2$ and $Rh/TiO_2/SiO_2$ catalysts [28].

Reaction with CD_3CN/CH_3OH (1:6) in $CDCl_3$ at 22 °C yielded methyl acetate and $Os_3(CO)_{11}NCCD_3$, probably via $Os_3(CH_2CO)(CO)_{10}NCCD_3$ [28, 45]; see also [24].

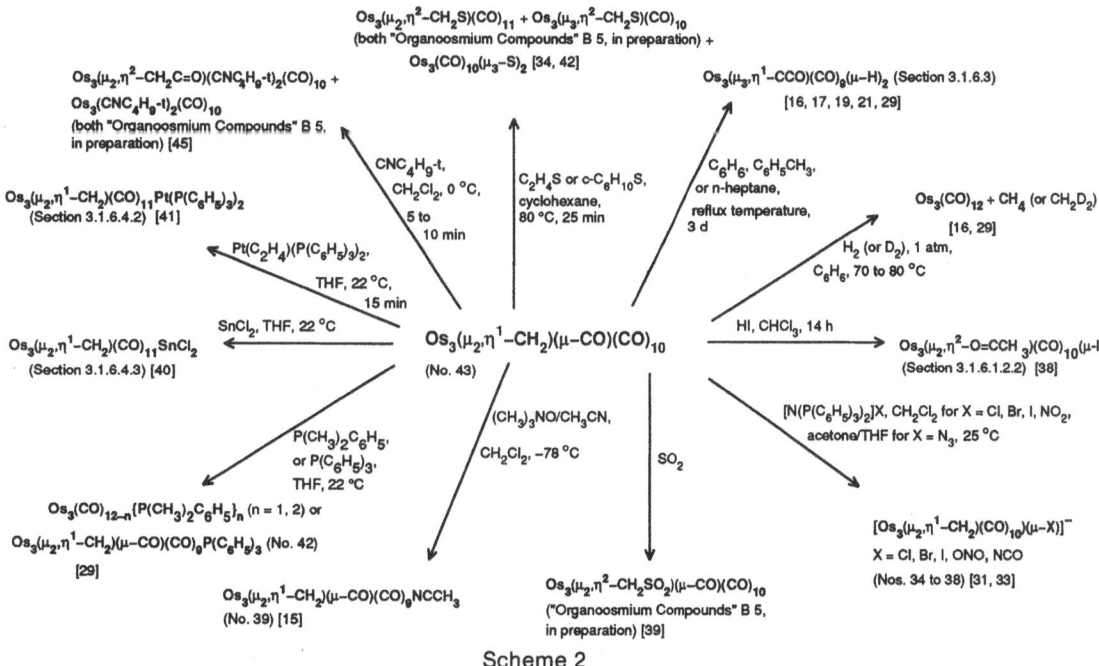

Scheme 2

$Os_3(\mu_2,\eta^1-CHSi(CH_3)_3)(\mu-CO)(CO)_{10}$ (Table **1**, No. 45) crystallizes in the monoclinic space group $P2_1/n - C_{2h}^5$ (No. 14) with a = 15.463(9), b = 9.352(3), c = 16.757(7) Å, β = 107.38(4)°; Z = 4, D_c = 2.77 g/cm³. The complex, see **Fig. 11**, has approximate C_s symmetry, both throughout the $Os_3(CO)_{10}$ core and also within the bridging ligands. Average bond distances of Os–CO and C≡O are 1.920 and 1.149 Å [15, 22].

References on pp. 34/6

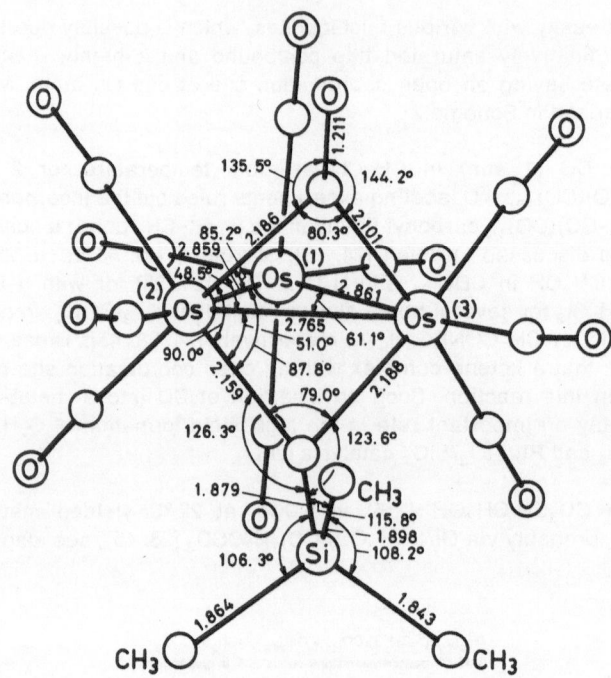

Fig. 11. Molecular structure of Os$_3$(μ_2,η^1-CHSi(CH$_3$)$_3$)(μ-CO)(CO)$_{10}$ (No. 45) with selected bond distances (in Å) and angles [23].

References:

[1] Bryan, E. G.; Jackson, W. G.; Johnson, B. F. G.; Kelland, J. W.; Lewis, J.; Schorpp, K. T. (J. Organomet. Chem. **108** [1976] 385/91).

[2] Churchill, M. R.; DeBoer, B. G.; Shapley, J. R.; Keister, J. B. (J. Am. Chem. Soc. **98** [1976] 2357/8).

[3] Deeming, A. J.; Hasso, S. (J. Organomet. Chem. **112** [1976] C 39/C 42).

[4] Calvert, R. B.; Shapley, J. R. (J. Am. Chem. Soc. **99** [1977] 5225/6).

[5] Churchill, M. R.; DeBoer, B. G. (Inorg. Chem. **16** [1977] 1141/6).

[6] Calvert, R. B.; Shapley, J. R.; Schultz, A. J.; Williams, J. M.; Suib, S. L.; Stucky, G. D. (J. Am. Chem. Soc. **100** [1978] 6240/1).

[7] Calvert, R. B.; Shapley, J. R. (J. Am. Chem. Soc. **100** [1978] 7726/7).

[8] Shapley, J. R.; Tachikawa, M.; Churchill, M. R.; Lashewycz, R. A. (J. Organomet. Chem. **162** [1978] C 39/C 42).

[9] Churchill, M. R.; Lashewycz, R. A. (Inorg. Chem. **18** [1979] 848/52).

[10] Jordan, R. F.; Norton, J. R. (J. Am. Chem. Soc. **101** [1979] 4853/8).

[11] Schultz, A. J.; Williams, J. M.; Calvert, R. B.; Shapley, J. R.; Stucky, G. D. (Inorg. Chem. **18** [1979] 319/23).

[12] Orpen, A. G. (J. Chem. Soc. Dalton Trans. **1980** 2509/16).

[13] Busetto, L.; Green, M.; Howard, J. A. K.; Hessner, B.; Jeffery, J. C.; Mills, R. M.; Stone, F. G. A.; Woodward, P. (J. Chem. Soc. Chem. Commun. **1981** 1101/3).

[14] Green, M.; Hankey, D. R.; Murray, M.; Orpen, A. G.; Stone, F. G. A. (J. Chem. Soc. Chem. Commun. **1981** 689/91).

[15] Shapley, J. R.; Sievert, A. C.; Churchill, M. R.; Wasserman, H. J. (J. Am. Chem. Soc. **103** [1981] 6975/7).

[16] Steinmetz, G. R.; Geoffroy, G. L. (J. Am. Chem. Soc. **103** [1981] 1278/9).

[17] Arce, A. J.; Deeming, A. J. (J. Chem. Soc. Chem. Commun. **1982** 364/5).

[18] Churchill, M. R.; Wasserman, H. J. (Inorg. Chem. **21** [1982] 825/7).

[19] Johnson, B. F. G.; Lewis, J.; Raithby, P. R.; Sankey, S. W. (J. Organomet. Chem. **231** [1982] C 65/C 67).

[20] Oxton, I. A.; Powell, D. B.; Sheppard, N.; Burgess, K.; Johnson, B. F. G. (J. Chem. Soc. Chem. Commun. **1982** 719/21).

[21] Sievert, A. C.; Strickland, D. S.; Shapley, J. R.; Steinmetz, G. R.; Geoffroy, G. L. (Organometallics **1** [1982] 214/5).

[22] Busetto, L.; Green, M.; Hessner, B.; Howard, J. A. K.; Jeffery, J. C.; Stone, F. G. A. (J. Chem. Soc. Dalton Trans. **1983** 519/25).

[23] Churchill, M. R.; Wasserman, H. J. (J. Organomet. Chem. **248** [1983] 365/73).

[24] Morrison, E. D.; Steinmetz, G. R.; Geoffroy, G. L.; Fultz, W. C.; Rheingold, A. L. (J. Am. Chem. Soc. **105** [1983] 4104/5).

[25] Shapley, J. R.; Cree-Uchiyama, M. E.; St. George, G. M.; Churchill, M. R.; Bueno, C. (J. Am. Chem. Soc. **105** [1983] 140/2).

[26] Deeming, A. J.; Manning, P. J. (J. Organomet. Chem. **265** [1984] 87/94).

[27] Green, M.; Orpen, A. G.; Schaverien, C. J. (J. Chem. Soc. Chem. Commun. **1984** 37/90).

[28] Morrison, E. D.; Steinmetz, G. R.; Geoffroy, G. L.; Fultz, W. C.; Rheingold, A. L. (J. Am. Chem. Soc. **106** [1984] 4783/9).

[29] Steinmetz, G. R.; Morrison, E. D.; Geoffroy, G. L. (J. Am. Chem. Soc. **106** [1984] 2559/64).

[30] Farrugia, L. J.; Green, M.; Hankey, D. R.; Murray, M; Orpen, A. G.; Stone, F. G. A. (J. Chem. Soc. Dalton Trans. **1985** 177/90).

[31] Morrison, E. D.; Geoffroy, G. L.; Rheingold, A. L. (J. Am. Chem. Soc. **107** [1985] 254/5).

[32] Morrison, E. D.; Geoffroy, G. L. (J. Am. Chem. Soc. **107** [1985] 3541/5).

[33] Morrison, E. D.; Geoffroy, G. L.; Rheingold, A. L.; Fultz, W. C. (Organometallics **4** [1985] 1413/8).

[34] Adams, R. D.; Babin, J. E.; Tasi, M. (Organometallics **5** [1986] 1920/2).

[35] Bower, D. K.; Keister, J. B. (J. Organomet. Chem. **312** [1986] C 33/C 36).

[36] Cree-Uchiyama, M.; Shapley, J. R.; St. George, G. M. (J. Am. Chem. Soc. **108** [1986] 1316/7).

[37] Keiter, R. L.; Strickland, D. S.; Wilson, S. R.; Shapley, J. R. (J. Am. Chem. Soc. **108** [1986] 3846/7).

[38] Morrison, E. D.; Bassner, S. L.; Geoffroy, G. L. (Organometallics **5** [1986] 408/11).

[39] Morrison, E. D. (Diss. Pennsylvania State Univ., Univ. Park, Pa. 1985, pp. 1/201; Diss. Abstr. Int. B **46** [1986] 2301/2).

[40] Viswanathan, N.; Morrison, E. D.; Geoffroy, G. L.; Geib, S. J.; Rheingold, A. L. (Inorg. Chem. **25** [1986] 3100/2).

[41] Williams, G. D.; Lieszkovszky, M.-C.; Mirkin, C. A.; Geoffroy, G. L.; Rheingold, A. L. (Organometallics **5** [1986] 2228/33).

[42] Adams, R. D.; Babin, J. E.; Tasi, M. (Organometallics **6** [1987] 1717/27).

[43] Bassner, S. L.; Morrison, E. D.; Geoffroy, G. L. (Organometallics **6** [1987] 2207/14).

[44] Adams, R. D.; Tanner, J. T. (Organometallics **7** [1988] 2241/3).

[45] Bassner, S. L.; Geoffroy, G. L.; Rheingold, A. L. (Polyhedron **7** [1988] 791/805).

[46] Green, M.; Orpen, A. G.; Schaverien, C. J. (J. Chem. Soc. Dalton Trans. **1989** 1333/40).

[47] Adams, R. D.; Chen, G.; Tanner, J. T. (Organometallics **9** [1990] 1530/8).

[48] Bower, D. K.; Keister, J. B. (Organometallics **9** [1990] 2321/7).

[49] Fehlner, T. P. (Polyhedron **9** [1990] 1955/63).

[50] Kane-Maguire, L. A. P.; Ghazy, T. (Appl. Organomet. Chem. **4** [1990] 475/80).

[51] Rosenberg, E.; Kabir, S. E.; Hardcastle, K. I.; Day, M.; Wolf, E. (Organometallics **9** [1990] 2214/7).

[52] Arce, A. J.; Manzur, J.; Marquez, M.; De Sanctis, Y.; Deeming, A. J. (J. Organomet. Chem. **412** [1991] 177/93).

[53] Day, M. W.; Hajela, S.; Kabir, S. E.; Irving, M.; McPhillips, T.; Wolf, E.; Hardcastle, K. I.; Rosenberg, E.; Milone, L.; Gogetto, R.; Osella, D. (Organometallics **10** [1991] 2743/51).

[54] Deeming, A. J.; Powell, N. I.; Whittaker, C. (J. Chem. Soc. Dalton Trans. **1991** 1875/80).

[55] Kabir, S. E.; Day, M.; Irving, M.; McPhillips, T.; Minassian, H.; Rosenberg, E.; Hardcastle, K. I. (Organometallics **10** [1991] 3997/4004).

[56] Adams, R. D.; Tanner, J. T. (Appl. Organomet. Chem. **6** [1992] 449/62).

3.1.6.2.2 Compounds with Other μ_2,η^1-Bridging ^{1}L Ligands

The compounds dealt with in this section generally consist of an $Os_3(\mu_2,\eta^1-{}^1L)(CO)_nD$ skeleton bearing additional μ-hydrido and/or other bridging units. Only Nos. 46 to 50 have two $\mu_2,\eta^1-{}^1L$ ligands.

The structure of most of the compounds can be illustrated by the general Formulas I to V; the remaining complexes in Table 2 are presented by individual figures. Generally, structures that are not confirmed by X-ray analysis, can only be considered as a proposal based on spectroscopic and analytical data. In particular, the positions of bridging hydrides which mostly can not be observed crystallographically are often discussed controversially (see [21, 29, 30]).

V

Most of the compounds can be considered as zwitterions with the positive charge located at the μ_2,η^1-bridging ligand, and the negative charge at the Os_3 skeleton, based on the nucleophilic behavior of the Os atoms toward protonation [31].

Differences in the nature of stabilization of the ^1L ligand have been observed between the alkyl- and heteroatom substituted alkylidene complexes. In alkyl-substituted alkylidene compounds stabilization is achieved by a donor interaction between the formally saturated 18e-$Os(CO)_4$ group and the carbene C. In amino-substituted alkylidene complexes (see Formulas I, III, IV, and related compounds) the electron deficiency at the carbene carbon is partially compensated by π donation from the nitrogen into the vacant unhybridized p orbital. The substantial double-bond character is confirmed by the short C-N bond lengths based on X-ray analysis as well as by the restricted rotation about the C-NRR' bond, as evidenced by the diastereotopic methylene protons in R, or R', giving rise to a sharp AB pattern in ^1H NMR spectra (for example, see No. 15, p. 44) [12, 15, 37, 45].

In alkoxy-substituted complexes (see Formula II) the μ_2,η^1-COR moiety is best described as a bridging carbyne ligand with some remaining double-bond character in μ-COR. This was indicated by the planar nature of the $Os_2(\mu_2,\eta^1$-COR) unit, the restricted rotation about the μ_2,η^1-C-O bond leading to isomeric product mixtures, and by X-ray analysis revealing a C-O bond length of 1.33 Å (for No. 22, p. 62) which is between the values expected for single (1.40 Å) and double (1.20 Å) bonds [40]. For the possible flipping of the alkyl group a mechanism was discussed suggesting an inversion at the oxygen via a linear transition state [5, 12, 15, 40]. For $(\mu$-H)$Os_3(\mu_2,\eta^1$-COCH$_3$)(CO)$_9$(CNC$_4$H$_9$-t), bearing an additional terminal isocyanide ligand, see "Organoosmium Compounds" B 5, Section 3.1.4.3, in preparation.

In the case of $(\mu$-H)$Os_3(\mu_2,\eta^1$-CCHC(CH$_3$)$_2$)(CO)$_{10}$, No. 1, a dimethylvinyl-substituted alkylidene complex, the lack of electron donation from the $Os(CO)_4$ moiety is partially compensated for by π interactions of the carbene C with the vinyl unit [37].

The bridging phenyl group in compounds corresponding to Formula V is best described as a three-center two-electron ligand. The large differences in the Os-$(\mu$-C)-Os bond lengths may be associated with steric effects [1].

References on pp. 72/3

Some compounds listed in Table 2 were prepared by the following methods:

Method I: Preparation of $(\mu\text{-H})_2Os_3(\mu_2,\eta^1\text{-C=NR}_2)(CO)_8(\mu\text{-SR'})$ (Formula III; R = alkyl, R' = alkyl, aryl) from compounds with terminal carbene ligands ("Organoosmium Compounds" B 5, Section 3.1.5.1, in preparation).

a. Decarbonylation of $(\mu\text{-H})Os_3(=CHNR_2)(CO)_9(\mu\text{-SR'})$ by UV irradiation (λ ca. 313 to 578 nm) in cyclohexane at room temperature for 30 to 45 min, followed by separation by TLC with CH_2Cl_2/hexane (30:70 or 40:60) as eluant [21, 26, 29].

b. From $(\mu\text{-H})_2Os_3(=CHNR_2)(CO)_8(\mu_3,\eta^2\text{-SR''})$ (R'' = C_6H_4, $C_6H_3CH_3$-4) by refluxing in hexane for 4 h [29] or in $CHCl_3$ for 2 h [28, 34]; work-up by TLC as before [28, 29, 34].

Method II: Starting from compounds with terminal isocyanide ligands ("Organoosmium Compounds" B 5, Section 3.1.4.2, in preparation).

a. $(\mu\text{-H})Os_3(\mu_2,\eta^1\text{-C=NHR})(CO)_{10}$ (Formula I; R = alkyl, aryl) were prepared from $(\mu\text{-H})HOs_3(CNR)(CO)_{10}$ [6, 41]. Catalytic conversion in the presence of $N(C_2H_5)_3$ in $CHCl_3$ or THF at room temperature for 15 to 20 min, followed by addition of pentane and precipitation at $-78\,^\circ$C, gave yields of 95% [6]. Thermolysis in C_6H_6 at 80 °C under vacuum yielded 15 to 30%, along with $(\mu\text{-H})Os_3(\mu_2,\eta^2\text{-CH=NR})(CO)_{10}$ and/or $(\mu\text{-H})_2Os_3(CNR)(CO)_9$; the products were not isolated [41].

b. $Os_3(\mu_2,\eta^1\text{-C=NHR})(\mu_2,\eta^2\text{-O=CNHR'})(CO)_9D$ (Formula IV; R = n-C_3H_7, i-C_3H_7, $CH_2C_6H_5$, C_6H_5; R' = n-C_3H_7, i-C_3H_7, $CH_2C_6H_5$; D = CO, NH_2R') were prepared from $Os_3(CNR)(CO)_{11}$ with NH_2R' at room temperature. For D = CO the reactions are complete within 3 to 30 min, whereas for D = NH_2R' a reaction time of 6 to 18 h, and an excess of the amine were necessary. Work-up was performed either by TLC with hexane as eluant (D = CO) or by recrystallization from CH_2Cl_2/hexane (D = NH_2R') [45, 47].

The products with D = NH_2R' formed via complexes with D = CO [45, 47].

Method III: $(\mu\text{-H})Os_3(\mu_2,\eta^1\text{-COR})(CO)_{10}$ (Formula II; R = D, CH_3, C_2H_5) were prepared from $[(\mu\text{-H})Os_3(\mu\text{-CO})(CO)_{10}]M$ (M = $N(C_2H_5)_4$, $N(P(C_6H_5)_3)_2$) and an excess of RSO_3F without a solvent [5, 8] or in CH_2Cl_2 at room temperature in 2 to 4 h [12]. The products were either purified by sublimation at 353 K/0.1 Torr [5], or by TLC with cyclohexane as eluant [12].

Attempted preparation of $(\mu\text{-H})Os_3(\mu_2,\eta^1\text{-COH})(CO)_{10}$ (see No. 19) from $[(\mu\text{-H})Os_3(\mu\text{-CO})(CO)_{10}][N(P(C_6H_5)_3)_2]$ and HSO_3F in CH_2Cl_2 gave only $(\mu\text{-H})HOs_3(CO)_{11}$, but the monodeuterated $(\mu\text{-H})Os_3(\mu_2,\eta^1\text{-COD})(CO)_{10}$ could be obtained with DSO_3F in CD_2Cl_2 at $-80\,^\circ$C due to a very large isotopic effect in the ligand-to-metal H transfer [36].

The preparations can also be performed with $[(\mu\text{-H})Os_3(\mu\text{-CO})(CO)_{10}]$-$[NH(C_2H_5)_3]$ in situ generated from $Os_3(CO)_{12}$ and $N(C_2H_5)_3$ [12].

Method IV: $Os_3(\mu_2,\eta^1\text{-}C_6H_4R\text{-}4)(\mu_3,\eta^2\text{-P}(C_6H_4R\text{-}4)C_6H_3R\text{-}4)(CO)_8(\mu_2,\eta^1\text{-P}(C_6H_4R\text{-}4)_2)$ (Formula V; R = H, CH_3) were prepared from $Os_3(CO)_{12}$ and $P(C_6H_4R\text{-}4)_3$ (molar ratio 1:2) in refluxing xylene for 1h. The reaction mixtures of nine products for R = H, and three products for R = CH_3 were separated by fractional crystallization and column chromatography on alumina for several times [1, 2].

Table 2

Compounds with Other μ_2,η^1-Bridging ^{1}L Ligands.

An asterisk preceding the compound number indicates further information at the end of the table, pp. 57/72.

Explanations, abbreviations, and units on p. X

No. compound	method of preparation (yield in %) properties and remarks

compounds of the type $(\mu\text{-H})Os_3(\mu_2,\eta^1\text{-CCRR'})(CO)_{10}$

1

from $(\mu\text{-H})_2Os_3(CO)_{10}$ and $3,3\text{-}(CH_3)_2C_3H_2\text{-c}$ in hexane at ca. 25 °C in 16 h; separated by repeated recrystallization (ca. 10%; along with 80% $(\mu\text{-H})Os_3(\mu_3,\eta^1\text{-CCH}_2CH(CH_3)_2)(CO)_{10}$, Section 3.1.6.3); starting with $(\mu\text{-D})_2Os_3(CO)_{10}$ no deuterated product was formed suggesting a reaction via $[Os_3(CO)_{10}]$, a mechanism was discussed [14, 37]

from $(\mu\text{-H})_2Os_3(\mu_2,\eta^1\text{-CH}_2)(CO)_{10}$ (Section 3.1.6.2.1) with $3,3\text{-}(CH_3)_2C_3H_2\text{-c}$ in refluxing hexane [14, 37]

yellow crystals from hexane/ether at -20 °C [14, 37]

^{1}H NMR (CDCl$_3$): -16.26 (s, μ-H), 2.05, 2.13 (2 s's, both CH$_3$), 8.42 (s, CH) [14, 37]

^{13}C NMR (CD$_2$Cl$_2$): 20.6, 28.8 (both CH$_3$), 148.9 (C-3), 152.5 (d, CH, ^{1}J(H,C) = 150), 169.8 (s, 2 CO), 174.1 (d, 2 CO; ^{2}J(H,C) = 9.8), 175.7 (s, 4 CO), 176.6 (s, 2 CO), 332.3 (μ_2,η^1-C) [14, 37]; the spectrum indicated C$_s$ symmetry for the complex; the electron deficiency at C-1 is compensated by π interaction with the vinyl moiety, although there is free rotation about the (C-1)-(C-2) bond on the low-temperature NMR time scale [37]

IR (hexane): 1989, 2015, 2021, 2058, 2105 (all CO) [14, 37]

2

see also "Organorhenium Compounds", Part 2, 1989, p. 352

as a by-product in the preparation of the acyl complex $Os_3\{\mu_3,\eta^2\text{-C=C(C}_6H_5)\text{-}(C=C(C_6H_5)Re(CO)_4)C=O\}(\mu\text{-CO})(CO)_8$ from $(\mu\text{-H})Os_3(\mu_3,\eta^1\text{-CC}_6H_5)(CO)_{10}$ (Section 3.1.6.3) and $Re(C\equiv CC_6H_5)(CO)_5$ in CH$_2$Cl$_2$ for 20 d, followed TLC with hexane/benzene, 2:1 as eluant (impure product) [32]

^{1}H NMR (CDCl$_3$; measured in the presence of the μ-acyl complex): -15.76 (μ-H) [23, 32]

References on pp. 72/3

Table 2 (continued)

No. compound	method of preparation (yield in %) properties and remarks

2 (continued)

intermediate in the formation of the μ-acyl complex indicated by the facile conversion by loss of a CO group [18, 32]

earlier formulated as zwitterionic $[(\mu-H)Os_3(\mu_2,\eta^1-C=C(C_6H_5)C\equiv CC_6H_5)-(CO)_{10}][Re(CO)_5]$ [18], but the allenyl-substituted carbene structure seems to be more propable, based on a comparison with the spectroscopic data and the X-ray analysis of No. 3 [32]

*3 $(\mu-H)Os_3(\mu_2,\eta^1-CC(C_6H_5)=C=C(C_6H_5)Re(CO)_4P(CH_3)_2C_6H_5)(CO)_{10}$

see also "Organorhenium Compounds", Part 2, 1989, p. 328

by treatment of the μ-acyl complex $Os_3\{\mu_3,\eta^2-C=C(C_6H_5)-(C=C(C_6H_5)Re(CO)_4)C=O\}(\mu-CO)(CO)_8$ with $P(CH_3)_2C_6H_5$ (ca. 1:1) in CH_2Cl_2 for 2 d, followed by TLC with hexane/CH_2Cl_2 (2:1) as eluant (85%) [32]

bright orange crystalline solid [32]

^1H NMR (CD_2Cl_2): −15.94 (s, μ-H; $^1J(^{187}Os,$ μ-H) = 31.4), 1.48, 1.49 (both CH_3; $^2J(^{31}P,H)$ = 9.3 or 8.3), 6.9 to 7.6 (m, C_6H_5) [32]; ($CDCl_3$): −15.94 (μ-H) [18]

^{13}C {^1H} NMR (no medium given; −40 °C): 170.8, 171.5 (br s's, each 1 CO), 172.2, 173.5 (d's, each 1 CO; $^2J(C,H)$ = 10.1, 11.3), 174.9, 176.1 (br s's, each 1 CO of $Os(CO)_4$), 178.4, 178.6 (s's, each 1 CO), 178.7, 180.1 (s's, each 1 CO of $Os(CO)_4$); the signals of the $Os(CO)_4$ moiety are temperature-dependent [23]; see also [18]

^{31}P NMR ($CDCl_3$): −30.51 [32]

4

from $Os_3(\mu_3,\eta^2-CH=CH)(\mu-CO)(CO)_9$ and an excess of $P(CH_3)_2C_6H_5$ in $CDCl_3$ at room temperature for 16 h; purification by TLC with $CHCl_3$/pentane, 1:2 (60%) [11]; see also [3]

yellow crystals [11]

^1H NMR ($CDCl_3$, 27 °C): −16.04 (dd, μ-H; J(H,P) = 2.5), 1.95, 2.08 (2 d's, both CH_3; J(H,P) = 13.5), 6.60 (dd, CH; J(μ-H,H) = 1.3, J(H,P) = 40.7), 7.6 (m, C_6H_5) [3, 11]; the diastereotopic methyl groups give sharp signals indicating the absence of rotation about the C=C bond [11]

Table 2 (continued)

No. compound	method of preparation (yield in %) properties and remarks

IR (cyclohexane): 1947, 1966, 1975, 1983, 1992, 2007, 2037, 2044, 2090 (all CO) [3, 11]

*5

from $(\mu\text{-H})Os_3(\mu_2,\eta^2\text{-C=CC}_6H_5)(CO)_{10}$ as described for No. 4 (91%) [11]; see also [3]

bright yellow crystals [11]

^{1}H NMR (CDCl$_3$, 27 °C): -16.35 (d, μ-H; J(μ-H,P)=3.2), 1.76, 2.49 (2 d's, CH$_3$; J(H,P)=12.4), 7.3, 7.8 (2 m's, both C$_6$H$_5$) [3, 11]; the diastereotopic methyl groups give sharp signals indicating the absence of rotation about the C=C bond [11]

IR (cyclohexane): 1949, 1967, 1973, 1979, 1986, 1990, 2008, 2037, 2046, 2089 (all CO) [3, 11]

compounds of the type $(\mu\text{-H})Os_3(\mu_2,\eta^1\text{-C=NRR'})(CO)_9D$ (Formula I)

6 $(\mu\text{-H})Os_3(\mu_2,\eta^1\text{-C=NH}_2)(CO)_{10}$

from [$(\mu\text{-H})HOs_3(CO)_{10}CN][N(P(C_6H_5)_3)_2]$ (pregenerated in situ from $(\mu\text{-H})_2Os_3(CO)_{10}$ and [N(P(C$_6$H$_5$)$_3$)$_2$]CN at -78 °C) and CF$_3$CO$_2$H in CH$_2$Cl$_2$ at 25 °C for 20 h, followed by addition of NaHCO$_3$, stirring for 4 h and purification by TLC with pentane/CH$_2$Cl$_2$ as eluant (30%) [31]

by treatment of a CDCl$_3$ solution of [$(\mu\text{-H})_2Os_3(\mu_2,\eta^1\text{-C=NH}_2)(CO)_{10}$]$^+$ (No. 18) with a few drops of H$_2$O in CH$_2$Cl$_2$; work-up by TLC [31]

yellow crystals from CH$_2$Cl$_2$/hexane [31]

^{1}H NMR (CDCl$_3$): -16.52 (s, μ-H), 8.65 (s, NH$_2$); using a 90% ^{13}C enriched sample the μ-H signal became a doublet, whereas with a 98.8% ^{15}N enriched sample the NH split into a doublet [31]

^{13}C NMR (CDCl$_3$): 297.4 (dt, μ-C; J(μ-C,H)= 2.6, J(μ-C,μ-H)=5.8) [31]

^{15}N NMR (CDCl$_3$; CD$_3$NO$_2$): -206.5 (t, NH$_2$; J(H,N)=91.5) [31]

IR (cyclohexane): 1950, 1976, 1982, 1991, 2005, 2018, 2050, 2057, 2101 (all CO) [31]

IR (Nujol or hexachlorobutadiene): 1482 (νCN), 1642 (δNH$_2$), 3346, 3427 (both νNH); resonances for the ^{13}C and the ^{15}N enriched samples also given [31]

mass spectrum: [M]$^+$ [31]

protonation with CF$_3$CO$_2$H in CDCl$_3$ yielded cationic No. 18, based on ^{1}H NMR spectra [31]

References on pp. 72/3

Table 2 (continued)

No. compound	method of preparation (yield in %) properties and remarks

6 (continued)

treatment with neat CH_3OD resulted in $(\mu\text{-H})Os_3(\mu_2,\eta^1\text{-C=ND}_2)(CO)_{10}$ upon evaporation of the mixture to dryness, IR: 2437, 2578 (ND); in $CDCl_3$ with a tenfold excess of CH_3OD the H/D exchange is slow at 25 °C being far from complete after 1 h; similarly in the presence of $N(C_2H_5)_3$ or CF_3CO_2H rapid exchange to equilibria conditions in less than 1 min, with the base catalysing the reversible deprotonation of NH_2, while the acid is thought to protonate the nitrogen or osmium atoms, the most obvious nucleophilic centers [31]

reaction with $[N(C_4H_9)_4]OH$ in CH_2Cl_2 yielded $[HOs_3(CO)_{11}][N(C_4H_9)_4]$ [31]

7 $(\mu\text{-H})Os_3(\mu_2,\eta^1\text{-C=NHCH}_3)(CO)_{10}$

IIa (95%) [41], IIa (34%) [6]
light yellow, m. p. 138 to 141 °C [6]
^{1}H NMR (acetone-d_6): -16.67 (s, μ-H), 3.67 (d, CH_3; $^3J(H,H) = 4.6$), 10.45 (s, NH) [6]
IR ($CDCl_3$): 1557, 1990, 2003, 2020, 2052, 2060, 2100 (all CO), 3327 (NH) [6]
hydrogenation yielded $(\mu\text{-H})_4Os_4(CNCH_3)(CO)_{11}$ [9]

*8 $(\mu\text{-H})Os_3(\mu_2,\eta^1\text{-C=NHC}_4H_9\text{-t})(CO)_{10}$

IIa (60%) [6]
light yellow crystals, m. p. 134 to 136 °C from hexane at -20 °C [6, 7]
^{1}H NMR (acetone-d_6): -16.85 (s, μ-H), 1.51 (s, CH_3); no resonance for NH given [6]
IR ($CDCl_3$): 1980, 2002, 2015, 2050, 2057, 2098 (all CO), 3295 (NH) [6]

9 $(\mu\text{-H})Os_3(\mu_2,\eta^1\text{-C=NHC}_6H_5)(CO)_{10}$

IIa (unstable, not isolated) [6]
^{1}H NMR (acetone-d_6): -16.21 (d, μ-H; $J(\mu\text{-H,NH}) = 2.7$), 7.2 (m, C_6H_5), 12.52 (br s, NH) [6]
IR (($n\text{-C}_4H_9)_2O$): 1582, 1974, 1996, 2008, 2042, 2050, 2087 (all CO), 3300 (NH) [6]

10 $(\mu\text{-H})Os_3(\mu_2,\eta^1\text{-C=NHC}_6H_4CH_3\text{-2})(CO)_{10}$

IIa (15%) [41]
^{1}H NMR (C_6D_6, 28 °C): -16.68 (s, μ-H), 1.78 (s, CH_3), 6.59 (m, C_6H_4); no resonance for NH given [41]

References on pp. 72/3

Table 2 (continued)

No. compound	method of preparation (yield in %) properties and remarks

11　$(\mu-H)Os_3(\mu_2,\eta^1-C\!\doteq\!NHC_6H_4CH_3-4)(CO)_{10}$

IIa (30%) [41]

^{1}H NMR (C_6D_6, 28 °C): -16.60 (s, μ-H), 1.98 (s, CH_3), 6.70 (m, C_6H_4); no resonance for NH given [41]

12　$(\mu-H)Os_3(\mu_2,\eta^1-C\!\doteq\!NHC_6H_3(CH_3)_2-2,6)(CO)_{10}$

IIa (23%) [41]

^{1}H NMR (C_6D_6, 28 °C): -16.90 (s, μ-H), 1.80 (s, 2 CH_3), 6.84 (m, C_6H_3); no resonance for NH given [41]

13　$(\mu-H)Os_3(\mu_2,\eta^1-C\!\doteq\!NH(CH_2)_3Si(OC_2H_5)_3)(CO)_{10}$

by conversion of $(\mu-H)HOs_3(CN(CH_2)_3Si-(OC_2H_5)_3)(CO)_{10}$ in CH_2Cl_2 at 35 °C within 20 min (compare Preparation Method IIa, but in the absence of $N(C_2H_5)_3$), or at 0 °C in the presence of an oxide such as Al_2O_3, TiO_2, SiO_2, ZnO, or MgO within 2 min; work-up as in Preparation Method IIa [10]

^{1}H NMR ($CDCl_3$, 30 °C): -16.6 (s, μ-H), 0.8 (t, $SiCH_2$), 1.22 (t, CH_3), 1.9 (m, CCH_2C), 3.9 (q, OCH_2; t, NCH_2), 9.84 (NH) [10]

IR (cyclohexane): 1969, 1978, 1986, 2001, 2015, 2047, 2055, 2098 (all CO); spectrum of oxide supported species differed slightly in line width [10]

in oxide suspensions, some C_2H_5OH was eliminated indicating substitution of ethoxide groups by surface oxygens [10]

*14　$(\mu-H)Os_3(\mu_2,\eta^1-C\!\doteq\!N(CH_3)_2)(CO)_{10}$

from $Os_3(CO)_{12}$ and $N(CH_3)_3$ in nonane for 7 h; purified by chromatography (7%, along with 3% of $(\mu-H)Os_3(\mu_3,\eta^2-CH\!=\!NCH_3)(CO)_9$ and 6% of $(\mu-H)Os_3(CO)_{10}(\mu-OH))$ [4]

from $Os_3(CO)_{12}$ and $N(CH_3)_3$ in hexane in a sealed tube at 170 °C for 9 h under vacuum (ca. 4%, along with small amounts of unreacted $Os_3(CO)_{12}$, $(\mu-H)Os_3(\mu_2,\eta^2-CH\!=\!NCH_3)(CO)_{10}$ and $(\mu-H)Os_3(CO)_{10}-(\mu-NH_2))$ [4]

from $(\mu-H)Os_3(\mu_2,\eta^2-CH_2N(CH_3)_2)(CO)_{10}$ ("Organoosmium Compounds" B 5, Section 3.1.6.1.2.7, in preparation) in refluxing heptane in 20 min, followed by TLC with CH_2Cl_2/hexane (1:9) as eluant (60%, along with 28% of $(\mu-H)_2Os_3(\mu_3,\eta^2-CHN(CH_3)_2)-(CO)_9$, Section 3.1.6.2.4) [33]

Table 2 (continued)

No. compound	method of preparation (yield in %) properties and remarks
*14 (continued)	from $(\mu-H)_2Os_3(\mu_3,\eta^2-CHN(CH_3)_2)(CO)_9$ (Section 3.1.6.2.4) in refluxing octane for 5 h under CO atmosphere; work-up as before (80%) [33] by pyrolysis of $(\mu-H)Os_3(=CHN(CH_3)_2)-$ $(CO)_9(\mu-SC_6H_5)$ at 200 °C under vacuum for 5 min, followed by dissolving in CH_2Cl_2 and TLC with CH_2Cl_2/hexane (25:75) as eluant (14%, along with 19% of $Os_3(\mu_2,\eta^1-$ $C=N(CH_3)_2)(CO)_{10}(\mu-SC_6H_5)$, No. 25, 22% of $(\mu-H)_2Os_3(CO)_9(\mu_3-S)$, and $Os_6(\mu_2,\eta^1-$ $C=N(CH_3)_2)(CO)_{15}(\mu_3-S)(\mu-SC_6H_5)$ in 21% yield) [27] light yellow crystals from CH_2Cl_2/hexane [27, 33] ^1H NMR ($CDCl_3$, 27 °C): -16.75 (s, $\mu-H$), 3.84 (s, CH_3) [4] IR (cyclohexane): 1980, 1987, 1990, 2002, 2019, 2048, 2057, 2100 (all CO) [4] mass spectrum: $[M]^+$ [4]
15 $(\mu-H)Os_3(\mu_2,\eta^1-C=N(CH_3)CH_2C_6H_5)(CO)_{10}$	
	from $Os_3(CO)_{12}$ and $(CH_3)_2NCH_2C_6H_5$ in decane or decahydronaphthalene for 48 h under CO atmosphere, followed by TLC of the residue on silica with pentane as eluant (6% in decane, and 13% in decahydronaphthalene); similar reaction in decahydronaphthalene but under N_2 atmosphere gave 15%; by-products were $(\mu-H)Os_3(\mu_2,\eta^2-$ $C(C_6H_5)=NCH_3)(CO)_{10}$ (traces or 11%), $(\mu-H)Os_3(\mu_3,\eta^2-CH=NCH_3)(CO)_9$, and a mononuclear complex in 6 to 8 % (see "Organoosmium Compounds" A 3, Section 1.1.7, in preparation) [4] yellow solid [4] ^1H NMR ($CDCl_3$, 27 °C): -16.58 (s, $\mu-H$), 3.68 (s, CH_3), 5.15, 5.35 (AB quartet, CH_2; $J(H,H) = 14.2$); 7.12 to 7.48 (m, C_6H_5); the sharp AB quartet of the benzylic hydrogens is due to the hindered rotation about the C=N double bond [4] IR (cyclohexane): 1980, 1987, 1991, 2001, 2019, 2048, 2057, 2100 (all CO)] [4] mass spectrum: $[M]^+$ [4]

References on pp. 72/3

Table 2 (continued)

No. compound	method of preparation (yield in %) properties and remarks

16 $(\mu\text{-H})Os_3(\mu_2,\eta^1\text{-C=N(CH}_3)_2)(CO)_9NCCH_3$

by treatment of No. 14 with $(CH_3)_3NO \cdot 2\,H_2O$ (dropwise addition) in CH_3CN at ca. 25 °C, followed by removal of the solvent [24]

yellow crystals from CH_3CN; unstable in solution [24]

^1H NMR ($CDCl_3$): -16.08 (s, μ-H), 2.58 (s $NCCH_3$), 3.79 (s, $=N(CH_3)_2$) [24]

IR (cyclohexane): 1933, 1954, 1962, 1973, 1980, 1990, 1999, 2005, 2023, 2040, 2081 (all CO) [24]

the nitrile ligand is probably coordinated to the nonbridged Os atom in an axial position [24]

reaction with $As(C_6H_5)_3$ in $CDCl_3$ yielded No. 17 [24]

*17 $(\mu\text{-H})Os_3(\mu_2,\eta^1\text{-C=N(CH}_3)_2)(CO)_9As(C_6H_5)_3$

exists in solution as an equilibrium mixture of two isomers with the $As(C_6H_5)_3$ ligand being coordinated equatorial to the nonbridged (Isomer 1) and to a bridged Os atom (Isomer 2; see Scheme 1, p. 61) [24]

by addition of a slight excess of $As(C_6H_5)_3$ to a solution of No. 16 in $CDCl_3$; purification by TLC using CH_2Cl_2/cyclohexane (ca. 1:10) as eluant [24]

solid from $CH_3OH/i\text{-}C_3H_7OH$ [24]

^1H NMR ($CDCl_3$):

Isomer 1: -16.04 (s, μ-H), 3.78 (s, CH_3); the methyl resonances are not resolved, indicating that the $As(C_6H_5)_3$ ligand is bonded to the unbridged Os atom [24]

Isomer 2: -16.36 (s, μ-H), 2.89 (s, CH_3), 3.59 (s, CH_3) [24]

IR (cyclohexane): 1939, 1961, 1975, 1980, 1991, 1999, 2009, 2019, 2044, 2057, 2078, 2086 (all CO) [24]

$[(\mu\text{-H})_2Os_3(\mu_2,\eta^1\text{-C=NH}_2)(CO)_{10}]^+$

18

by protonation of $(\mu\text{-H})Os_3(\mu_2,\eta^1\text{-C=NH}_2)(CO)_{10}$ (No. 6) with a tenfold excess of CF_3CO_2H in $CDCl_3$ (quantitative; not isolated) [31]

^1H NMR ($CDCl_3$, 27 °C): -20.14, -15.73 (2 d's, both μ-H; J=1) [31]

reconversion to No. 6 occurred upon treatment with H_2O in CH_2Cl_2 solution [31]

References on pp. 72/3

Table 2 (continued)

No. compound	method of preparation (yield in %) properties and remarks

compounds of the type $(\mu\text{-H})Os_3(\mu_2,\eta^1\text{-COR})(CO)_9D$ (Formula II)

19 $(\mu\text{-H})Os_3(\mu_2,\eta^1\text{-COD})(CO)_{10}$

III (not isolated), attempted preparation of the undeuterated species failed [36]
^{1}H NMR (CD$_2$Cl$_2$, $-80\,°$C): -16.3 (s, μ-H) [36]
^{13}C NMR (CD$_2$Cl$_2$, $-93\,°$C): 168.8 (2 CO), 172.8 (1 CO), 173.0 (1 CO), 174.6 (2 CO), 179.0 (4 CO), 346.7 (COD) [36]

*20 $(\mu\text{-H})Os_3(\mu_2,\eta^1\text{-COCH}_3)(CO)_{10}$

III (81%) [12], III (50%) [5, 8]
by carbonylation of $(\mu\text{-H})_3Os_3(\mu_3,\eta^1\text{-COCH}_3)(CO)_9$ (Section 3.1.6.3) in decane under 14 to 70 atm of CO at 100 °C for 4 to 5 d (45%) [15]
from $[(\mu\text{-H})Os_3(CO)_{11}NCCH_3]^+$ and KCN at 25 °C for 1 h (30%) [35]
yellow solid from CH$_3$OH [12]; soluble in most organic solvents [5, 12]
^{1}H NMR (CD$_2$Cl$_2$): -16.2 (μ-H), 4.6 (CH$_3$) [5]; similar in CDCl$_3$ [12]
^{13}C NMR (CH$_2$Cl$_2$, $-80\,°$C): 168.7, 169.4 (2 CO-b,b'), 173.9, 174.5 (2 CO-c,c'; J(H,C) = 9.3), 174.6, 175.4 (2 CO-d,d'), 178.6, 178.7 (2 CO-a,a'), 179.1 (1 CO-f), 180.6 (1 CO-e), 352.2 (μ_2,η^1-C); for assignment, see Scheme 2, p. 63 [5]
IR (hexane): 1456 (COCH$_3$) [12], 1979, 1990, 1997, 2009, 2023, 2056, 2064, 2106 (all CO) [5]; similar in cyclohexane [12]
mass spectrum: [M]$^+$ [5, 12]

*21 $(\mu\text{-H})Os_3(\mu_2,\eta^1\text{-COC}_2H_5)(CO)_{10}$

III [8, 12]
from $[(\mu\text{-H})Os_3(\mu\text{-CO})(CO)_{10}][N(C_2H_5)_4]$ and $[(C_2H_5)_3O]BF_4$ in hexane at room temperature (40%) [5]
yellow solid; soluble in most organic solvents [5, 12]
^{1}H NMR (CD$_2$Cl$_2$): -16.2 (μ-H), 1.66 (t, CH$_3$; J(H,H) = 7), 4.68 (q, CH$_2$; J(H,H) = 7) [5]; similar in CDCl$_3$ [12]
^{13}C NMR (CD$_2$Cl$_2$, $-80\,°$C): 169.3, 170.0 (2 CO-b,b'), 174.4, 175.0 (2 CO-c,c'; J(H,C) = 9.3), 175.2, 176.0 (2 CO-d,d'), 179.3 (2 CO-a,a'), 179.7 (1 CO-f), 181.2 (1 CO-e), 349.7 (μ_2,η^1-C); for assignment, see Scheme 2, p. 63 [5]

References on pp. 72/3

Table 2 (continued)

No. compound	method of preparation (yield in %) properties and remarks

IR (hexane): ca. 1680 (COC_2H_5), 1977, 1989, 1994, 2007, 2022, 2055, 2062, 2105 (all CO) [5]; identical in cyclohexane [12]

mass spectrum: $[M]^+$ [5]

*22 $(\mu\text{-H})Os_3(\mu_2,\eta^1\text{-COCH}_2Cl)(CO)_{10}$

from $(\mu\text{-H})_2Os_3(\mu_3,\eta^1\text{-CCO})(CO)_9$ (Section 3.1.6.3) and BCl_3 in CH_2Cl_2 [40]

light yellow rods from CH_2Cl_2 at $-15\,°C$ [40]

*23 $(\mu\text{-H})Os_3(\mu_2,\eta^1\text{-COCH}_3)(CO)_9P(C_6H_5)_3$

equilibrium of anti (more stable) and syn isomers (Isomers 1 and 2, Scheme 2, p. 63)

from $(\mu\text{-H})Os_3(\mu_2,\eta^1\text{-COCH}_3)(CO)_{10}$ (No. 20, contaminated with $HOs_3(CO)_{10}(\mu\text{-OCH}_3)$) and $P(C_6H_5)_3$ in CH_3CN at room temperature in the presence of $(CH_3)_3NO\cdot2\,H_2O$, followed by TLC with 10% CH_2Cl_2 in hexane (ca. 10%, along with $HOs_3(CO)_9P(C_6H_5)_3(\mu\text{-OCH}_3)$ as a by-product); recrystallized from CH_3OH [15]

^1H NMR ($CDCl_3$, $-20\,°C$):

Isomer 1: -15.62 (d, μ-H; J(H,P)$=6.7$), 4.16 (s, anti-CH_3), 7.4 (m, C_6H_5); ($CDCl_3$, 25 °C): -15.61 (d, μ-H; J(H,P)$=7$), 4.17 (s, anti-CH_3) [15]

Isomer 2: -15.72 (d, μ-H; J(H,P)$=6.7$), 3.63 (s, syn-CH_3), 7.4 (m, C_6H_5); ($CDCl_3$, 25 °C): -15.68 (d, μ-H; J(H,P)$=7$), 3.69 (br s, syn-CH_3) [15]

IR (cyclohexane): 1950, 1962, 1972, 1982, 1989, 2001, 2012, 2049, 2090 (all CO) [15]

mass spectrum: $[M]^+$ [15]

*24 $(\mu\text{-H})Os_3(\mu_2,\eta^1\text{-COCH}_3)(CO)_9As(C_6H_5)_3$

equilibrium of anti (more stable) and syn isomers (Isomers 1 and 2, Scheme 2, p. 63), the ratio is ca 1.14:1 in $CDCl_3$ at room temperature (?) [15]

from $(\mu\text{-H})Os_3(\mu_2,\eta^1\text{-COCH}_3)(CO)_{10}$ (No. 20, contaminated with $HOs_3(CO)_{10})(\mu\text{-OCH}_3)$ and $As(C_6H_5)_3$ in decane at 110 °C for 5 h, followed by TLC with 10% CH_2Cl_2 in hexane (47%), along with unreacted starting material, $Os_3(CO)_{12}$, $Os_3(CO)_{11}As(C_6H_5)_3$, and $HOs_3(CO)_9As(C_6H_5)_3(\mu\text{-OCH}_3)$ [15]

^1H NMR ($CDCl_3$):

Isomer 1: -15.70 (br s, μ-H), 3.86 (br s, anti-CH_3), 7.4 (m, C_6H_5) [15]

Isomer 2: 4.17 (br s, syn-CH_3) [15]

IR (cyclohexane): 1946, 1957, 1972, 1982, 1990, 2001, 2008, 2014, 2050, 2092 (all CO) [15]

                                       References on pp. 72/3

Table 2 (continued)

No. compound	method of preparation (yield in %) properties and remarks

Os$_3$(μ_2,η^1-C=N(CH$_3$)$_2$)(CO)$_{10}$(μ-SC$_6$H$_5$)

*25

by pyrolysis of (μ-H)Os$_3$(=CHN(CH$_3$)$_2$)-(CO)$_9$(μ-SC$_6$H$_5$) as described for No. 14, see p. 44 (19%) [27]

yellow crystals from CH$_2$Cl$_2$/hexane at -20 °C [27]

^{1}H NMR (CD$_2$Cl$_2$): 3.86 (s, CH$_3$), 7.25 (m, C$_6$H$_5$) [27]

IR (hexane): 1967, 1987, 2003, 2012, 2046, 2060, 2095 (all CO) [27]

compounds of the type (μ-H)$_2$Os$_3$(μ_2,η^1-C=NR$_2$)(CO)$_8$(μ-SR') (Formula III)

26 (μ-H)$_2$Os$_3$(μ_2,η^1-C=N(CH$_3$)$_2$)(CO)$_8$(μ-SCH$_3$)

Ia (68%) [26]

yellow solid [26]

^{1}H NMR (CDCl$_3$): -15.95, -13.91 (2 s's, both μ-H), 2.58 (s, SCH$_3$), 3.87, 3.99 (2 s's, both NCH$_3$) [26]

IR (hexane): 1951, 1974, 1990, 2012, 2018, 2048, 2085 (all CO) [26]

refluxing in octane gave (μ-H)$_2$Os$_6$(μ_2,η^1-C=N(CH$_3$)$_2$)$_2$(CO)$_{12}$(μ_3,η^1-SCH$_3$)$_2$ [26]

*27 (μ-H)$_2$Os$_3$(μ_2,η^1-C=N(CH$_3$)$_2$)(CO)$_8$(μ-SC$_6$H$_5$)

Ia (37%) [21, 29], Ib (88%) [29], Ib (60%, along with (μ-H)Os$_3$(CO)$_8$(η^2-C$_6$H$_5$CH=N(CH$_3$)$_2$)(μ_3-S) in 20% yield) [28, 34]

air-stable crystals from C$_6$H$_6$ at 25 °C [21, 29]

^{1}H NMR (CD$_2$Cl$_2$): -15.84, -13.71 (2 s's, both μ-H), 3.85, 4.01 (2 s's, both CH$_3$), 7.30 (m, C$_6$H$_5$) [21, 29]

IR (hexane): 1953, 1975, 1990, 1999, 2012, 2020, 2050, 2086 (all CO) [21, 29]

*28 (μ-H)$_2$Os$_3$(μ_2,η^1-C=N(CH$_3$)$_2$)(CO)$_8$(μ-SC$_6$F$_5$)

Ia (53%) [29]

solid [29]

^{1}H NMR (CD$_2$Cl$_2$): -15.61, -13.34 (2 s's, both μ-H), 3.84, 3.97 (2 s's, both CH$_3$) [29]

IR (hexane): 1953, 1975, 1996, 2001, 2018, 2029, 2054, 2090 (all CO) [29]

References on pp. 72/3

Table 2 (continued)

No. compound	method of preparation (yield in %) properties and remarks

*29 $(\mu-H)_2Os_3(\mu_2,\eta^1-C=N(CH_3)_2)(CO)_8(\mu-SC_6H_4CH_3-4)$

Ia (35%), Ib [29]
solid [29]
^1H NMR (CD_2Cl_2): -15.83, -13.85 (s's, both
  $\mu-H$), 2.11 (s, CH_3-4), 3.84, 4.04 (s's, both
  NCH_3), 7.05 (m, C_6H_4) [29]
IR (hexane): 1952, 1974, 1990, 1999, 2011, 2020,
  2050, 2086 (all CO) [29]

*30 $(\mu-H)_2Os_3(\mu_2,\eta^1-C=N(C_2H_5)_2)(CO)_8(\mu-SC_6H_4CH_3-4)$

Ia (21%), Ib [29]
solid [29]
^1H NMR (CD_2Cl_2): -15.95, -13.92 (s's, both
  $\mu-H$), 1.04, 1.35 (2 t's, both CH_3 of C_2H_5;
  $J(H,H)=7.3$), 2.06 (s, CH_3-4), 3.75, 4.03 (q's,
  both NCH_2; $^2J(H,H)=7.3$), 7.11 (m, C_6H_4) [29]
IR (hexane): 1950, 1973, 1989, 1997, 2011, 2019,
  2049, 2085 (all CO) [29]

compounds of the type $(\mu-H)Os_3(\mu_2,\eta^1-C=N(CH_3)_2)(\eta^2-{}^1L)(CO)_7(\mu-SC_6H_5)$

*31

from $(\mu-H)_2Os_3(\mu_2,\eta^1-C=N(CH_3)_2)(CO)_8-$
  $(\mu-SC_6H_5)$ (No. 27) and an excess of
  $CH_2(N(CH_3)_2)_2$ in refluxing heptane for 2 h;
  purified by TLC with CH_2Cl_2/hexane, ca.
  25:75 (27%, along with 37% of the isomer
  No. 32) [30]
from $(\mu-H)Os_3(CO)_{10}(\mu-SC_6H_5)$ and an excess
  of $CH_2(N(CH_3)_2)_2$ (11%, along with 17% of
  No. 32) [30]
yellow crystals from C_6H_6 [30]
^1H NMR (CD_2Cl_2): -15.99 (t, $\mu-H$; $J(H,H)=1.6$),
  2.35, 3.82 (2 t's, each 1 H of CH_2; $J(H,H)=$
  1.8), 3.23, 3.51, 3.64, 3.75 (4 s's, all CH_3), 7.33
  (m, C_6H_5); by hydride decoupling, CH_2
  formed two doublets with $J=1.95$ [30]
IR (hexane): 1932, 1954, 1959, 1975, 1992, 2032,
  2066 (all CO) [30]

converted into No. 32 (in 86% yield) upon refluxing in octane for 3 h in the presence of
  CO, the transformation proceeded by
  $\mu_2,\eta^1-C=N(CH_3)_2$ ligand migration initially induced by CO addition at Os-1 resulting in an
  intermediate containing $\mu_3,\eta^1-C-N(CH_3)_2$ or
  terminal $\equiv C-N(CH_3)_2$, followed by elimination
  of a CO group from Os-2; similar treatment
  with $CH_2(N(CH_3)_2)_2$ for 1 h led to No. 32 (in

Table 2 (continued)

No. compound	method of preparation (yield in %) properties and remarks

*31 (continued)

18% yield), probably via the corresponding intermediates [30]

*32

for formation, see No. 31

yellow crystals from CH_2Cl_2/hexane at $-20\,°C$ [30]

^{1}H NMR (CD_2Cl_2): -10.25 (s, μ-H), 2.05, 4.55 (2 d's, each 1 H of CH_2; J(H,H) = 11.5), 2.83, 3.04, 3.79, 3.84 (4 s's, all CH_3), 7.30 (m, C_6H_5) [30]

IR (hexane): 1946, 1970, 1990, 2002, 2012, 2062 (all CO) [30]

thermolysis in toluene yielded No. 50; a mechanism was discussed [34]

compounds of the type $Os_3(\mu_2,\eta^1\text{-}C=NHR)(\mu_2,\eta^2\text{-}O=CNHR')(CO)_9D$ (Formula IV)

33 $Os_3(\mu_2,\eta^1\text{-}C=NHC_3H_7\text{-}n)(\mu_2,\eta^2\text{-}O=CNHC_3H_7\text{-}i)(CO)_{10}$

IIb (91%) [45, 47]

in traces upon column chromatographic work-up of No. 37 on silica gel besides 73% of the isonitrile complex $(\mu\text{-}H)Os_3(\mu_2,\eta^2\text{-}O=CNHC_3H_7\text{-}i)$-$(CNC_3H_7\text{-}n)(CO)_9$ ("Organoosmium Compounds" B 5, Section 3.1.4.3, in preparation) [47]

microcrystalline solid [45, 47]

^{1}H NMR ($CDCl_3$): 1.06 (d, CH_3 of C_3H_7-i), 1.08 (t, CH_3 of C_3H_7-n), 1.87 (m, $CH_2\mathbf{CH_2}CH_3$), 3.59 (m, $\mathbf{CH_2}CH_2CH_3$), 3.96 (m, CH), 5.75 (d, NHC_3H_7-i), 9.00, 9.16 (2 br s's, both NHC_3H_7-n, corresponding to the two isomers observed in a ratio of 3:2) [45, 47]

^{13}C NMR ($CDCl_3$): 11.3 (CH_3 of C_3H_7-n), 21.8, 22.3, 22.5, 22.9 (all CH_3 of C_3H_7-i), 42.8 ($CH_2\mathbf{CH_2}CH_3$), 62.6, 62.8 (CH, $\mathbf{CH_2}CH_2CH_3$), 171.5, 172.2, 172.3, 172.9, 176.6, 177.1, 177.4, 178.7, 180.1, 180.4, 181.5, 182.2, 183.0, 184.9, 186.5 (all CO, corresponding to the two isomers), 212.9, 213.1 (both O=C, corresponding to the two isomers), 271.6, 272.2 (both C=N, corresponding to the two isomers) [45, 47]

IR (CH_2Cl_2): 1958, 1975, 2003, 2037, 2053, 2090 (all CO) [45, 47]

FAB mass spectrum: $[M]^+$, $[M-x\ CO]^+$, $x = 1$ to 6 [45, 47]

References on pp. 72/3

Table 2 (continued)

No. compound	method of preparation (yield in %) properties and remarks
	reaction with $NH_2C_3H_7$-i resulted in No. 37; see Preparation Method IIb [45, 47] treatment with excess $P(C_6H_5)_3$ in CH_2Cl_2 for 7 d gave No. 42 [47]
34 $Os_3(\mu_2,\eta^1$-C=NHC$_3$H$_7$-i)(\mu_2,\eta^2$-O=CNHC$_3H_7$-n)(CO)$_{10}$	IIb (85%) [45, 47] microcrystalline solid [45] ^{1}H NMR (CDCl$_3$): 0.87 (t, CH$_3$ of C$_3$H$_7$-n), 1.41 (d, CH$_3$ of C$_3$H$_7$-i), 1.44 (m, CH$_2$**CH$_2$**CH$_3$), 3.09 (m, **CH$_2$**CH$_2$CH$_3$), 3.99 to 4.20 (m, CH), 5.93 (br s, NHC$_3$H$_7$-n), 8.95, 9.09 (br d's, both NHC$_3$H$_7$-i, corresponding to the two isomers observed in a ratio of 3:2) [45, 47] ^{13}C NMR (CDCl$_3$): 11.3 (CH$_3$ of C$_3$H$_7$-n), 21.4, 22.4 (both CH$_3$ of C$_3$H$_7$-i), 43.0 (CH$_2$**CH$_2$**CH$_3$), 64.1, 64.2 (NHCH and NHCH$_2$), 171.3, 172.2, 172.7, 176.8, 177.3, 177.5, 178.9, 180.0, 180.4, 181.6, 182.3, 183.1, 184.1, 184.9, 186.3 (all CO, corresponding to the two isomers), 213.5 (both O=C), 264.8, 265.6 (both C=N, corresponding to the two isomers) [45, 47] IR (CH$_2$Cl$_2$): 1957, 1978, 2002, 2036, 2053, 2090 (all CO) [45, 47] reaction with $NH_2C_3H_7$-n resulted in No. 38; see Preparation Method IIb [45, 47]
*35 $Os_3(\mu_2,\eta^1$-C=NHCH$_2$C$_6$H$_5$)(\mu_2,\eta^2$-O=CNHC$_3H_7$-i)(CO)$_{10}$	IIb (85%) [47] by carbonylation of No. 39 in CH$_2$Cl$_2$ at room temperature for 1 h (quantitative), based on IR spectroscopy [47] in traces upon column chromatographic work-up of No. 39 on silica gel besides 61% of the isonitrile complex $(\mu$-H)Os$_3(\mu_2,\eta^2$-O=CNHC$_3$-H$_7$-i)(CNCH$_2$C$_6$H$_5$)(CO)$_9$ ("Organoosmium Compounds" B 5, Section 3.1.4.3, in preparation) [47] crystals [47] ^{1}H NMR (CDCl$_3$): 1.06, 1.10 (d's, both CH$_3$), 4.00 (m, CH), 4.72, 4.79 (d's, **CH$_2$**C$_6$H$_5$, corresponding to the two isomers observed in a ratio of 7:3), 5.77 (br d, NHC$_3$H$_7$-i), 7.36 to 7.49 (m, C$_6$H$_5$), 9.11, 9.26 (2 br s's, both NHCH$_2$C$_6$H$_5$, corresponding to the two isomers observed in a ratio of 7:3) [47]

References on pp. 72/3

Table 2 (continued)

No. compound	method of preparation (yield in %) properties and remarks

*35 (continued)

IR (CH$_2$Cl$_2$): 1954, 1973, 2004, 2037, 2053, 2090 (all CO) [47]

reaction with NH$_2$C$_3$H$_7$–i resulted in No. 39; see Preparation Method IIb [47]

36 Os$_3$(μ_2,η^1-C=NHC$_6$H$_5$)(μ_2,η^2-O=CNHC$_3$H$_7$-i)(CO)$_{10}$

IIb (89%) [45, 47]

in traces upon column chromatographic work-up of No. 40 on silica gel besides 75% of the isonitrile complex (μ–H)Os$_3$(μ_2,η^2-O=CNHC$_3$H$_7$–i)(CNC$_6$H$_5$)(CO)$_9$ ("Organo-osmium Compounds" B 5, Section 3.1.4.3, in preparation) [47]

microcrystalline solid [45, 47]

^{1}H NMR (CDCl$_3$): 1.04 (d, CH$_3$), 3.96 (m, CH), 5.73 (d, NHC$_3$H$_7$–i), 7.34 to 7.43 (m, C$_6$H$_5$), 10.85, 11.00 (2 br s's, both NHC$_6$H$_5$, corresponding to the two isomers observed in a ratio of 1:1) [45, 47]

IR (CH$_2$Cl$_2$): 1960, 1975, 2004, 2039, 2055, 2091 (all CO) [45, 47]

reaction with NH$_2$C$_3$H$_7$–i resulted in No. 40; see Preparation Method IIb [45, 47]

37 Os$_3$(μ_2,η^1-C=NHC$_3$H$_7$-n)(μ_2,η^2-O=CNHC$_3$H$_7$-i)(CO)$_9$NH$_2$C$_3$H$_7$-i

IIb [45], IIb (82%) [47]

solid [45, 47]

^{1}H NMR (CDCl$_3$): 1.01 (t, CH$_3$ of C$_3$H$_7$–n), 1.09 (dd, CH$_3$ of NHC$_3$H$_7$–i), 1.25 (dd, CH$_3$ of NH$_2$C$_3$H$_7$–i), 1.81 (m, CH$_2$**CH$_2$**CH$_3$), 3.02 (br s, NH$_2$), 3.16 (m, CH of NH$_2$C$_3$H$_7$–i), 3.60 (m, **CH$_2$**CH$_2$CH$_3$), 3.97 (m, CH of NHC$_3$H$_7$–i), 5.76 (br s, NHC$_3$H$_7$–i), 8.58, 8.77 (2 br s's, both NH of NHC$_3$H$_7$–n, corresponding to the two isomers observed in a ratio of 2:3) [45, 47]

IR (CH$_2$Cl$_2$): 1902, 1945, 1957, 1983, 1992, 2027, 2069 (all CO) [45, 47]

both column chromatographic purification on silica gel and protonation by CH$_3$CO$_2$H in CH$_2$Cl$_2$ at 25 °C yielded (μ–H)Os$_3$(μ_2,η^2-O=CNHC$_3$H$_7$–i)(CNC$_3$H$_7$-n)(CO)$_9$ ("Organo-osmium Compounds" B 5, Section 3.1.4.3, in preparation) [45, 47]; the contact with silica gel in addition led to traces of No. 33 [47]

38 Os$_3$(μ_2,η^1-C=NHC$_3$H$_7$-i)(μ_2,η^2-O=CNHC$_3$H$_7$-n)(CO)$_9$NH$_2$C$_3$H$_7$-n

IIb [45], IIb (80%) [47]

solid [45, 47]

References on pp. 72/3

Table 2 (continued)

No. compound	method of preparation (yield in %) properties and remarks

^{1}H NMR (CDCl$_3$): 0.86 (t, CH$_3$ of NH$_2$C$_3$H$_7$-n), 0.94 (t, CH$_3$ of NHC$_3$H$_7$-n), 1.37 (d, CH$_3$ of C$_3$H$_7$-i), 1.52 (m, NH$_2$CH$_2$**CH$_2$**CH$_3$), 1.66 (m, NHCH$_2$**CH$_2$**CH$_3$), 2.28 (br s, NH$_2$), 3.07 (m, NH**CH$_2$**CH$_2$CH$_3$, NH$_2$**CH$_2$**CH$_2$CH$_3$), 4.12 (m, CH), 5.94 (br s, NHC$_3$H$_7$-n), 8.47, 8.65 (2 br s's, both NHC$_3$H$_7$-i, corresponding to the two isomers observed in a ratio of 3:2 [45, 47]
IR (CH$_2$Cl$_2$): 1902, 1942, 1955, 1961, 1988, 2027, 2069 (all CO) [45, 47]

39 Os$_3$(μ_2,η^1-C=NHCH$_2$C$_6$H$_5$)(μ_2,η^2-O=CNHC$_3$H$_7$-i)(CO)$_9$NH$_2$C$_3$H$_7$-i

IIb (70 to 80%; easily contaminated with the starting complex No. 35) [47]
^{1}H NMR (CDCl$_3$): 1.11 (dd, CH$_3$ of NHC$_3$H$_7$-i), 1.26 (dd, CH$_3$ of NH$_2$C$_3$H$_7$-i), 3.11 (br s, NH$_2$), 3.26 (m, CH of NH$_2$C$_3$H$_7$-i), 4.01 (m, CH of NHC$_3$H$_7$-i), 4.77 (d, **CH$_2$**C$_6$H$_5$), 5.79 (br d, NHC$_3$H$_7$-i), 7.31 to 7.41 (m, C$_6$H$_5$), 8.74, 8.86 (br s's, both NHCH$_2$, corresponding to the two isomers observed in a ratio of 3:2) [47]
IR (CH$_2$Cl$_2$): 1900, 1940, 1956, 1981, 1992, 2026, 2069 (all CO) [47]
carbonylation in CH$_2$Cl$_2$ gave No. 35 in quantitative yield [47]
column chromatographic work-up on silica gel yielded the isonitrile complex (μ-H)Os$_3$(μ_2,η^2-O=CNHC$_3$H$_7$-i)(CNCH$_2$C$_6$H$_5$)-(CO)$_9$ ("Organoosmium Compounds" B 5, Section 3.1.4.3, in preparation), along with traces of No. 35 [47]

*40 Os$_3$(μ_2,η^1-C=NHC$_6$H$_5$)(μ_2,η^2-O=CNHC$_3$H$_7$-i)(CO)$_9$NH$_2$C$_3$H$_7$-i

IIb (83%) [45, 47]
crystalline solid from CH$_2$Cl$_2$/hexane [45, 47]
^{1}H NMR (CDCl$_3$): 1.11 (dd, CH$_3$ of NHC$_3$H$_7$-i), 1.27 (dd, CH$_3$ of NH$_2$C$_3$H$_7$-i), 3.08 (NH$_2$), 3.23 (m, CH of NH$_2$C$_3$H$_7$-i), 4.02 (m, CH of NHC$_3$H$_7$-i), 5.75 (d, NHC$_3$H$_7$-i), 7.35 to 7.43 (m, C$_6$H$_5$), 10.43, 10.63 (2 br s's, NHC$_6$H$_5$, corresponding to the two isomers observed in a ratio of 9:1 [45, 47]
IR (KBr): 1436 (O=C) [45, 47]; (CH$_2$Cl$_2$): 1902, 1943, 1958, 1982, 1992, 2029, 2071 (all CO) [45, 47]

References on pp. 72/3

Table 2 (continued)

No. compound	method of preparation (yield in %) properties and remarks

*40 (continued)

both column chromatographic purification on silica gel and protonation by CH_3CO_2H in CH_2Cl_2 at 25 °C yielded $(\mu-H)Os_3(\mu_2,\eta^2-O=CNHC_3H_7-i)(CNC_6H_5)(CO)_9$ ("Organoosmium Compounds" B 5, Section 3.1.4.3, in preparation) [45, 47]; the contact with silica gel in addition led to traces of No. 36 [47]

41 $Os_3(\mu_2,\eta^1-C=NHC_6H_5)(\mu_2,\eta^2-O=CNHCH_2C_6H_5)(CO)_9NH_2CH_2C_6H_5$

IIb (75%) [47]

^{1}H NMR ($CDCl_3$): 3.45 (br s, NH_2), 4.15 (m, $NH_2\mathbf{CH_2}C_6H_5$), 4.37 (d, $NHCH_2C_6H_5$), 6.17 (br s, $CONHCH_2C_6H_5$), 7.28 to 7.47 (m, C_6H_5), 10.46, 10.67 (br s's, NHC_6H_5, corresponding to the two isomers observed in a ratio of 3:7) [47]

IR (CH_2Cl_2): 1906, 1954, 1958, 1993, 2030, 2072 [47]

mass spectrum: $[M]^+$, $[M-CO]^+$, $[M-C_6H_5]^+$, $[M-CO]^+$ [47]

42 $Os_3(\mu_2,\eta^1-C=NHC_3H_7-n)(\mu_2,\eta^2-O=CNHC_3H_7-i)(CO)_9P(C_6H_5)_3$

by treatment of No. 33 with an excess of $P(C_6H_5)_3$ in CH_2Cl_2 at room temperature for 7 d, followed by TLC with hexane as eluant (21%) [47]

yellow microcrystals [47]

^{1}H NMR ($CDCl_3$): 0.82 (t, CH_3 of C_3H_7-n), 1.12 (d, CH_3 of C_3H_7-i), 1.87 (m, $CH_2\mathbf{CH_2}CH_3$), 3.57 to 3.84 (m, $\mathbf{CH_2}CH_2CH_3$, CH), 5.65 (br d, NHC_3H_7-i), 7.26 to 7.91 (m C_6H_5), 8.87 (br s, NHC_3H_7-n) [47]

IR (CH_2Cl_2): 1908, 1947, 1991, 2032, 2071 (all CO) [47]

$Os_3(\mu_2,\eta^1-C_6H_5)(\mu_2,\eta^2-O=CC_6H_5)(CO)_8(\mu_3-Se)_2$

*43

by decarbonylation of $Os_3(CO)_{10}(\mu-SeC_6H_5)_2$ in refluxing cyclohexane for 8 h, followed by work-up by TLC with light petroleum/CH_2Cl_2 (4:1) as eluant (23%, along with 6% of $Os_2(CO)_6(\mu-SeC_6H_5)_2$ and 47% of $Os_3(CO)_{10}(\mu-SeC_6H_5)_2$) [42, 44]; similar in refluxing octane (13%) [42, 44]

yellow crystals [42, 44]

^{1}H NMR (CD_2Cl_2): 7.22, 7.37 (2 m's, each 2 H-meta of C_6H_5), 7.45, 7.50 (2 m's, each

References on pp. 72/3

Table 2 (continued)

No. compound	method of preparation (yield in %) properties and remarks

1 H-para of C_6H_5), 7.61, 8.26 (2 m's, each
2 H-ortho of C_6H_5) indicated inequivalent
C_6H_5 groups [42, 44]
IR (cyclohexane): 1966, 1990, 2006, 2018, 2022,
2027, 2080, 2102 (all CO) [42, 44]

compounds of the type $Os_3(\mu_2,\eta^1-C_6H_4R-4)(\mu_3,\eta^2-P(C_6H_4R-4)C_6H_3R-4)(CO)_8(\mu_2,\eta^1-P(C_6H_4R-4)_2)$ (Formula V)

*44 $Os_3(\mu_2,\eta^1-C_6H_5)(\mu_3,\eta^2-P(C_6H_5)C_6H_4)(CO)_8(\mu_2,\eta^1-P(C_6H_5)_2)$
IV (12%) [1, 2]
by thermolysis of $Os_3(CO)_{10}(P(C_6H_5)_3)_2$ in refluxing xylene for 1 h; the mixture of six
products was separated by column chromatography on alumina (10%) [2]
orange, m. p. 195 °C [1, 2]
IR (CCl_4): 1942, 1973, 1993, 2000, 2023, 2046,
2074 (all CO) [2]
mass spectrum: $[M-x\ CO]^+$ (x = 1 to 8), $[M-x\ CO]^{2+}$ (x = 2 to 8), and fragments of $[M-8\ CO]^+$ [1, 2]

45   $Os_3(\mu_2,\eta^1-C_6H_4CH_3-4)(\mu_3,\eta^2-P(C_6H_4CH_3-4)C_6H_3CH_3-4)(CO)_8(\mu_2,\eta^1-P(C_6H_4CH_3-4)_2)$
IV [1], IV (14%) [2]
orange, m.p. 204 °C [1, 2]
IR (CCl_4): 1938, 1970, 1991, 1998, 2020, 2045,
2063 (all CO) [2]

compounds with two μ_2,η^1-bridging ^1L ligands

*46

from $Os_3(CO)_{11}NCCH_3$ and an excess of $CNCF_3$
(by condensation at −196 °C) in CH_2Cl_2 at
25 °C for ca. 5 min, followed by washing with
CH_2Cl_2/pentane, ca. 2:1 (35%); No. 48 (4%)
could be isolated from the solution by TLC
[43, 46]
yellow crystals from CH_2Cl_2/hexane at 4 °C
[43, 46]
^{19}F NMR (acetone-d_6, −34 °C): −59.95,
−59.71 (s's); the signals of the inequivalent
$CNCF_3$ ligands coalesce at 10 °C giving a
broad resonance at $\delta = -60.2$ ppm at 23 °C
[43, 46]
IR (CH_2Cl_2): 1588, 1625 (both CN), 1996, 2037,
2059, 2073, 2121 (all CO) [43, 46]
decarbonylation in refluxing benzene yielded
No. 48 [43, 46]

References on pp. 72/3

Table 2 (continued)

No. compound	method of preparation (yield in %) properties and remarks

*47

as a by-product in the preparation of No. 48
(17%) [46]
orange crystals [46]
^{1}H NMR (CD$_2$Cl$_2$): 2.92 (s, CH$_3$) [46]
^{19}F NMR (CD$_2$Cl$_2$): -61.4, -60.1 (s's, both
CF$_3$) [46]
IR (CH$_2$Cl$_2$): 1588, 1617 (all CN), 1971, 2012,
2032, 2045, 2110, 2129 (all CO) [46]
thermolysis in CH$_2$Cl$_2$ at reflux temperature for
40 min, followed by TLC with CH$_2$Cl$_2$/hexane
(1:1) as eluant gave Nos. 48 and 49 in 11
and 23% yield, respectively [46]

*48

from Os$_3$(CO)$_{10}$(NCCH$_3$)$_2$ and an excess of
CNCF$_3$ as described at No. 46; isolated by
TLC with CH$_2$Cl$_2$/hexane (1:1) as eluant
(45%); by-products were Nos. 47 and 49 in
17 and 2% yield, respectively [46]
from No. 49 by treatment with CO (1 atm) in
CH$_2$Cl$_2$ at room temperature for 5h, followed
by TLC separation (42%) [46]
by decarbonylation of No. 46 in refluxing benzene for 5 min; purification by TLC with
CH$_2$Cl$_2$/hexane (1:4) as eluant (87%) [43,
46], or by thermolysis of No. 47 in CH$_2$Cl$_2$
(11%) [46]
as a by-product in the preparation of No. 46
(4%) [43, 46]
purple crystals from CH$_2$Cl$_2$/hexane at 4 °C
[43, 46]
^{19}F NMR (acetone-d$_6$): -59.28 (s, 2 CF$_3$)
[43, 46]
IR (hexane): 1661, 1696, 1718 (all CN), 2006,
2033, 2059, 2079, 2123 (all CO) [43, 46]
reacted with (CH$_3$)$_3$NO/CH$_3$CN to give No. 49
[46]

*49

by thermolysis of No. 47 in refluxing CH$_2$Cl$_2$
(23%) [46]
by treatment of No. 48 with an excess of
(CH$_3$)$_3$NO/CH$_3$CN in CH$_2$Cl$_2$/CH$_3$CN (2:1) at
room temperature for 20 min (18%) [46]
as a by-product in the preparation of No. 48
(2%) [46]
purple crystals [46]
^{1}H NMR (CD$_2$Cl$_2$): 2.90 (s, CH$_3$) [46]

References on pp. 72/3

Table 2 (continued)

No. compound	method of preparation (yield in %) properties and remarks
	^{19}F NMR (CD_2Cl_2): -59.1, -58.5 (s's, both CF_3); bridging $CNCF_3$ are inequivalent due to the presence of the $NCCH_3$ ligand [46] IR (hexane): 1636, 1692, 1705, 2215 (all CN), 1986, 2001, 2018, 2029, 2057, 2065, 2100 (all CO) [46] stirring in CH_2Cl_2 under 1 atm of CO resulted in No. 48 [46]

*50 | obtained as an inseparable mixture of two slowly interconverting isomers, based on ^1H NMR [34] by thermolysis of No. 32 in refluxing toluene for 4 h; purification by TLC using CH_2Cl_2/hexane (ca. 30:70) as eluant (70%) [34] crystals from CH_2Cl_2/hexane at $-25\,°C$ [34] ^1H NMR (toluene-d_8): Isomer 1: -14.55 (s, μ-H), 3.24, 3.25, 3.38, 3.39 (4 s's, all CH_3), 7.45 (m, C_6H_5) [34] Isomer 2: -13.30 (s, μ-H), 2.53, 3.02, 3.42, 3.43 (4 s's, all CH_3), 7.45 (m, C_6H_5); the ratio of Isomer 1 to Isomer 2 was evaluated as 3:1, with Isomer 1 probably being the one observed in the solid state [34]; for variable temperature ^1H NMR, see p. 72 IR (hexane): 1930, 1942, 1978, 1988, 2007, 2067 (all CO) [34]

*Further information:

$(\mu\text{-H})Os_3(\mu_2,\eta^1\text{-CC}(C_6H_5)\text{=C=C}(C_6H_5)Re(CO)_4P(CH_3)_2C_6H_5)(CO)_{10}$ (Table 2, No. 3) crystallizes in the triclinic space group $P\bar{1}-C_i^1$ (No. 2) with a = 11.914(3), b = 12.844(2), c = 14.170(3) Å, α = 85.18(2)°, β = 86.48(2)°, γ = 76.50(2)°; Z = 2, R = 0.031. The structure determination (see Fig. 12) showed an allenyl-substituted μ-carbene moiety [23] refuting earlier descriptions as a zwitterion [18]. The hydride ligand was not observed crystallographically but is believed to bridge the Os(1)–Os(2) bond. The Os–CO and Re–CO distances are in the range of 1.85 to 1.96 and of 1.90 to 2.03 Å, respectively [23].

$(\mu\text{-H})Os_3(\mu_2,\eta^1\text{-C=C}(C_6H_5)P(CH_3)_2C_6H_5)(CO)_{10}$ (Table 2, No. 5) crystallizes in the triclinic space group $P\bar{1}-C_i^1$ (No. 2) with a = 14.075(3), b = 12.210(3), c = 8.724(2) Å, α = 100.86(3)°, β = 95.41(2)°, γ = 101.51(3)°; Z = 2, D_c = 2.534 g/cm³, R = 0.040. The structure is shown in Fig. 13. The short (μ_2,η^1-C)–C distance indicates a double bond; the planarity of these two C atoms with the substituents at them is consistent with a sp^2 hybridization of these atoms. The average bond distances of Os–CO and C≡O are 1.925 and 1.132 Å.

$(\mu\text{-H})Os_3(\mu_2,\eta^1\text{-C=NHC}_4H_9\text{-t})(CO)_{10}$ (Table 2, No. 8) crystallizes in the monoclinic space group $P2_1/n-C_{2h}^5$ (No. 14) with a = 13.651(4), b = 9.156(4), c = 18.275(5) Å, β = 111.4(2)°; Z = 4,

References on pp. 72/3

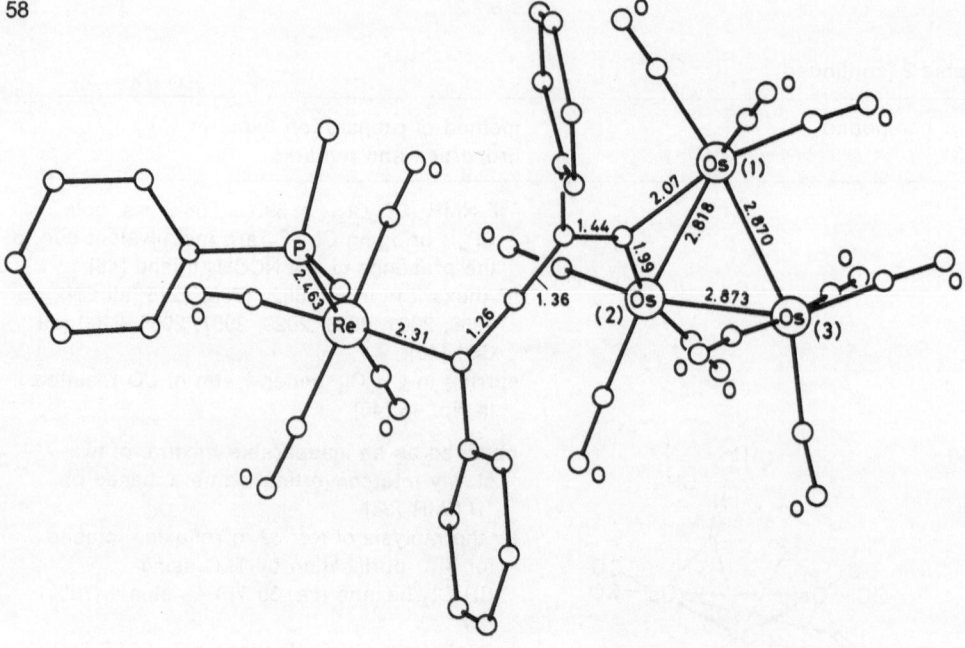

Fig. 12. Molecular structure of (μ-H)-$Os_3\{\mu_2,\eta^1-CC(C_6H_5)=C=C(C_6H_5)-Re(CO)_4P(CH_3)_2C_6H_5\}(CO)_{10}$ (No. 3) with selected bond distances (in Å) [23].

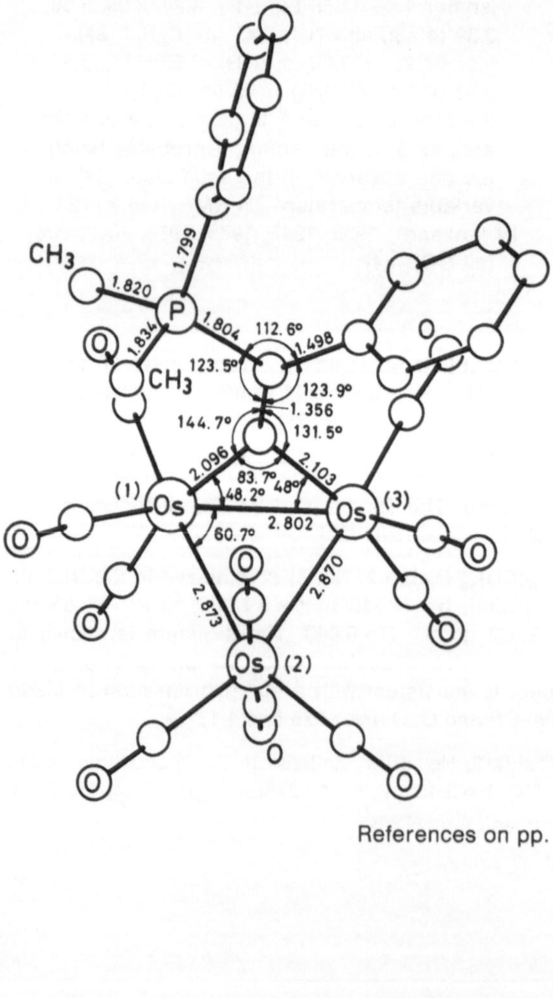

Fig. 13. Molecular structure of (μ-H)-$Os_3(\mu_2,\eta^1-C=C(C_6H_5)P(CH_3)_2C_6H_5)-(CO)_{10}$ (No. 5) with selected bond distances in (Å) and bond angles [11].

References on pp. 72/3

$D_c = 2.92$ g/cm³, R = 0.049. The structure is shown in **Fig. 14**. The symmetrically coordinated μ_2,η^1-iminyl ligand is probably strongly trans directing as was concluded from the observation that the Os-C bond distances to the carbonyl ligands trans to the iminyl ligand were unusually long (Os(2)-CO = 1.956 Å, and Os(3)-CO = 1.948 Å). The bridging hydride ligand was not crystallographically observed, but thought to occupy the opposite side of the Os(2)-Os(3) bond. The angle μ_2,η^1-C-N-C(CH₃)₃ is significantly large, which is probably due to the steric repulsion effects of the t-C₄H₉ group. The other average bond distances of Os-CO and C≡O are 1.90 and 1.15 Å [7].

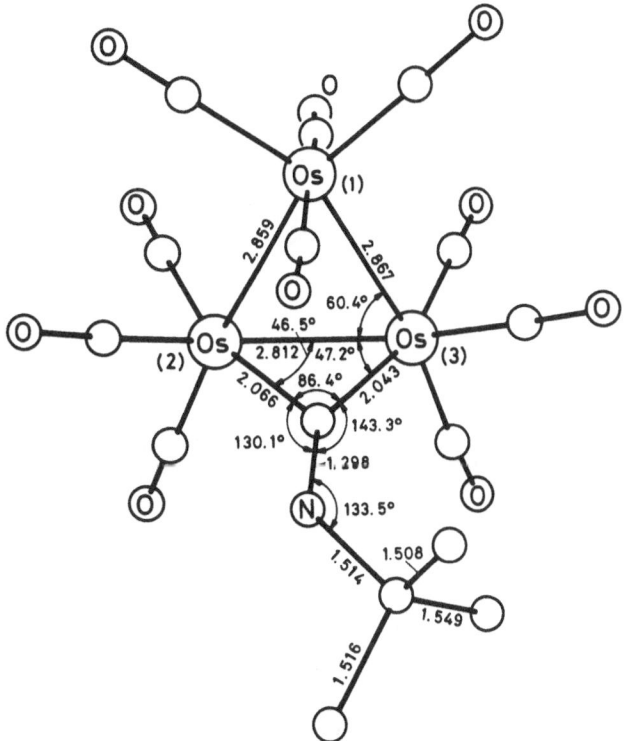

Fig. 14. Molecular structure of $(\mu\text{-H})Os_3(\mu_2,\eta^1\text{-C=NHC}_4\text{H}_9\text{-t})(CO)_{10}$ (No. 8) with selected bond distances (in Å) and angles [7].

$(\mu\text{-H})Os_3(\mu_2,\eta^1\text{-C=N(CH}_3)_2)(CO)_{10}$ (Table 2, No. **14**) crystallizes in the triclinic space group P$\bar{1}$ − C$_i^1$ (No. 2) with a = 9.147(2), b = 13.036(2), c = 8.969(2) Å, α = 104.42(2)°, β = 109.13(2)°, γ = 90.07(2)°; Z = 2, D_c = 3.09 g/cm³, R = 0.0419. The short C-N distance (see **Fig. 15**) is due to a CN double bond. The hydride ligand is not observed crystallographically but it is believed to bridge the Os(1)-Os(2) bond. Average bond distances of Os-CO and C≡O are 1.92 and 1.13 Å [33].

Reaction with H₂S [38] or CH≡CCO₂CH₃ [39] in refluxing n-octane yielded $(\mu\text{-H})_2$-Os₃(=CHN(CH₃)₂)(CO)₈(μ₃-S) ("Organoosmium Compounds" B 5, Section 3.1.5.1, in preparation) and $(\mu\text{-H})Os_3\{\mu_3,\eta^3\text{-(CH}_3)_2\text{NCCCHCO}_2\text{CH}_3\}(CO)_9$ (Formula VI), respectively [38, 39]. Upon treatment with (CH₃)₃NO·2 H₂O in CH₃CN, No. 16 was obtained [24].

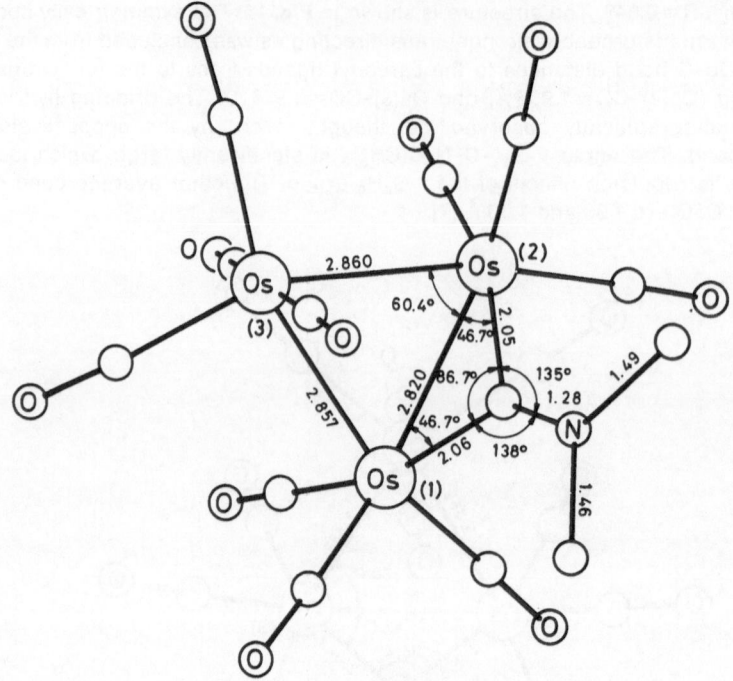

Fig. 15. Molecular structure of $(\mu\text{-H})Os_3(\mu_2,\eta^1\text{-C=N}(CH_3)_2)(CO)_{10}$ (No. 14) with selected bond distances (in Å) and bond angles [33].

$(\mu\text{-H})Os_3(\mu_2,\eta^1\text{-C=N}(CH_3)_2)(CO)_9As(C_6H_5)_3$ (Table **2**, No. **17**). An equilibrium of two isomers, bearing the $As(C_6H_5)_3$ ligand equatorial at the nonbridged (Isomer 1) or at one of the bridged Os atoms (Isomer 2) was observed in solution and evaluated by ^1H NMR spectroscopy (see Scheme 1, p. 61). Isomer 1 is formed first as the kinetically controlled product by reaction of $As(C_6H_5)_3$ with No. 16. The equilibrium constant [Isomer 1]/[Isomer 2] was 4.4(0.2) in $CDCl_3$ at 21 °C. The first order rate constants k_{obs} for isomerization to equilibrium increased with the temperature from $0.51(0.04) \times 10^{-4}\,s^{-1}$ at 27 °C to $3.14(0.05) \times 10^{-4}\,s^{-1}$ at 49 °C. Forward and reverse rate constants, k_f and k_r, respectively, were calculated as: $k_f = 0.42(0.07) \times 10^{-4}\,s^{-1}$ and $k_r = 0.09(0.01) \times 10^{-4}\,s^{-1}$ at 27 °C, and $k_f = 2.60(0.03) \times 10^{-4}\,s^{-1}$ and $k_r = 0.54(0.03) \times 10^{-4}\,s^{-1}$ at 49 °C. Activation parameters for rearrangement amounted to $\Delta H^+ = 15.5 \pm 1.3$ kcal/mol and $\Delta S^+ = -27 \pm 5$ cal·mol^{-1}·K^{-1}. The isomerization is an intramolecular process including the opening of the hydride and methylidyne bridges resulting in an intermediate with only terminally coordinated ligands, which then moved and closed the bridges to different metal atoms; for a detailed discussion, see [24].

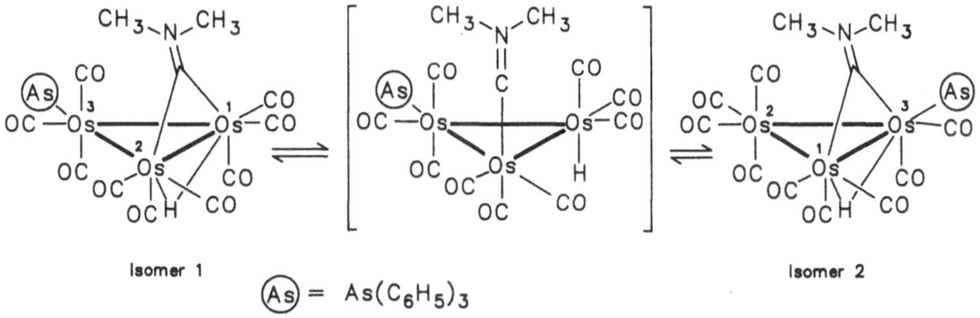

Isomer 1 ⇌ [...] ⇌ Isomer 2

$$(As) = As(C_6H_5)_3$$

Scheme 1

(μ-H)Os$_3$(μ$_2$,η^1-COR)(CO)$_{10}$ (R = CH$_3$, C$_2$H$_5$; Table 2, Nos. 20 and 21). Variable temperature ^{13}C NMR spectroscopy of Nos. 20 and 21 gave evidence of alkyl group 'flipping' and polytopic rearrangement of the Os(CO)$_4$ and Os(CO)$_3$ units; see Scheme 2, p. 63 (R = CH$_3$ or C$_2$H$_5$, D = CO). Low temperature ^{13}C NMR spectra showed for every carbonyl group a separate signal because of the lack of any symmetry. Coalescence of the a,a' resonances began at −80 °C; upon warming to 30 °C the signals of b,b', c,c', and, finally, d,d' collapsed into one signal for each pair. The approximate free energy barrier for the flipping of the alkyl group was calculated to be ΔG$^{\pm}$ = 13.4 kcal/mol for R = CH$_3$ and 13.6 kcal/mol for R = C$_2$H$_5$. Coalescence of the CO signals of d,d', e, and f, which are assigned to a rearrangement of the Os(CO)$_4$ group, started between 60 and 90 °C to give one sharp signal at 130 °C; ΔG$^{\pm}$ = 17 kcal/mol [13]. For No. 21 the coalescence process proceeded in two steps. At 80 °C the d,d' and e signals have collapsed via a trigonal twist mechanism, whereas the CO signal due to f is still sharp; it began to broaden at 90 °C. Finally at 170 °C, the degenerated CO resonances of the Os(CO)$_3$ collapsed due to a polytopic rearrangement of the Os(CO)$_3$ groups [5]; see also Nos. 23 and 24, p. 63.

(μ-H)Os$_3$(μ$_2$,η^1-COCH$_3$)(CO)$_{10}$ (No. 20) reacted with donor compounds D such as P(C$_6$H$_5$)$_3$ and As(C$_6$H$_5$)$_3$ to yield (μ-H)Os$_3$(μ$_2$,η^1-COCH$_3$)(CO)$_9$D (Nos. 23 and 24). The reaction with an excess of As(C$_6$H$_5$)$_3$ in decane was of pseudo first order in cluster concentration and of zero order in As(C$_6$H$_5$)$_3$ concentration; k = (19 to 22) × 10^{-6} s^{-1} at 90 °C, (66 ± 2) × 10^{-6} s^{-1} at 100 °C, and (230 ± 10) × 10^{-6} s^{-1} at 109.9 °C. The activation parameters for the disappearence of No. 20 were ΔH$^{\pm}$ = 34.1 ± 0.7 kcal/mol and ΔS$^{\pm}$ = + 13 ± 2 cal·mol^{-1}·K^{-1}. The substitution was inhibited by CO, revealing a mechanism involving CO dissociation [15].

Hydrogenation of No. 20 between 60 and 120 °C in decane produced (μ-H)$_3$Os$_3$(μ$_3$,η^1-COCH$_3$)(CO)$_9$ (Section 3.1.6.3) [8, 12, 15] and (μ-H)$_2$Os$_3$(CO)$_{10}$ [15]. Monitoring the progress of the hydrogenation by IR spectroscopy revealed that in an open system (bubbling H$_2$ at 1 atm for 1 h, sweeping out released CO [8, 12, 15]), the rate of disappearance of (μ-H)Os$_3$(μ$_2$,η^1-COCH$_3$)(CO)$_{10}$ increased with increasing conversion, while in a closed system (50 atm H$_2$ for 3 d [15]) the rate of the disappearence decreased with increasing conversion. The observations led to the conclusion that CO dissociation is the first step followed by reversible oxidative addition of molecular hydrogen to the intermediate [HOs$_3$(μ$_3$,η^1-COCH$_3$)(CO)$_9$], and that the released CO inhibited the hydrogenation by recombination with the intermediate. The limiting rate constant at 90 °C was 1 × 10^{-4} s^{-1} for an open system and 1 × 10^{-5} s^{-1} for a closed system [15]. Aspects of the hydrogenation reaction in comparison to the iron and ruthenium analogs, and/or other μ_3-CX groups were discussed [8, 12, 15].

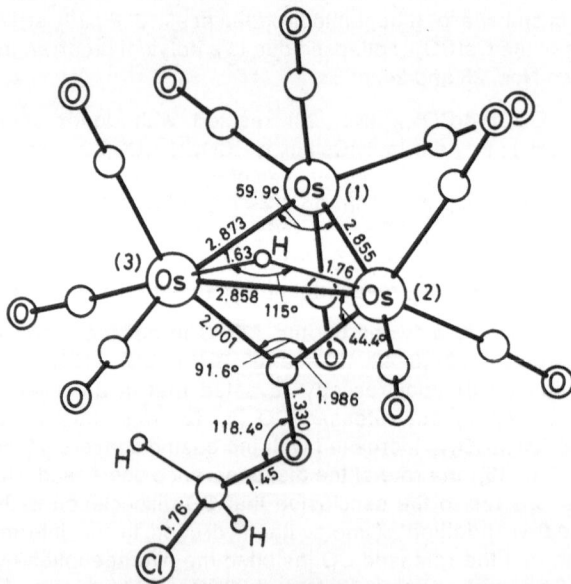

VII

No. 20 is a versatile starting material for other Os_3 clusters. Refluxing with an equimolar quantity of CNC_4H_9-t in hexane gave $(\mu\text{-H})Os_3(\mu_2,\eta^1\text{-COCH}_3)(CNC_4H_9\text{-t})(CO)_9$ ("Organoosmium Compounds" B 5, Section 3.1.4.3, in preparation) [5]. Reaction with $LiBH(C_2H_5)_3$ yielded $[(\mu\text{-H})Os_3(\mu_2,\eta^1\text{-CHOCH}_3)(CO)_{10}]^-$ (Section 3.1.6.2.1) [13]. Treatment with diphenylacetylene in toluene at 95 °C for 106 h resulted in $(\mu\text{-H})Os_3(\mu_3,\eta^3\text{-CH}_3OCC(C_6H_5)CC_6H_5)$-$(CO)_9$ (Formula VII); the reaction is assumed to be a model for Fischer-Tropsch chain growth [16]. Reaction with C_6H_5Li at 0 °C in ether followed by addition of $CF_3SO_3CH_3$ gave $(\mu\text{-H})Os_3(\mu_3,\eta^1\text{-CC}_6H_5)(CO)_{10}$ and $Os_3(\mu_3,\eta^1\text{-CC}_6H_5)(\mu_3,\eta^1\text{-COCH}_3)(CO)_9$ (both Section 3.1.6.3) at 35 °C after 60 to 72 h [20, 25] or $(\mu\text{-H})Os_3(\mu_3,\eta^1\text{-CC}_6H_5)(=C(OCH_3)_2)(CO)_9$ ("Organoosmium Compounds" B 5, Section 3.1.5.2, in preparation) at 0 °C after 30 min [19]. Reaction with $(C_2H_4)_2PtP(C_6H_{11}\text{-c})_3$ at room temperature yielded $Os_3(\mu_5\text{-C})$-$(CO)_9(\mu\text{-OCH}_3)(\mu\text{-H})Pt_2(\mu\text{-CO})(P(C_6H_{11}\text{-c})_3)_2$ (Section 3.1.6.4.2) [17]. With $AuCH_3P(C_6H_5)_3$ at 90 °C in toluene formed $Os_3(\mu_2,\eta^1\text{-COCH}_3)(CO)_{10}AuP(C_6H_5)_3$ (Section 3.1.6.4.3) [22].

$(\mu\text{-H})Os_3(\mu_2,\eta^1\text{-COCH}_2Cl)(CO)_{10}$ (Table 2, No. 22) crystallizes in the monoclinic space group $P2_1/c - C_{2h}^5$ (No. 14) with $a = 8.028(1)$, $b = 17.602(4)$, $c = 13.805(3)$ Å, $\beta = 104.88(1)°$; $Z = 4$, $D_c = 3.270$ g/cm³, $R = 0.021$. The structure is shown in **Fig. 16**. Average bond distances of Os-CO and $C\equiv O$ are 1.922 and 1.14 Å [40].

Fig. 16. Molecular structure of $(\mu\text{-H})Os_3(\mu_2,\eta^1\text{-COCH}_2Cl)(CO)_{10}$ (No. 22) with selected bond distances (in Å) and bond angles [40].

References on pp. 72/3

(μ-H)Os$_3$(μ$_2$,η^1-COCH$_3$)(CO)$_9$D (D = P(C$_6$H$_5$)$_3$, As(C$_6$H$_5$)$_3$; Table **2**, Nos. **23** and **24**) were obtained as mixtures of anti and syn isomers (Scheme 2, Isomers 1 and 2; R = CH$_3$) due to the restricted rotation about the C–OCH$_3$ bond (see also p. 37), evaluated by ^{1}H NMR spectroscopy. Isomers 1 and 2 differ in the relative orientation between R and D; D itself is assumed to occupy a position trans to the Os(CO)$_4$ unit and cis to the bridging hydride [15].

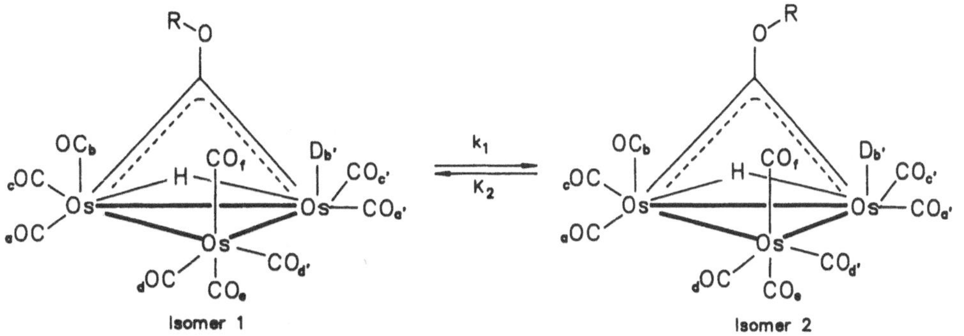

Scheme 2

Rate constants for the isomeric conversion of No. 23 in CDCl$_3$, estimated by variable temperature ^{1}H NMR [15]:

T (°C)	k(1 → 2) (s^{-1})	k(2 → 1) (s^{-1})
25	—	7
35	11	32
44	29	54

The free energy difference amounts to 0.4 kcal at 25 °C. The syn rotamer is for steric reasons less stable than the anti one. The average free activation energy ΔG$^+$ for the conversion 2 → 1 was evaluated to be 16.1 ± 0.2 kcal/mol at 35 °C; the significantly higher value compared to Nos. 20 and 21 (p. 61) is associated with steric interactions between the C$_6$H$_5$ and CH$_3$ groups [15].

Os$_3$(μ$_2$,η^1-C=N(CH$_3$)$_2$)(CO)$_{10}$(μ-SC$_6$H$_5$) (Table **2**, No. **25**) crystallizes in the monoclinic space group P2$_1$/c – C$_{2h}^5$ (No. 14) with a = 8.856(3), b = 8.945(4), c = 30.848(7) Å, β = 92.22(2)°; Z = 4, D$_c$ = 2.76 g/cm^3, R = 0.0403. The structure is shown in **Fig. 17**. The nonbonding Os(1)–Os(2) separation is bridged by the SC$_6$H$_5$ ligand and by the CN(CH$_3$)$_2$ unit; the C–N bond has a partial multiple bond character. A noncrystallographically imposed mirror plane passes through Os(3), μ$_2$,η^1-C, N, S and the C$_6$H$_5$ group. Average bond distances of Os–CO and C≡O are 1.93 and 1.12 Å [27].

(μ-H)$_2$Os$_3$(μ$_2$,η^1-C=NR$_2$)(CO)$_8$(μ-SR') (R = CH$_3$, C$_2$H$_5$, R' = C$_6$H$_5$, C$_6$F$_5$, C$_6$H$_4$CH$_3$-4; Table **2**, Nos. **27** to **30**). (μ-H)$_2$Os$_3$(μ$_2$,η^1-C=N(CH$_3$)$_2$)(CO)$_8$(μ-SC$_6$H$_5$) (No. **27**) crystallizes in the triclinic space group P$\bar{1}$ – C$_i^1$ (No. 2) with a = 9.278(1), b = 32.668(5), c = 7.644(1) Å, α = 90.26(1)°, β = 99.99(1)°, γ = 96.07(1)°; Z = 4, D$_c$ = 2.82 g/cm^3, R = 0.0530. The structure of one of the symmetry-independent molecules is shown in **Fig. 18**. The two inequivalent bridging hydrides were not observed crystallographically. Average bond distances of Os–CO and C≡O are 1.89 and 1.16 Å [21, 29].

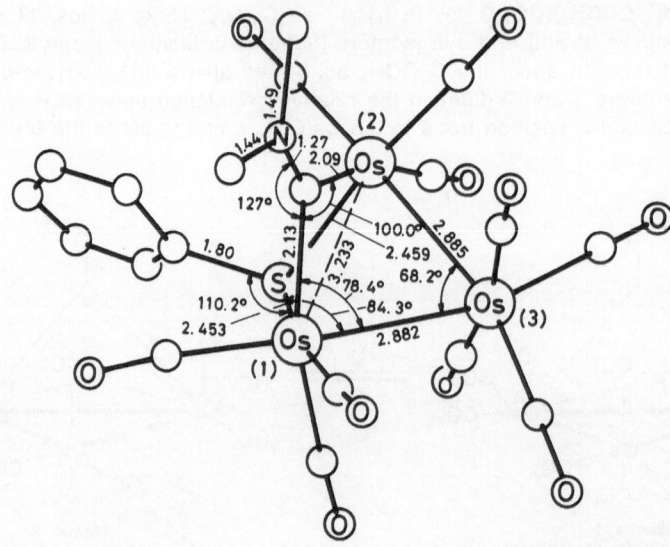

Fig. 17. Molecular structure of
$Os_3(\mu_2,\eta^1-C=N(CH_3)_2)(CO)_{10}$-
$(\mu-SC_6H_5)$ (No. 25) with
selected interatomic distances
(in Å) and bond angles [27].

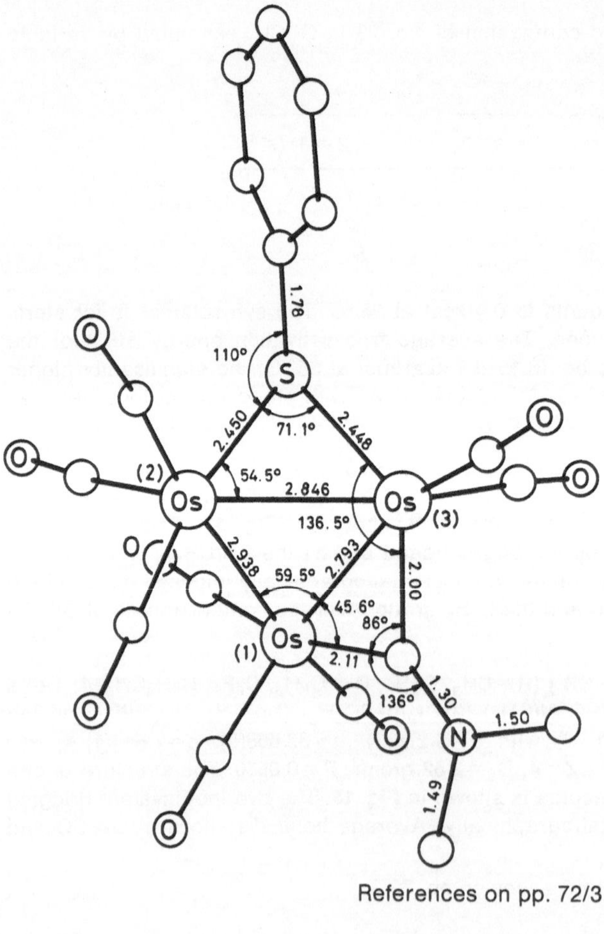

Fig. 18. Molecular structure of
one of the symmetry–independent molecules of
$(\mu-H)_2Os_3(\mu_2,\eta^1-C=N(CH_3)_2)$-
$(CO)_8(\mu-SC_6H_5)$ (No. 27) with
average selected bond distances (in Å) and bond angles
[21, 29].

References on pp. 72/3

Pyrolysis of Nos. 27 to 30 in octane gave compounds of the type $(\mu\text{-}H)_2Os_3(=CR'NR_2)$-$(CO)_8(\mu_3\text{-}S)$ ("Organoosmium Compounds" B 5, Section 3.1.5.1, in preparation); a mixture of Nos. 27 and 30 yielded no mixed complexes indicating that the sigmatropic R' shift was intramolecular [29].

Reaction of No. 27 with $CH_2(N(CH_3)_2)_2$ in refluxing heptane gave Nos. 31 and 32. Probably No. 31 was formed first via intermediate $[(\mu\text{-}H)_2Os_3(\mu_2,\eta^1\text{-}C=N(CH_3)_2)(CO)_7(\eta^1\text{-}N(CH_3)_2CH_2N(CH_3)_2)(\mu\text{-}SC_6H_5)]$ and then converted into No. 32 by shifting of the bridging $C=N(CH_3)_2$ ligand [30].

$(\mu\text{-}H)Os_3(\mu_2,\eta^1\text{-}C=N(CH_3)_2)(\eta^2\text{-}CH_2N(CH_3)_2)(CO)_7(\mu\text{-}SC_6H_5)$ (Table 2, No. 31) crystallizes in the triclinic space group $P\bar{1} - C_i^1$ (No. 2) with a = 13.509(2), b = 12.075(3), c = 9.925(2) Å, α = 102.18(2)°, β = 103.39(1)°, γ = 94.89(2)°; Z = 2, D_c = 2.33 g/cm³, R = 0.049. The structure is shown in **Fig. 19**. The methylene hydrogens and the hydride ligand were not observed crystallographically. Average bond distances of Os-CO and C≡O are 1.89 and 1.14 Å [30].

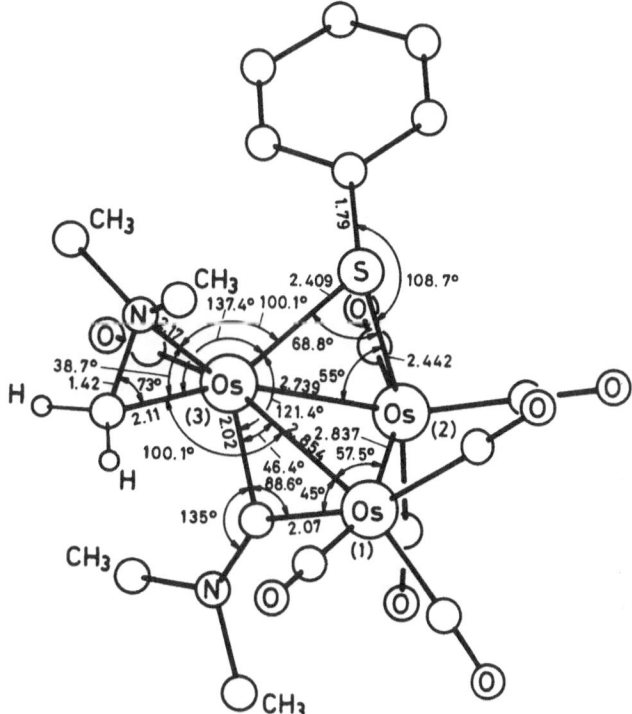

Fig. 19. Molecular structure of $(\mu\text{-}H)Os_3(\mu_2,\eta^1\text{-}C=N(CH_3)_2)(\eta^2\text{-}CH_2N(CH_3)_2)(CO)_7(\mu\text{-}SC_6H_5)$ (No. 31) with selected bond distances (in Å) and bond angles; the methylene hydrogens are shown in idealized positions [30].

$(\mu\text{-}H)Os_3(\mu_2,\eta^1\text{-}C=N(CH_3)_2)(\mu_2,\eta^2\text{-}CH_2N(CH_3)_2)(CO)_7(\mu\text{-}SC_6H_5)$ (Table 2, No. 32) crystallizes in the monoclinic space group $P2_1/c - C_{2h}^5$ (No. 14) with a = 9.986(1), b = 24.909(5), c = 10.027(1) Å, β = 97.81°; Z = 4, D_c = 2.66 g/cm³, R = 0.038. The structure is shown in **Fig. 20**. The hydride ligand was not observed crystallographically, but it is thought to bridge either the Os(1)-Os(2) or the Os(2)-Os(3) bond. Average bond distances of Os-CO and C≡O are 1.88 and 1.15 Å [30].

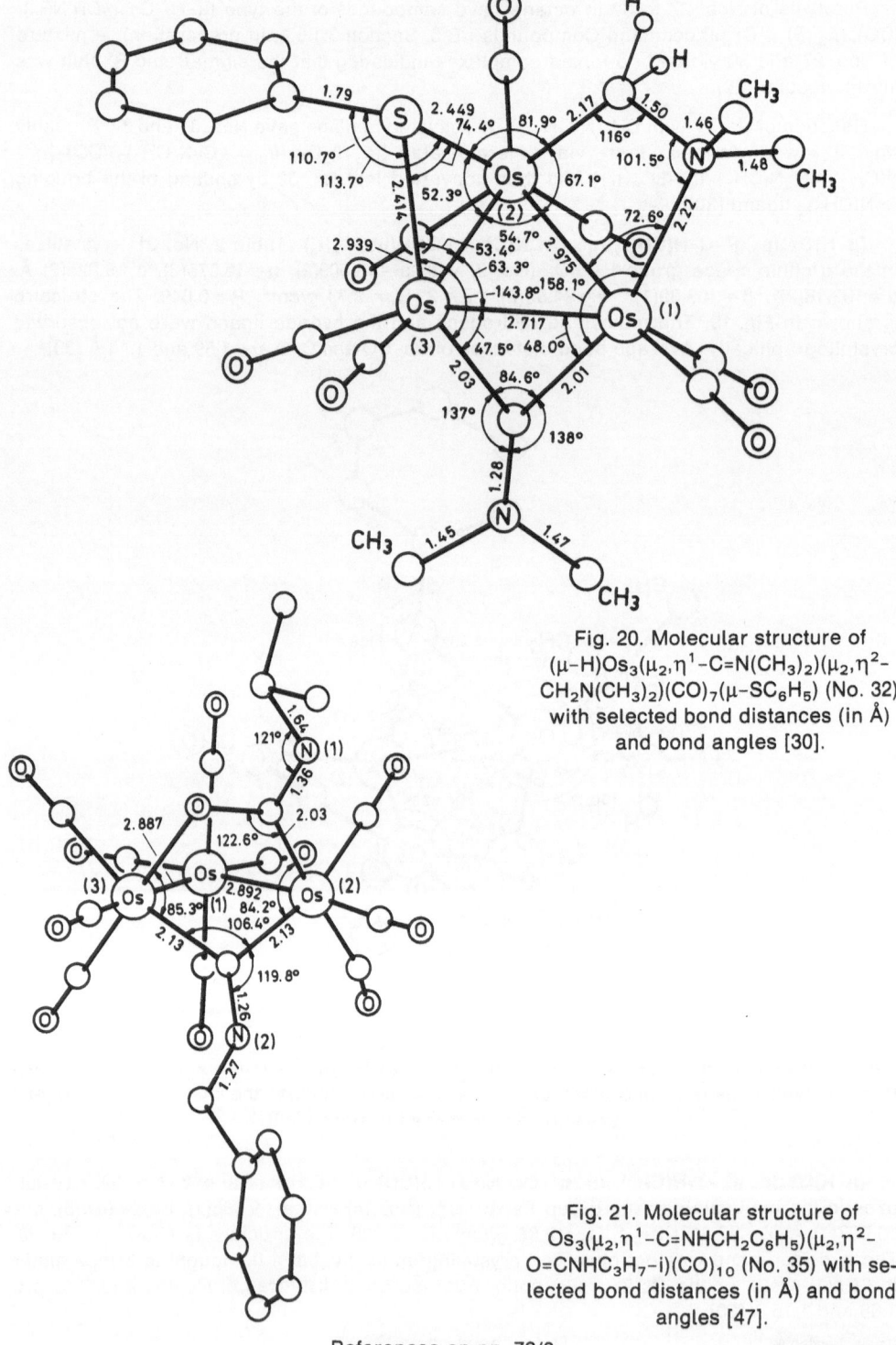

Fig. 20. Molecular structure of
$(\mu-H)Os_3(\mu_2,\eta^1-C=N(CH_3)_2)(\mu_2,\eta^2-CH_2N(CH_3)_2)(CO)_7(\mu-SC_6H_5)$ (No. 32)
with selected bond distances (in Å)
and bond angles [30].

Fig. 21. Molecular structure of
$Os_3(\mu_2,\eta^1-C=NHCH_2C_6H_5)(\mu_2,\eta^2-O=CNHC_3H_7-i)(CO)_{10}$ (No. 35) with selected bond distances (in Å) and bond
angles [47].

References on pp. 72/3

$Os_3(\mu_2,\eta^1\text{-}C\text{=}NHCH_2C_6H_5)(\mu_2,\eta^2\text{-}O\text{=}CNHC_3H_7\text{-}i)(CO)_{10}$ (Table **2**, No. **35**) crystallizes in the orthorhombic space group Pbca $-$ D_{2h}^{15} (No. 61) with a $=$ 11.9975(14), b $=$ 17.797(4), c $=$ 25.819(3) Å; Z $=$ 8, $D_c =$ 2.542 g/cm^3, R $=$ 0.056. Os(2) and Os(3) of the isosceles Os$_3$ triangle are bridged by both an aminocarbene group and a carboxamido ligand; see **Fig. 21**. The elongated nonbonding Os(2)-Os(3) separation of 3.4074 Å reveals that the initial metal-metal bond in the starting material has been cleaved. Restricted rotation about the C-N bond and the short C-N distance in the aminocarbene ligand are indicative for a C-N multiple bond. The Os atoms at the Os(CO)$_3$ groups adopt a pseudooctahedral geometry. The Os-CO bonds range from 1.72 to 2.07 Å (average: 1.95 Å), and the C≡O bonds from 0.89 to 1.36 Å (average: 1.12 Å) [47].

$Os_3(\mu_2,\eta^1\text{-}C\text{=}NHC_6H_5)(\mu_2,\eta^2\text{-}O\text{=}CNHC_3H_7\text{-}i)(CO)_9NH_2C_3H_7\text{-}i$ (Table **2**, No. **40**) crystallizes in the triclinic space group P$\bar{1}$ $-$ C$_i^1$ (No. 2) with a $=$ 8.704(1), b $=$ 13.100(2), c $=$ 13.546(4) Å, $\alpha =$ 93.46(2)°, $\beta =$ 89.91(2)°, $\gamma =$ 106.39(1)°; Z $=$ 2, $D_c =$ 2.376 g/cm^3, R $=$ 0.038. The structure, shown in **Fig. 22**, revealed that the metal-metal bond in the doubly bridged Os(1)-Os(2) moiety has been cleaved, indicated by the longer interatomic distance of 3.356 Å compared to the other distances in the Os$_3$ triangle [45, 47].

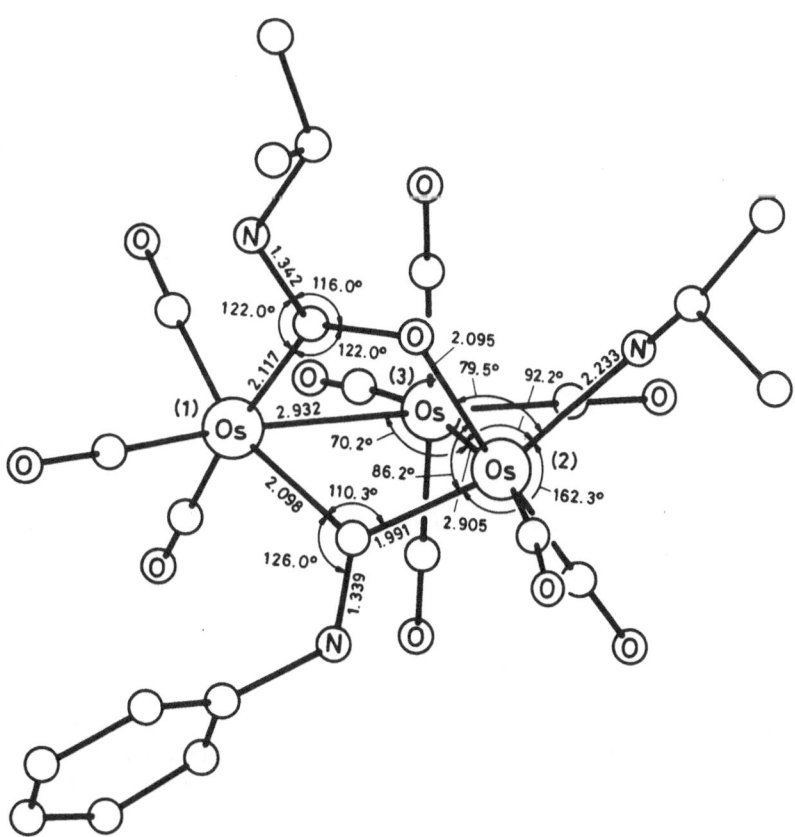

Fig. 22. Molecular structure of $Os_3(\mu_2,\eta^1\text{-}C\text{=}NHC_6H_5)(\mu_2,\eta^2\text{-}O\text{=}CNHC_3H_7\text{-}i)(CO)_9NH_2C_3H_7\text{-}i$ (No. 40) with selected bond distances (in Å) and bond angles [45, 47].

References on pp. 72/3

Os$_3$(μ_2,η^1-C$_6$H$_5$)(μ_2,η^2-O=CC$_6$H$_5$)(CO)$_9$(μ_3-Se)$_2$ (Table **2**, No. **43**) crystallizes in the triclinic space group P$\bar{1}$ − C$_i^1$ (No. 2) with a = 8.808(3), b = 15.346(4), c = 20.108(5) Å, α = 105.23(2)°, β = 99.52(2)°, γ = 102.59(2)°; Z = 4, D$_c$ = 3.03 g/cm^3, R = 0.0513. One of the two independent but closely similar molecules in the unit cell is shown in **Fig. 23**. No metal–metal bonds were observed; the Os–Os distances range between 3.195 and 3.785 Å. The C$_6$H$_5$CO ligand can be regarded as a normal three-electron donating acyl bridge formed by Se–C$_6$H$_5$ bond cleavage and C$_6$H$_5$ migration to CO. The bridging C$_6$H$_5$ group, coordinated to Os(3) by a σ bond and to Os(1) by long-range π-interactions, can also be considered as a three-electron donor. The three metal centers are all octrahedral diamagnetic d^6 Os^{II} atoms [42].

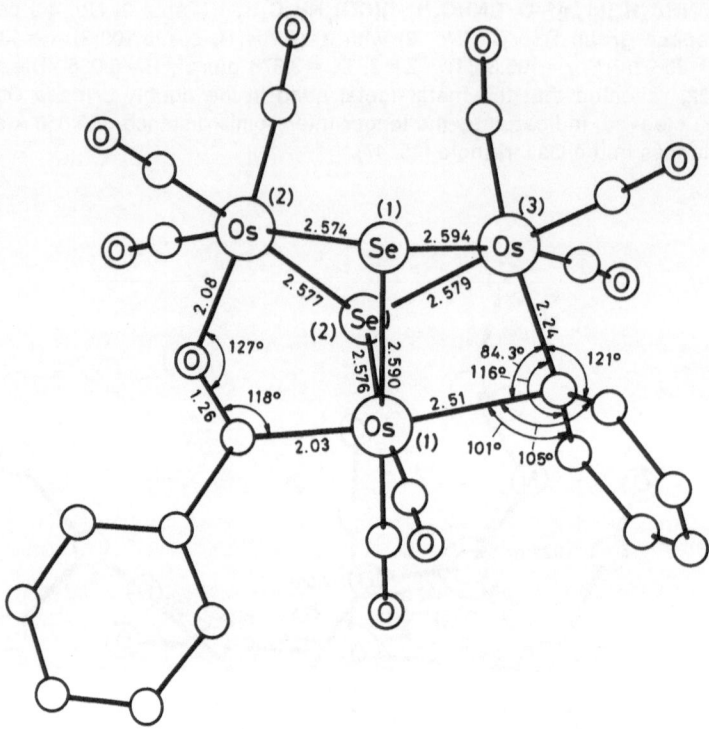

Fig. 23. Molecular structure of one of the two independent molecules of Os$_3$(μ_2,η^1-C$_6$H$_5$)(μ_2,η^2-O=CC$_6$H$_5$)(CO)$_8$(μ_3-Se)$_2$ (No. 43) with selected bond distances (in Å) and bond angles [42].

Os$_3$(μ_2,η^1-C$_6$H$_5$)(μ_3,η^2-P(C$_6$H$_5$)C$_6$H$_4$)(CO)$_9$(μ_2,η^1-P(C$_6$H$_5$)$_2$) (Table **2**, No. **44**) crystallizes in the tetragonal space group I4$_1$/a − C$_{4h}^6$ (No. 88) with a = b = 18.818, c = 41.191 Å; Z = 16, R = 0.049. The plane of the bridging phenyl group (see **Fig. 24**) is orthogonal to the Os(1)–Os(3) bond direction with an Os(3)–C–Os(1) bond angle of 85.0°. The significant difference in the bond lengths between the bridging carbon of C$_6$H$_5$ and Os(1) and Os(3), respectively, is due to the strong trans influence exerted by the μ-P atom at Os(3), in contrast to that of the CO group which is positioned trans to the Os(1)–C bond of the μ-C$_6$H$_5$ ligand [1].

References on pp. 72/3

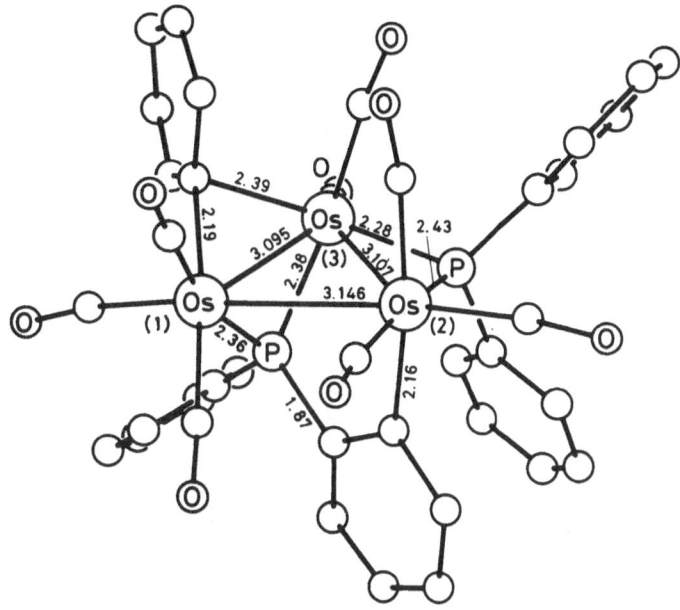

Fig. 24. Molecular structure of $Os_3(\mu_2,\eta^1\text{-}C_6H_5)(\mu_3,\eta^2\text{-}P(C_6H_5)C_6H_4)(CO)_8(\mu_2,\eta^1\text{-}P(C_6H_5)_2)$ (No. 44) with selected bond distances (in Å) [1].

$Os_3(\mu_2,\eta^1\text{-}C\text{=}NCF_3)_2(CO)_{10}D$ (D = CO, NCCH$_3$; Table 2, Nos. **46** and **47**). No. 46 crystallizes in the orthorhombic space group Pca2$_1$ – C$_{2v}^5$ (No. 29) with a = 14.456(2), b = 11.077(2), c = 14.301(4) Å; Z = 4, D$_c$ = 3.10 g/cm^3, R = 0.029 [43, 46], whereas No. 47 crystallizes in the triclinic space group P$\bar{1}$ – C$_i^1$ (No. 2) with a = 10.390(3), b = 15.416(3), c = 8.079(1) Å; α = 99.16(2)°, β = 104.61(2)°, γ = 94.40(2)°; Z = 2, D$_c$ = 2.97 g/cm^3, R = 0.024 [46]. Nos. 46 and 47 are structurally analogous with the exception that one CO group at the central Os(2) in No. 46 is substituted by an NCCH$_3$ ligand (No. 47; see **Fig. 25**). Both molecules consist of an open triosmium cluster with the μ_2,η^1-CNCF$_3$ ligands bridging each of the two metal-metal bonds [43, 46]. Selected bond distances and angles for Nos. 46 and 47 are [46]:

atoms	distance (Å)		atoms	angle (°)	
	No. 46	No. 47		No. 46	No. 47
Os(1)–Os(2)	2.847(1)	2.823(9)	Os(1)–Os(2)–Os(3)	140.39(4)	137.06(2)
Os(1)–Os(3)	2.859(1)	2.863(1)	C(1)–N(1)–C(3)	117 (2)	121.0 (9)
Os(1)–μ-C(1)	2.01(2)	2.08(1)	C(2)–N(2)–C(4)	120 (2)	120 (1)
Os(2)–μ-C(1)	2.10(2)	2.06(1)			
Os(2)–μ-C(2)	2.13(2)	2.13(1)			
Os(3)–μ-C(2)	2.07(2)	2.06(1)			
C(1)–N(1)	1.35(2)	1.27(2)			
C(2)–N(2)	1.27(2)	1.27(1)			

$Os_3(\mu_2,\eta^1\text{-}C\text{=}NCF_3)_2(CO)_9D$ (D = CO, NCCH$_3$; Table 2, Nos. **48** and **49**). No. 48 crystallizes in the orthorhombic space group Pbcn – D$_{2h}^{14}$ (No. 60) with a = 26.940(6), b = 11.361(3), c =

References on pp. 72/3

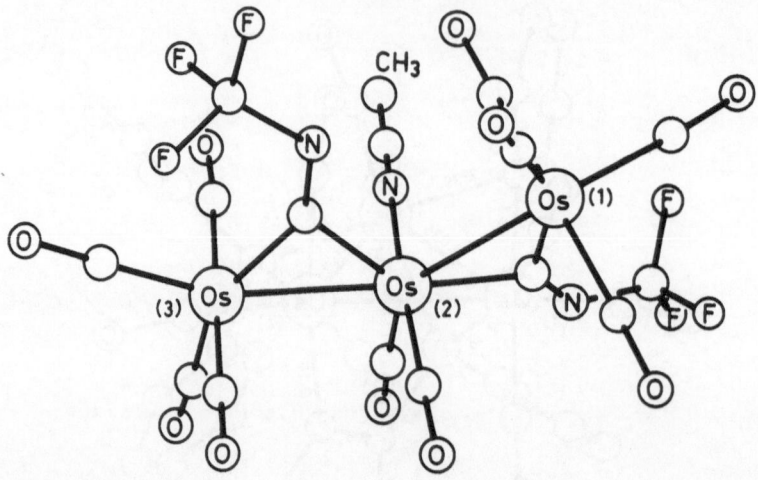

Fig. 25. Molecular structure of $Os_3(\mu_2,\eta^1-C=NCF_3)_2(CO)_{10}NCCH_3$ (No. 47) [46].

14.450(2) Å; Z = 8, D_c = 3.13 g/cm³, R = 0.034 [43, 46], whereas No. 49 crystallizes in the monoclinic space group $P2_1/n - C_{2h}^5$ (No. 14) with a = 19.065(4), b = 13.080(2), c = 9.418(2) Å; β = 90.12°; Z = 4, D_c = 2.98 g/cm³, R = 0.036 [46]. Nos. 48 and 49 are structurally analogous with the exception that an axial CO group at the unbridged Os(2) atom in No. 48 is substituted by an NCCH₃ ligand (No. 49; in **Fig. 26**). In contrast to Nos. 46 and 47, Nos. 48 and 49 consist of a closed Os₃ triangle having three metal-metal bonds; the two μ_2,η^1-CNCF₃ ligands bridge the same edge of the cluster. The doubly-bridged Os-Os bond is significantly shorter than the unbridged metal-metal bonds [43, 46]. Selected bond distances and angles for Nos. 48 and 49 are [46]:

atoms	distance (Å)		atoms	angle (°)	
	No. 48	No. 49		No. 48	No. 49
Os(1)-Os(2)	2.850(1)	2.736(1)	C(1)-N(1)-C(3)	126 (2)	120 (1)
Os(2)-Os(3)	2.850(1)	2.833(9)	C(2)-N(2)-C(4)	125 (2)	124 (2)
Os(1)-Os(3)	2.742(1)	2.738(9)			
Os(1)-μ-C(1)	2.15(2)	2.08(2)			
Os(3)-μ-C(1)	2.11(2)	2.13(2)			
Os(1)-μ-C(2)	2.11(2)	2.12(2)			
Os(3)-μ-C(2)	2.15(2)	2.09(2)			
C(1)-N(1)	1.22(2)	1.25(2)			
C(2)-N(2)	1.24(3)	1.21(2)			

(μ-H)Os₃(μ_2,η^1-C=N(CH₃)₂)₂(CO)₇(μ-SC₆H₅) (Table 2, No. **50**) crystallizes in the monoclinic space group $P2_1/c - C_{2h}^5$ (No. 14) with a = 16.891(2), b = 8.390(2), c = 16.966(2) Å, β = 98.26(1)°; Z = 4, D_c = 2.76 g/cm³, R = 0.042. The structure, shown in **Fig. 27**, is assumed to correspond to the major isomer (Isomer 1), based on ¹H NMR determinations in solution. One of the μ_2,η^1-carbene ligands and the benzenethiolato ligand lie on one side, while the second μ_2,η^1-carbene ligand is on the opposite side of the Os₃-plane. The Os(1)-Os(3) bond is extremely short, probably caused by steric interactions between the bridging ligands. The

References on pp. 72/3

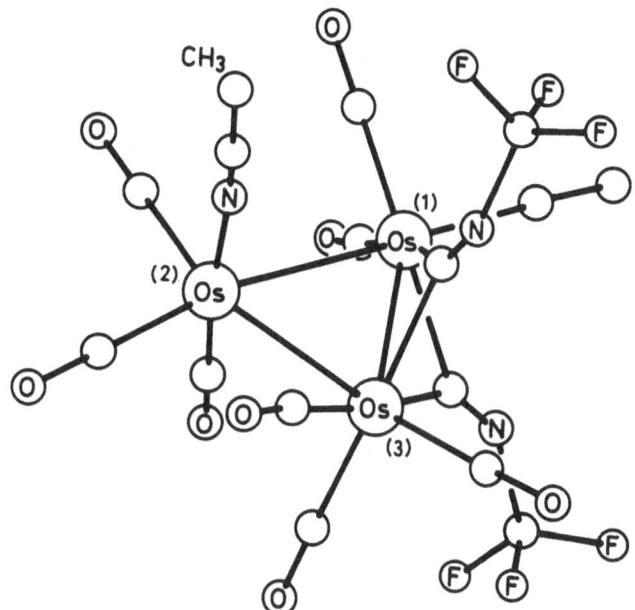

Fig. 26. Molecular structure of $Os_3(\mu_2,\eta^1\text{-}C=NCF_3)_2(CO)_9NCCH_3$ (No. 49) [46].

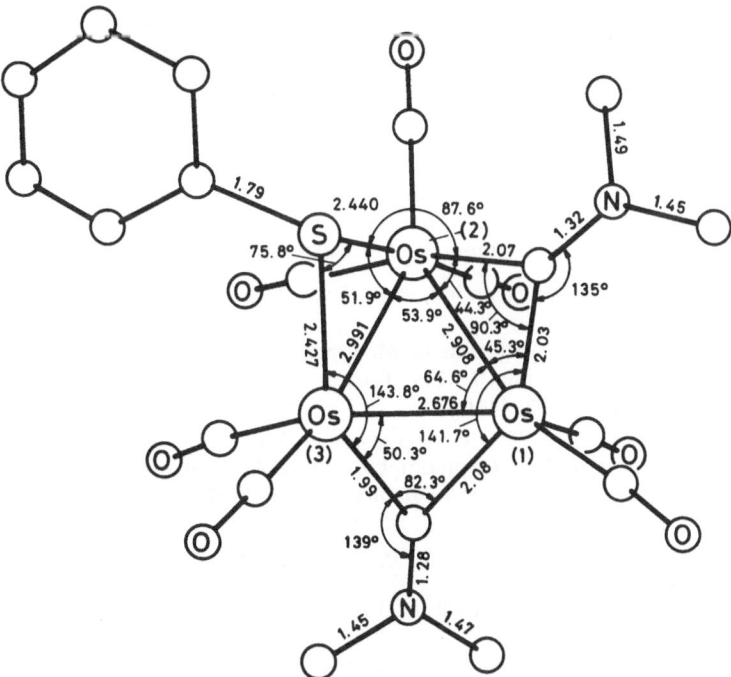

Fig. 27. Molecular structure of $(\mu\text{-}H)Os_3(\mu_2,\eta^1\text{-}C=N(CH_3)_2)_2(CO)_7(\mu\text{-}SC_6H_5)$ (No. 50) with selected bond distances (in Å) and bond angles [34].

hydride ligand is not observed crystallographically but it is believed to bridge the Os(2)–Os(3) bond, based on the displacements of the CO ligands and the elongation of the Os(2)–Os(3) bond. Average bond distances of Os–C and C≡O are 1.87 and 1.16 Å [34].

Variable temperature ^{1}H NMR investigations between 25 and 105 °C indicated an increasing interconversion rate with increasing temperature. At 60 °C, the resonances of Isomer 1 at $\delta = 3.24$ and 3.25 ppm were shifted toward the lower field resulting in broad, but slightly separated signals around 3.34 to 3.37 ppm; a strong broadening was observed at 84 °C followed by collapsing into an extremely broad singlet at 105 °C. The low–field resonances of Isomer 2 at $\delta = 3.42$ and 3.43 ppm merged at ca. 84 °C with the average of the low–field pair from Isomer 1, whereas the resonances at $\delta = 2.53$ and 3.02 ppm were almost completely collapsed into the baseline. Over the range of 84 to 105 °C the low–field average (ca. 3.39 ppm) of both isomers had sharpened [34].

The results of the ^{1}H NMR determinations led to the conclusion that the two isomers probably differ in the orientation of the phenyl ligand being achieved by an inversion of the configuration on the sulfur atom placing the C_6H_5 group away from (Isomer 1) or nearly over the face of the cluster (Isomer 2). The large separation of the higher–field CH_3 signals of the μ_2,η^1-C=N(CH$_3$)$_2$ bridge lying on the same side of the Os$_3$ face as μ-SC$_6$H$_5$ in Isomer 2 at 25 °C is probably induced by a ring current shielding effect, which is affected by the orientation of the C_6H_5 group over the Os$_3$ group [34].

References:

[1] Bradford, C. W.; Nyholm, R. S.; Gainsford, G. J.; Guss, J. M.; Ireland, P. R.; Mason, R. (J. Chem. Soc. Chem. Commun. **1972** 87/9).

[2] Bradford, C. W.; Nyholm, R. S. (J. Chem. Soc. Dalton Trans. **1973** 529/33).

[3] Deeming, A. J.; Hasso, S. (J. Organomet. Chem. **112** [1976] C 39/C 42).

[4] Yin, C. C.; Deeming, A. J. (J. Organomet. Chem. **133** [1977] 123/38).

[5] Gavens, P. D.; Mays, M. J. (J. Organomet. Chem. **162** [1978] 389/401).

[6] Adams, R. D.; Golembeski, N. M. (J. Am. Chem. Soc. **101** [1979] 2579/87).

[7] Adams, R. D.; Golembeski, N. M. (Inorg. Chem. **18** [1979] 2255/8).

[8] Keister, J. B. (J. Chem. Soc. Chem. Commun. **1979** 214/5).

[9] Churchill, M. R.; Hollander, F. J. (Inorg. Chem. **19** [1980] 306/10).

[10] Evans, J.; Gracey, B. P. (J. Organomet. Chem. **228** [1982] C 4/C 6).

[11] Henrick, K.; McPartlin, M.; Deeming, A. J.; Hasso, S.; Manning, P. (J. Chem. Soc. Dalton Trans. **1982** 899/906).

[12] Keister, J. B.; Payne, M. W.; Muscatella, M. J. (Organometallics **2** [1983] 219/25).

[13] Shapley, J. R.; Cree-Uchiyama, M. E.; St. George, G. M.; Churchill, M. R.; Bueno, C. (J. Am. Chem. Soc. **105** [1983] 140/2).

[14] Green, M.; Orpen, A. G.; Schaverien, C. J. (J. Chem. Soc. Chem. Commun. **1984** 37/9).

[15] Bavaro, L. M.; Keister, J. B. (J. Organomet. Chem. **287** [1985] 357/67).

[16] Beanan, L. R.; Keister, J. B. (Organometallics **4** [1985] 1713/21).

[17] Farrugia, L. J.; Miles, A. D.; Stone, F. G. A. (J. Chem. Soc. Dalton Trans. **1985** 2437/47).

[18] Koridze, A. A.; Kizas, O. A.; Kolobova, N. E.; Petrovskii, P. V. (J. Organomet. Chem. **292** [1985] C 1/C 3).

[19] Shapley, J. R.; Yeh, W.-Y.; Churchill, M. R.; Li, Y.-J. (Organometallics **4** [1985] 1898/1900).

[20] Yeh, W.-Y.; Shapley, J. R.; Li, Y.-J.; Churchill, M. R. (Organometallics **4** [1985] 767/72).

[21] Adams, R. D.; Babin, J. E.; Kim, H.-S. (Organometallics **5** [1986] 1924/5).

[22] Farrugia, L. J. (Acta Crystallogr. C **42** [1986] 680/2).

[23] Koridze, A. A.; Kizas, O. A.; Kolobova, N. E.; Yanovskii, A. I.; Struchkov, Yu. T. (J. Organomet. Chem. **302** [1986] 413/5).

[24] Shaffer, M. R.; Keister, J. B. (Organometallics **5** [1986] 561/6).

[25] Yeh, W.-Y.; Shapley, J. R.; Ziller, J. W.; Churchill, M. R. (Organometallics **5** [1986] 1757/63).

[26] Adams, R. D.; Babin, J. E. (Inorg. Chem. **26** [1987] 980/4).

[27] Adams, R. D.; Babin, J. E. (Organometallics **6** [1987] 2236/41).

[28] Adams, R. D.; Babin, J. E. (J. Am. Chem. Soc. **109** [1987] 6872/3).

[29] Adams, R. D.; Babin, J. E.; Kim, H.-S. (J. Am. Chem. Soc. **109** [1987] 1414/24).

[30] Adams, R. D.; Babin, J. E.; Kim, H.-S. (Organometallics **6** [1987] 749/54).

[31] Deeming, A. J.; Donovan-Mtunzi, S.; Kabir, S. E.; Arce, A. J.; De Sanctis, Y. (J. Chem. Soc. Dalton Trans. **1987** 1457/61).

[32] Koridze, A. A.; Kizas, O. A.; Kolobova, N. E.; Petrovskii, P. V. (Izv. Akad. Nauk SSSR, Ser. Khim. **1987** 1630/5; Bull. Acad. Sci. USSR Div. Chem. Sci. **36** [1987] 1508/12).

[33] Adams, R. D.; Babin, J. E. (Organometallics **7** [1988] 963/9).

[34] Adams, R. D.; Babin, J. E. (New J. Chem. **12** [1988] 641/8).

[35] Kabir, S. E. (Indian J. Chem. A **27** [1988] 677/9).

[36] Pribich, D. C.; Rosenberg, E. (Organometallics **7** [1988] 1741/5).

[37] Green, M.; Orpen, A. G.; Schaverien, C. J. (J. Chem. Soc. Dalton Trans. **1989** 1333/40).

[38] Adams, R. D.; Babin, J. E.; Kim, H.-S.; Tanner, J. T.; Wolfe, T. A. (J. Am. Chem. Soc. **112** [1990] 3426/35).

[39] Adams, R. D.; Babin, J. E.; Wolfe, T. A. (Organometallics **9** [1990] 440/6).

[40] Krause, J. A.; Workman, D. P.; Shore, S. G. (Acta Crystallogr. C **46** [1990] 2086/8).

[41] Anslyn, E. V.; Green, M.; Nicola, G.; Rosenberg, E. (Organometallics **10** [1991] 2600/5).

[42] Arce, A. J.; Arrojo, P.; Deeming, A. J.; De Sanctis, Y. (J. Chem. Soc. Chem. Commun. **1991** 1491/2).

[43] Adams, R. D.; Chi, Y.; DesMarteau, D. D.; Lentz, D.; Marschall, R. (J. Am. Chem. Soc. **114** [1992] 1909/10).

[44] Arce, A. J.; Arrojo, P.; De Sanctis, Y.; Deeming, A. J.; West, D. J. (Polyhedron **11** [1992] 1013/21).

[45] Lin, Y.-W.; Gau, H.-M.; Wen, Y.-S.; Lu, K.-L. (Organometallics **11** [1992] 1445/7).

[46] Adams, R. D.; Chi, Y.; DesMarteau, D. D.; Lentz, D.; Marschall, R.; Scherrmann, A. (J. Am. Chem. Soc. **114** [1992] 10822/6).

[47] Lu, K.-L.; Chen, C.-J.; Lin, Y.-W.; Gau, H.-M.; Hong, F.-E.; Wen, Y.-S. (Organometallics **12** [1993] 2188/96).

3.1.6.2.3 Compounds with μ_3,η^2-Bridging ^{1}L Ligands of the CR'C(R)=O Type

The compounds dealt with in this section all have μ_3,η^2-bridging ^{1}L ligands, where a carbon atom bridges two Os atoms and a keto oxygen atom interacts with the third Os atom of the Os$_3$ triangle; see Formula IVa (bridging hydrides omitted). The most frequent compound types are represented by Formulas I to III; other types having μ_3,η^2-bridging ^{1}L ligands are presented by individual figures in Table 3.

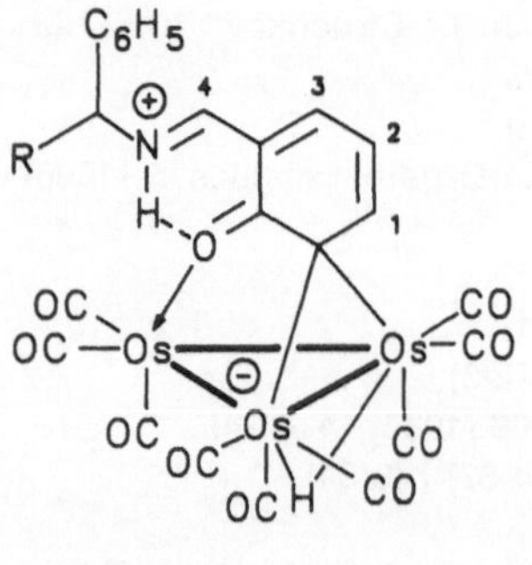

I 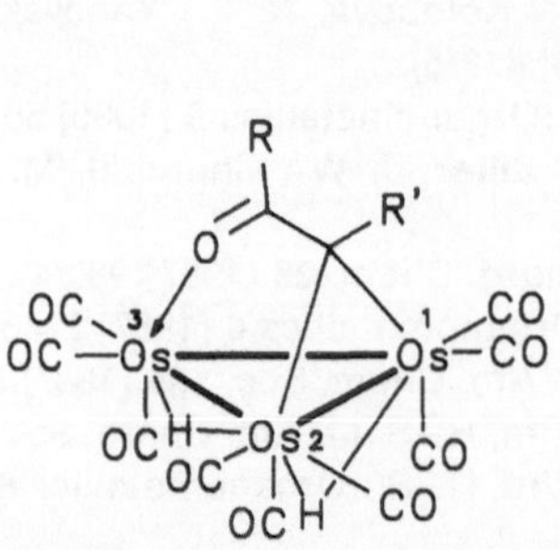   II 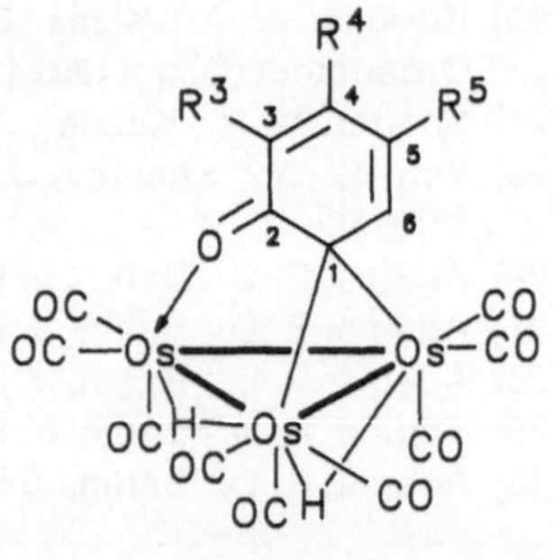   III

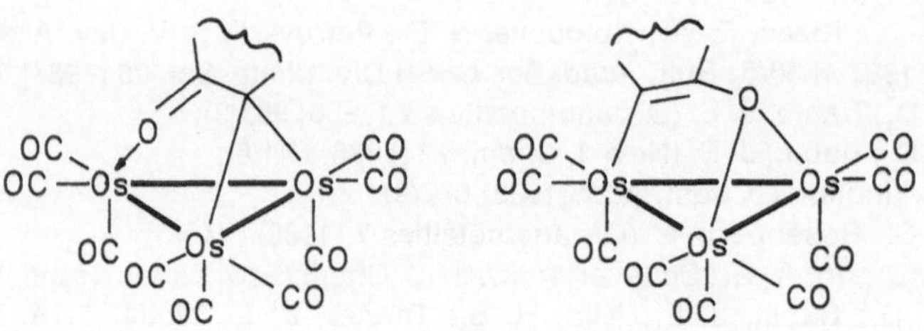

a   IV   b

Earlier, Nos. 4, 8 to 10, 13, 14, 16, 23, and 24 were incorrectly formulated as valence tautomers of type IVb (bridging hydrides omitted) having a bridging oxygen and a nonbridging σ bonded carbon [1, 2]. But detailed NMR and IR spectroscopic reinvestigations, as well as some X-ray analyses, indicated structures corresponding to type IVa for Nos. 4, 8 to 10, and 13 [3 to 6, 8, 9]. Although Nos. 14, 16, 23, and 24 [1] were not reinvestigated, they were also assigned in Table 3 to structure type IVa based on their spectra in comparison to the reexamined compounds [5, 6]. Alternative coordination possibilities shown in Formulas V to VII were found in starting materials or in additional products in the preparation of Nos. 2 and 3 [8].

The compounds of type I (Nos. 2 and 3) are best described by a zwitterionic structure as their ν(CO) absorptions are at ca. 35 to 45 cm^{-1} lower wavenumbers than, for example, in No. 13 [8].

The complexes of type III show hydride exchange at rates essentially insensitive to the ring substituents. All show similarly broad hydride signals at 30 °C which sharpened into doublets at lower temperatures; the mechanism of hydride exchange is unknown [6].

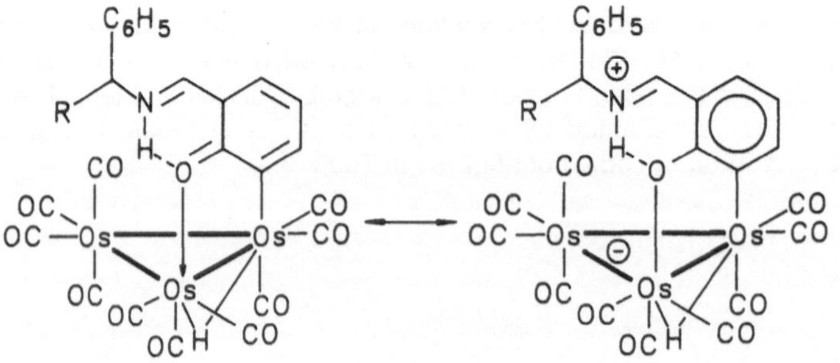

V

References on pp. 87/8

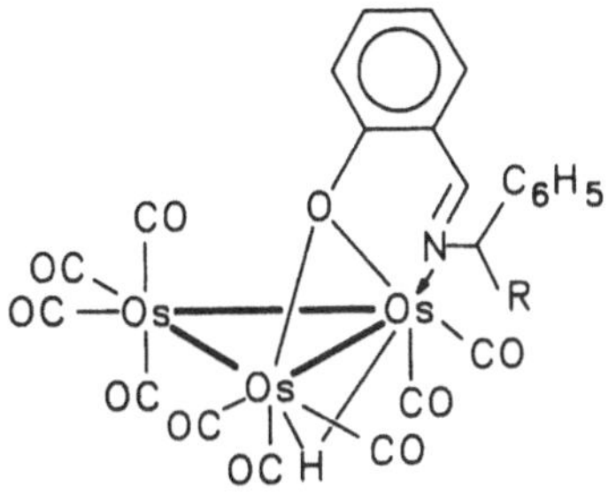

VI 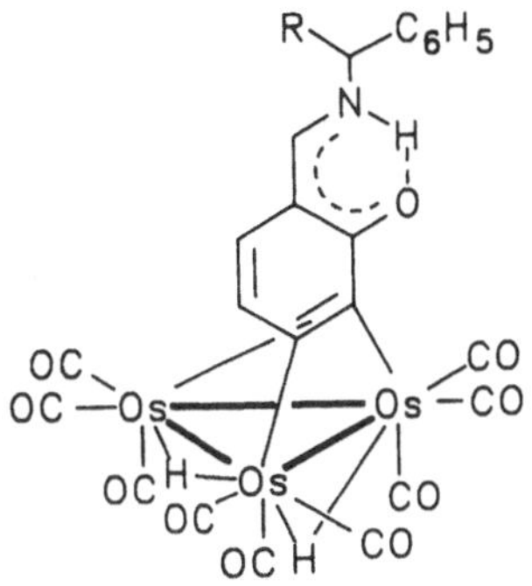   VII

Most compounds listed in Table 3 were prepared by the following methods:

Method I: $(\mu\text{-H})_2Os_3(\mu_3,\eta^2\text{-O=C}_6H_8)(CO)_9$ (No. 11), $(\mu\text{-H})_2Os_3(\mu_3,\eta^2\text{-O=C}_6H_{4-n}R_n)(CO)_9$
             (Nos. 13, 15 to 21), and $(\mu\text{-H})_2Os_3(\mu_3,\eta^2\text{-O=C}_{10}H_6\text{-1})(CO)_9$ (No. 24) were pre-
             pared from $Os_3(CO)_{12}$ with $O=C_6H_{10}R_nC_6H_{5-n}OH$, or $1\text{-C}_{10}H_7OH$ either directly
             at 170 to 185 °C [3] or in dry decane [4, 5], nonane [6], or xylene [1, 6] at
             reflux temperature for 4 to 16 h; work-up by TLC with pentane/toluene (ca.
             5:1) as eluant [1, 5].

             The cyclohexanone complex No. 11 was obtained in better yields with cyclohex-
             enone instead of cyclohexanone; the expected cyclohexenone cluster No. 12
             was not formed by this reaction [4, 5].

Method II: $(\mu\text{-H})_2Os_3(\mu_3,\eta^2\text{-O=CHCR})(CO)_9$ (Nos. 4, 8 to 10) were prepared by decarbony-
             lation/hydrogen transfer from $(\mu\text{-H})Os_3(\mu_2,\eta^2\text{-O=CCH}_2R)(CO)_{10}$ (R = H,
             n-C$_4$H$_9$, n-C$_5$H$_{11}$, C$_6$H$_5$; see "Organoosmium Compounds" B5, Section
             3.1.6.1.2.2, in preparation) in dry nonane at reflux temperature for 30 min
             to 8 h [2, 4, 5].

Method III: Nos. 1 to 3, 6, and 12 were prepared by thermal decarbonylation of $(\mu\text{-H})Os_3$-
             $\{\mu_2,\eta^2\text{-O=CHC=P(C}_6H_5)_3\}(CO)_{10}$, $(\mu\text{-H})Os_3(\mu_2,\eta^2\text{-O=C}_6H_3\text{=CHNHCHRC}_6H_5)$-
             $(CO)_{10}$ (see Formula V; R = H, CH$_3$), $(\mu\text{-H})Os_3\{\mu_2,\eta^2\text{-O=C(OCH}_3)CH_2\}(CO)_{10}$,
             or $(\mu\text{-H})Os_3(\mu_2,\eta^2\text{-O=C}_6H_7\text{-c})(CO)_{10}$ (all "Organoosmium Compounds" B 5,
             Section 3.1.6.1.2.3, in preparation) in refluxing heptane, octane, nonane, or
             toluene for 2 to 8 h in the dark [4, 5, 8, 11, 12, 13]; work-up by TLC with
             ether/light petroleum (7:4) as eluant [8].

Method IV: $(\mu\text{-H})_2Os_3(\mu_3,\eta^2\text{-O=CHCH})(CO)_9$ (No. 4), $(\mu\text{-H})_2Os_3(\mu_3,\eta^2\text{-O=C}_6H_3R)(CO)_9$
             (Nos. 13, 15, 17), and $(\mu\text{-H})_2Os_3(\mu_3,\eta^2\text{-O=C}_{10}H_6\text{-2})(CO)_9$ (No. 23) were pre-
             pared from $(\mu\text{-H})Os_3(CO)_{10}(\mu_2,\eta^1\text{-OR'})$ (R' = CH=CH$_2$, C$_6$H$_5$, C$_6$H$_4$F-4,
             C$_6$H$_4$OH-4, C$_{10}$H$_7$-2) by decarbonylation/hydrogen transfer in refluxing cyclo-
             hexane, octane or nonane for 9 to 65 h [1, 2, 5, 6, 7]; work-up by TLC with
             pentane or light petroleum/toluene (ca. 5:1) as eluant [1, 5].Decarbonylation/
             hydrogen-transfer reaction of $(\mu\text{-H})Os_3(CO)_{10}(\mu_2,\eta^1\text{-OCH=CH}_2)$ in refluxing
             cyclohexane gave No. 4 in low yields, additionally with $(\mu\text{-H})_3Os_3(\mu_3,\eta^1\text{-CH})$-
             $(CO)_9$ in 39% yield and $(\mu\text{-H})Os_3(\mu_2,\eta^2\text{-O=CCH}_2)(CO)_{10}$, whereas only No. 4
             and $(\mu\text{-H})_3Os_3(\mu_3,\eta^1\text{-CH})(CO)_9$ were observed in refluxing nonane (30 min).
             Only traces of No. 4 were formed under CO in refluxing cyclohexane after
             38 h; mainly unreacted startingmaterial, $Os_3(CO)_{12}$, and $(\mu\text{-H})_3Os_3(\mu_3,\eta^1\text{-CH})$-
             $(CO)_9$ were obtained [5].

References on pp. 87/8

Table 3
Compounds with μ_3,η^2-Bridging ^{1}L Ligands Bonded by a μ_2-C Atom and a Keto O Atom.
An asterisk preceding the compound number indicates further information at the end of
the table, pp. 84/7.
Explanations, abbreviations, and units on p. X

No. compound	method of preparation (yield in %) properties and remarks

compound of the type $(\mu\text{-H})Os_3\{\mu_3,\eta^2\text{-O=CHCP}(C_6H_5)_3\}(CO)_9$

*1

III [11, 13]
by decarbonylation of $(\mu\text{-H})Os_3\{\mu_2,\eta^2\text{-}$
O=CHC=P$(C_6H_5)_3\}(CO)_{10}$ in CH_2Cl_2 at 20 °C
in daylight for 3 d; isolated by TLC with
CH_2Cl_2/light petroleum (80%) [11, 13]
yellow-orange crystals from CH_2Cl_2/hexane
[13]
^{1}H NMR (CDCl$_3$): −12.71 (s, μ-H), 7.1 to 7.8
(m, C_6H_5), 10.65 (d, CH; J(H,P)=7.0) [11, 13]
IR (CH_2Cl_2): 1955, 1985, 2015, 2036, 2071 (all
CO) [11, 13]; due to the zwitterionic character the ν(CO) wavenumbers are, as expected,
ca. 20 to 30 cm^{-1} lower than for the closely
related No. 26 [13]
decarbonylation in heptane or toluene at 96 to
100 °C yielded
$(\mu\text{-H})_2Os_3\{\mu_3,\eta^3\text{-O=CHCP}(C_6H_5)_2C_6H_4\}(CO)_8$
[11, 13]

compounds of the type $(\mu\text{-H})Os_3(\mu_3,\eta^2\text{-O=}C_6H_3CH=NHCHRC_6H_5)(CO)_9$ (Formula I)

2 $(\mu\text{-H})Os_3(\mu_3,\eta^2\text{-O=}C_6H_3CH=NHCH_2C_6H_5)(CO)_9$

III (6%; along with
$(\mu\text{-H})Os_3(\mu_2,\eta^2\text{-}C_{14}H_{12}NO)(CO)_9$, Formula VI,
p. 75, and $(\mu\text{-H})_2Os_3(\mu_3,\eta^2\text{-}C_{14}H_{11}NO)(CO)_9$,
Formula VII, p. 75, R=H) [8]
by-product in the preparation of
$(\mu\text{-H})Os_3(\mu_2,\eta^2\text{-O=}C_6H_3CH=NHCH_2C_6H_5)$-
$(CO)_{10}$ (see Formula V, p. 74, R=H) from
$Os_3(CO)_{10}(NCCH_3)_2$ and
$C_6H_5CH_2N=CHC_6H_4OH\text{-}2$ in benzene for 1 h;
isolated by TLC with ether/light petroleum
(1:4) as eluant (ca. 3%) [8]
dark green crystals [8]
^{1}H NMR (CDCl$_3$, 30 °C): −12.37 (s, μ-H), 4.92
(br, CH$_2$), 6.31 (t, H^2; J(H,H)=7.2), 7.73 (dd,
H^3; J(H,H)=2.1, 7.3), 7.94 (dd, H^1; J(H,H)=
1.6, 7.2). 8.03 (d, H^4; J(H,NH)=11.2), 12.50
(br, NH) [8]; for numbering, see Formula I
IR (cyclohexane): 1942, 1951, 1970, 1988, 2018,
2042, 2073 (all CO) [8]

References on pp. 87/8

Table 3 (continued)

No. compound	method of preparation (yield in %) properties and remarks

3 $(\mu-H)Os_3\{\mu_3,\eta^2-O=C_6H_3CH=NHCH(CH_3)C_6H_5\}(CO)_9$

III (along with $(\mu-H)Os_3(\mu_2,\eta^2-C_{15}H_{14}NO)(CO)_9$, Formula VI, p. 75, and $(\mu-H)_2Os_3(\mu_3,\eta^2-C_{15}H_{13}NO)(CO)_9$, Formula VII, p. 75, R=CH$_3$) [8]

by-product in the preparation of $(\mu-H)Os_3-\{\mu_2,\eta^2-O=C_6H_3CH=NHCH(CH_3)C_6H_5\}(CO)_{10}$ ("Organoosmium Compounds" B 5, Section 3.1.6.1.2.3, in preparation; see Formula V, p. 74, R=CH$_3$) from $Os_3(CO)_{10}(NCCH_3)_2$ and $C_6H_5CH(CH_3)N=CHC_6H_4OH-2$ as described for No. 2 (only traces) [8]

dark green crystals [8]

IR (cyclohexane): 1940, 1952, 1970, 1988, 2020, 2044, 2074 (all CO) [8]

compounds of the type $(\mu-H)_2Os_3(\mu_3,\eta^2-O=CRCR')(CO)_9$ (Formula II) and $[(\mu-H)_3Os_3\{\mu_3,\eta^2-O=C(OCH_3)CH\}(CO)_9]BF_4$

4 $(\mu-H)_2Os_3(\mu_3,\eta^2-O=CHCH)(CO)_9$

II (along with $(\mu-H)_3Os_3(\mu_3,\eta^1-CH)(CO)_9$, Section 3.1.6.3) [2, 5], IV (3.5%) [2, 5, 7]

from $(\mu-H)Os_3(CO)_{10}(\mu-OCH=CH_2)$ upon UV irradiation in cyclohexane at 25 °C for 2.5 h; work-up by TLC with CH$_2$Cl$_2$/light petroleum as eluant (42%) [7, 9]

from $Os_3(\mu_3,\eta^2-O=CHCH)(\mu-CO)(CO)_9$ (No. 26) and H$_2$ (1 atm) in refluxing cyclohexane or octane for 45 or 120 min (56 to 76%); lower yields (18 to 38%) in refluxing octane for 1.5 h in the absence of H$_2$; work-up as before [7, 9]

pale yellow crystals [9]; brown solid [5]

^{1}H NMR (CDCl$_3$): −14.42 (d, μ-H), −12.55 (d, μ-H; J(H,H)=2.0), 4.39 (d, CH), 11.59 (d, CH=O; J(H,H)=4.0) [7, 9]; the formation of two isomers was deduced from ^{1}H NMR determinations in CDCl$_3$ and acetone-d$_6$ showing a pair of doublets for the μ_3,η^2-^{1}L ligand and a pair of hydride doublets; the relative intensities of the two sets varied with solvent and temperature [5]; ^{1}H NMR results in [5] are doubtful because No. 4 was obtained only in small amounts [7]

IR (KBr): 1500 (C=O) [5, 7, 9]; (cyclohexane): 1953, 1983, 2002, 2014, 2027, 2056, 2084, 2112 (all CO) [5, 7, 9]

References on pp. 87/8

Table 3 (continued)

No. compound	method of preparation (yield in %) properties and remarks

5 $(\mu-H)_2Os_3\{\mu_3,\eta^2-O=C(N(C_2H_5)_2)CH\}(CO)_9$

by pyrolysis of $(\mu-H)_3Os_3\{\mu_3,\eta^1-CC(O)-N(C_2H_5)_2\}(CO)_9$ (Section 3.1.6.3) in dry toluene at room temperature for 24 h in a sealed tube under 1 atm of CO, followed by heating to 125 °C for 10 min; isolated by TLC (77%) [12]

pale yellow solid [12]

^1H NMR (CDCl$_3$): -15.25 (d, Os(2)HOs(3); $^2J(H,H) = 1.5$), -13.84 (t, Os(1)HOs(2); J(H,H) = 1.5), 1.20 (t, CH$_3$), 3.13 (q, CH$_2$; $^3J(H,H) = 7.0$), 3.37 (q, CH$_2$; $^3J(H,H) = 7.0$), 4.38 (d, CH; $^3J(H,H) = 1.5$); for numbering, see Formula II; the signal pattern observed for the ethyl groups was assigned to a partial resolution of the four CH$_2$ and two CH$_3$ resonances and not to a hydride migration process producing a plane of symmetry [12]

IR (cyclohexane): 1527 (C≈O), 1948, 1961, 1975, 1995, 2003, 2013, 2017, 2045, 2074, 2100 (all CO) [12]

mass spectrum: $[M]^+$, $[M - x\ CO - y\ H]^+$ [12]

6 $(\mu-H)_2Os_3\{\mu_3,\eta^2-O=C(OCH_3)CH\}(CO)_9$

III (17%) [12]

by pyrolysis of $(\mu-H)_3Os_3(\mu_3,\eta^1-CCO_2CH_3)(CO)_9$ (Section 3.1.6.3) in dry toluene at 120 to 125 °C for 21 h in a sealed tube under 1 atm of CO; isolated by TLC (78%) [12]

pale yellow solid [12]

^1H NMR (CDCl$_3$): -15.13 (d, Os(2)HOs(3); $^2J(H,H) = 1.6$), -13.67 (t, Os(1)HOs(2); J(H,H) = 1.2), 3.53 (s, CH$_3$), 3.94 (d, CH; $^3J(H,H) = 1.2$) [12]; for numbering, see Formula II

IR (cyclohexane): 1531 (C≈O), 1952, 1965, 1983, 2001, 2009, 2022, 2053, 2081, 2107 (all CO) [12]

mass spectrum: $[M]^+$, $[M - x\ CO - y\ H]^+$ [12]

protonation with HBF$_4 \cdot$O(C$_2$H$_5$)$_2$ yielded No. 7 [12]

7 $[(\mu-H)_3Os_3\{\mu_3,\eta^2-O=C(OCH_3)CH\}(CO)_9]BF_4$

from No. 6 and an excess of HBF$_4 \cdot$O(C$_2$H$_5$)$_2$ in CD$_2$Cl$_2$ at 0 °C; precipitated with ether [12]

white solid [12]

References on pp. 87/8

Table 3 (continued)

No. compound	method of preparation (yield in %) properties and remarks
	^1H NMR (CD_3CN): -17.34 (d, Os(1)HOs(3) and Os(2)HOs(3); ^3J(H,H) = 1.5), -14.47 (br s, Os(1)HOs(2)), 3.70 (s, CH_3), 4.04 (br s, CH) [12]; for numbering, see Formula II IR (CH_2Cl_2): 1977, 1996, 2016, 2046, 2062, 2076, 2110, 2125, 2151 (all CO) [12]
8 $(\mu-H)_2Os_3(\mu_3,\eta^2-O=CHCC_4H_9-n)(CO)_9$	II (15%) [2, 5] solid [5] ^1H NMR ($CDCl_3$, 27 °C): -14.05, -12.18 (d's, both μ-H), 0.87 (m, CH_3), 1.24 (m, $CH_2-\beta,\gamma$), 2.12 (m, $CH_2-\alpha$), 11.36 (s, CH) [5] IR (KBr): 1490 (C=O); (cyclohexane): 1980, 2001, 2014, 2027, 2055, 2084, 2109 (all CO) [5]
9 $(\mu-H)_2Os_3(\mu_3,\eta^2-O=CHCC_5H_{11}-n)(CO)_9$	II (10%) [2, 5] solid [5] ^1H NMR ($CDCl_3$, 27 °C): -14.00, -12.06 (d's, both μ-H), 0.88 (m, CH_3), 1.24 (m, $CH_2-\beta,\gamma,\delta$), 2.08 (m, $CH_2-\alpha$), 11.44 (s, CH) [5] IR (KBr): 1498 (C=O); (cyclohexane): 1980, 2002, 2014, 2026, 2055, 2083, 2108 (all CO) [5]
10 $(\mu-H)_2Os_3(\mu_3,\eta^2-O=CHCC_6H_5)(CO)_9$	II (17%) [2, 4, 5] yellow crystals [5] ^1H NMR ($CDCl_3$, 27 °C): -13.88, -11.79 (d's, both μ-H), 6.9 to 7.7 (m, C_6H_5), 11.27 (s, CH) [5]; ($CDCl_3$, -50 °C): -13.88, -11.79 (d's, each μ-H) [4] IR (KBr): 1497 (C=O)/(C=C) [5]; (cyclohexane): 1985, 1990, 2003, 2017, 2027, 2059, 2087, 2113 (all CO) [4, 5]

compounds of the type $(\mu-H)_2Os_3(\mu_3,\eta^2-O=C_6H_n-c)(CO)_9$ (n = 6 or 8)

11	I (20% using cyclohexenone, 17% using cyclohexanone, in both cases along with not properly characterized $(\mu-H)Os_3(C_6H_{11}O)(CO)_{10})$ [4, 5] yellow crystals [5] ^1H NMR ($CDCl_3$, 27 °C): -14.03, -12.36 (d's, both μ-H), 1.3 to 1.9 (m, $CH_2-\beta,\gamma,\delta$), 2.30 (m, $CH_2-\alpha$) [4, 5] IR (Nujol): 1496 (C=O) [5]; (cyclohexane): 1970, 1998, 2010, 2020, 2023, 2051, 2080, 2105 (all CO) [4, 5]

References on pp. 87/8

Table 3 (continued)

No. compound	method of preparation (yield in %) properties and remarks

12

III (74%, along with compound VIII, p. 84) [4, 5]

yellow crystals [5]

^1H NMR (CDCl$_3$, 27 °C): -14.02, -12.27 (d's, each μ-H), 1.78, 2.44 (m's, both CH$_2$), 4.94 (dt, CH; J(H,H) = 4.3, 9.6), 5.81 (d, CH; J(H,H) = 9.6) [4, 5]

IR (Nujol mull): 1494, 1505, 1615 (C=O)/(C=C) [5]; (cyclohexane): 1982, 2000, 2012, 2024, 2053, 2082, 2107 (all CO) [4, 5]

compounds of the type (μ-H)$_2$Os$_3$(μ$_3$,η^2-O=C$_6$H$_{4-n}$R$_n$)(CO)$_9$ (Formula III) and [((μ-H)$_3$Os$_3$(μ$_3$,η^2-O=C$_6$H$_4$)(CO)$_9$]$^+$

13 (μ-H)$_2$Os$_3$(μ$_3$,η^2-O=C$_6$H$_4$)(CO)$_9$

I (along with (μ-H)Os$_3$(CO)$_{10}$(μ$_2$,η^1-OC$_6$H$_5$) and Os$_3$(CO)$_{10}$(μ$_2$,η^1-OC$_6$H$_5$)$_2$ as the main-product) [3], IV (48%) [1]

yellow, air-stable crystals [1, 6]

^1H NMR (CDCl$_3$, 27 °C): -14.06, -11.66 (d's, each μ-H; J(H,H) = 1.5 at -60 °C [1, 4, 6]), 6.04 (ddd, H^5; J(H^5,H^6) = 7.5, J(H^5,H^4) = 8.5, J(H^5,H^3) = 1.8), 6.42 (d, H^3; J(H^3,H^4) = 7.0), 7.24 (dd, H^6; J(H^4,H^6) = 1.5), 7.69 (ddd, H^4) [1, 6]; for numbering, see Formula III

^{13}C NMR (CDCl$_3$): 116.5 (C^5), 124.1 (C^3), 138.6 (C^4), 166.1 (C^6), 175.0, 178.3 (C^2, C^1) [3, 6]; for numbering, see Formula III

IR (cyclohexane): 1981, 1988, 2003, 2014, 2028, 2035, 2058, 2086, 2112 (all CO) [1, 4, 6]

mass spectrum: [M]$^+$, [M$-$x CO$-$y H]$^+$ [1]

protonation with CF$_3$CO$_2$H yielded No. 14 [1]

thermolysis in nonane at 146 °C under CO (sealed glass tube) gave (μ-H)Os$_3$(CO)$_{10}$(μ$_2$,η^1-OC$_6$H$_5$) [3]

14 [(μ-H)$_3$Os$_3$(μ$_3$,η^2-O=C$_6$H$_4$)(CO)$_9$]$^+$

from No. 13 and neat CF$_3$CO$_2$H for several days, (60%) [1]

^1H NMR (no medium given, 27 °C): -15.36 (s, 2 μ-H), -13.96 (s, 1μ-H) [1]

15 (μ-H)$_2$Os$_3$(μ$_3$,η^2-O=C$_6$H$_3$F-5)(CO)$_9$

I [6], IV [6]

yellow air-stable crystals [6]

^1H NMR (CDCl$_3$, 27 °C): -13.85, -11.41 (d's, each μ-H; J(H,H) = 1.9 measured at low temperature), 6.40 (dd, H^3; J(H,H) = 10.0, J(H,^{19}F) = 4.5), 6.86 (dd, H^6; J(H,H) = 3.5, J(H,^{19}F) = 9.0), 7.59 (ddd, H^4; J(H,H) = 10.0,

Table 3 (continued)

No. compound	method of preparation (yield in %) properties and remarks

3.5, J(H,^{19}F) = 7.9) [6]; for numbering, see
Formula III
^{13}C NMR (CD$_2$Cl$_2$): 125.2 (C^3; J(C,^{19}F) = 7.5),
129.7 (C^4; J(C,^{19}F) = 27.7), 145.0 (C^6;
J(C,^{19}F) = 15.1) [6]; for numbering, see For-
mula III
IR (cyclohexane): 1982, 1988, 2004, 2015, 2030,
2036, 2058, 2086, 2113 (all CO) [6]

16 (μ-H)$_2$Os$_3$(μ_3,η^2-O=C$_6$H$_3$OH-3)(CO)$_9$

I (7%, impure) [1]
yellow crystals [1]
^{1}H NMR (CDCl$_3$, 27 °C): −14.02, −11.15 (s's,
both μ-H), 5.62 (s, OH), 5.98, 6.72, 7.16 (t, d,
d, H^4 to H^6; J = 8.0) [1]
IR (cyclohexane): 1983, 1990, 2005, 2016, 2030,
2037, 2059, 2087, 2113 (all CO) [1]

17 (μ-H)$_2$Os$_3$(μ_3,η^2-O=C$_6$H$_3$OH-5)(CO)$_9$

I [6], IV [6]
yellow air-stable crystals [6]
^{1}H NMR (acetone-d$_6$, 27 °C): −13.67, −11.44
(d's, each μ-H; J(H,H) = 1.6; measured below
−20 °C), 6.41 (d, H^3; J(H,H) = 9.5), 6.75 (d,
H^6; J(H,H) = 3.0), 7.59 (dd, H^4; J(H,H) = 3.3,
9.5) [6]; for numbering, see Formula III
IR (cyclohexane): 1980, 1986, 2002, 2013, 2025,
2034, 2056, 2084, 2110 (all CO) [6]

18 (μ-H)$_2$Os$_3$(μ_3,η^2-O=C$_6$H$_2$(CH$_3$)$_2$-3,4)(CO)$_9$

I (60%) [6]
yellow air-stable crystals [6]
^{1}H NMR (CDCl$_3$, 27 °C): −14.14, −11.77 (d's,
both μ-H; J(H,H) = 1.8; measured at low temperature), 1.86 (s, CH$_3$-4), 1.97 (s, CH$_3$-3),
5.89 (d, H^5; J(H,H) = 8.7), 7.00 (d, H^6;
J(H,H) = 8.7) [6]; for numbering, see Formula
III
^{13}C NMR (CD$_2$Cl$_2$): 120.3 (C^5), 130.9 (C^3), 148.4
(C^4), 162.8 (C^6), 176.0, 176.6 (C^2, C^1) [6]; for
numbering, see Formula III
IR (cyclohexane): 1978, 1984, 2000, 2011, 2024,
2030, 2053, 2083, 2110 (all CO) [6]

19 (μ-H)$_2$Os$_3$(μ_3,η^2-O=C$_6$H$_2$(CH$_3$)$_2$-3,5)(CO)$_9$

I (75%) [6]
yellow air-stable crystals [6]
^{1}H NMR (CDCl$_3$, 27 °C): −14.00, −11.68 (d's,
each μ-H; J(H,H) = 1.8; measured at low tem-

References on pp. 87/8

Table 3 (continued)

No. compound	method of preparation (yield in %) properties and remarks

19 (continued)

perature), 2.01 (s, CH_3-3), 2.04 (s, CH_3-5), 6.81 (s, H^6), 7.38 (s, H^4) [6]; for numbering, see Formula III

^{13}C NMR (CD_2Cl_2): 124.8 (C^5), 132.9 (C^3), 141.1 (C^4), 161.9 (C^6), 175.9, 176.2 (C^2, C^1) [6]; for numbering, see Formula III

IR (cyclohexane): 1979, 1985, 2000, 2012, 2025, 2033, 2056, 2084, 2110 (all CO) [6]

20 $(\mu-H)_2Os_3(\mu_3,\eta^2-O=C_6H_3(C_3H_7-i)-3)(CO)_9$

I [3], I (28%) [6]

yellow air-stable crystals [6]

^{1}H NMR ($CDCl_3$, 27 °C): -14.02, -11.66 (d's, each μ-H; J(H,H)=2.0; measured a low temperature), 0.96 (d, CH_3; J(H,H)=7.2), 3.00 (m, CH), 6.02 (t, H^5; J(H,H)=8.0), 7.15 (dd, H^4; J(H,H)=1.8, 8.0), 7.54 (dd, H^6; J(H,H)=1.8, 8.0); for numbering, see Formula III; below 10 °C the CH_3 groups give two doublets with a separation of only $\delta=0.01$ ppm at 10 °C [6]; see also [3]

IR (cyclohexane): 1980, 1987, 2002, 2013, 2027, 2034, 2057, 2086, 2112 (all CO) [6]

*21 $(\mu-H)_2Os_3(\mu_3,\eta^2-O=C_6H_3CH_2C_6H_5-3)(CO)_9$

I [3], I (26%) [6]

yellow air-stable crystals [6]

^{1}H NMR ($CDCl_3$, 27 °C): -14.76, -11.90 (d, μ-H; J(H,H)=2.0; measured at low temperature), 3.72 (s, CH_2), 6.00 (t, H^5; J(H,H)=6.4), 6.2 (m, H^4, C_6H_5), 7.54 (d, H^6; J(H,H)=6.4); for numbering, see Formula III; the asymmetry of the complex resulting from the bridging hydrides did not effect diastereotopic methylene hydrogens as expected [6]

IR (cyclohexane): 1980, 1987, 2003, 2013, 2028, 2035, 2057, 2086, 2112 (all CO) [6]

compound of the type $(\mu-H)_2Os_3(\mu_3,\eta^2-O=C_7H_6)(CO)_9$

22

from $(\mu-H)_2Os_3(CO)_{10}$ and cyclohepta-2,4,6-trienone in refluxing nonane for 3.5 h; purified by TLC with light petroleum/CH_2Cl_2 as eluant (20%) [10]

from $(\mu-H)Os_3(CO)_{10}(\eta^3-OC_7H_7)$ in refluxing light petroleum 2 h; isolated as before (51%) [10]

from $(\mu-H)Os_3(\mu_2,\eta^2-O=C_7H_5)(CO)_{10}$ ("Organoosmium Compounds" B 5, Section 3.1.6.1.2.3,

References on pp. 87/8

Table 3 (continued)

No. compound	method of preparation (yield in %) properties and remarks

in preparation) and H_2 (1 atm) in refluxing light petroleum for 30 min; purified as before (32%) [10]

orange-yellow crystals [10]

^{1}H NMR ($CDCl_3$): -14.17, -12.07 (d's, each μ-H, J(H,H) = 1.8), 2.58, 2.67 (dd's, AB pattern, both H^7; J = 3.6, 10.2), 5.35 (dt, H^6; J = 3.6, 7.7), 5.54 (dd, H^4; J = 4.6, 10.7), 6.15 (d, H^3; J = 10.7), 6.26 (dd, H^5; J = 4.6, 7.7) [10]

IR (cyclohexane): 1980, 1999, 2010, 2023, 2029, 2052, 2080, 2106 (all CO) [10]

compounds of the type $(μ\text{-}H)_2Os_3(μ_3,η^2\text{-}O{=}C_{10}H_6)(CO)_9$ and $[(μ\text{-}H)_3Os_3(μ_3,η^2\text{-}O{=}C_{10}H_6\text{-}2)(CO)_9]^+$

23

IV (45%) [1]

yellow crystals [1]

^{1}H NMR ($CDCl_3$, 27 °C): -13.34, -10.69 (d's, each μ-H; J(H,H) = 1.5 at -60 °C), 6.48 (d, H^3), 7.42 (m, H^5 to H^8), 8.06 (d, H^4; J(H^4,H^3) = 9.5) [1]

IR (cyclohexane): 1980, 1988, 2002, 2013, 2022, 2032, 2055, 2083, 2108 (all CO) [1]

mass spectrum: $[M]^+$, $[M-x\,CO-y\,H]^+$ [1]

24

I (9%, along with $(μ\text{-}H)_2Os_3(CO)_9(μ_2,η^2\text{-}OC_{10}H_6)$, Formula IX, p. 84, in 38% yield) [1]

yellow crystals [1]

^{1}H NMR ($CDCl_3$, 27 °C): -13.80, -11.54 (d's, both μ-H; J(H,H) = 1.5 at -60 °C), 6.27, 6.86 (d's, H^3, H^4; J(H^3,H^4) = 8.2), 7.46, 7.76, 8.16 (m's, H^5 to H^8) [1]

IR (cyclohexane): 1980, 1998, 2012, 2025, 2032, 2055, 2083, 2108 (all CO) [10]

mass spectrum: $[M]^+$, $[M-x\,CO-y\,H]^+$ [1]

protonation with CF_3COOH gave No. 25 [1]

25 $[(μ\text{-}H)_3Os_3(μ_3,η^2\text{-}O{=}C_{10}H_6\text{-}2)(CO)_9]^+$

from No. 24 and CF_3CO_2H for several days [1]

^{1}H NMR (no medium given, 27 °C): -15.48 (s, 2 μ-H), -13.28 (s, 1 μ-H) [1]

References on pp. 87/8

Table 3 (continued)

No. compound	method of preparation (yield in %) properties and remarks

compound of the type Os$_3$(μ_3,η^2-O=CHCH)(μ-CO)(CO)$_9$

*26

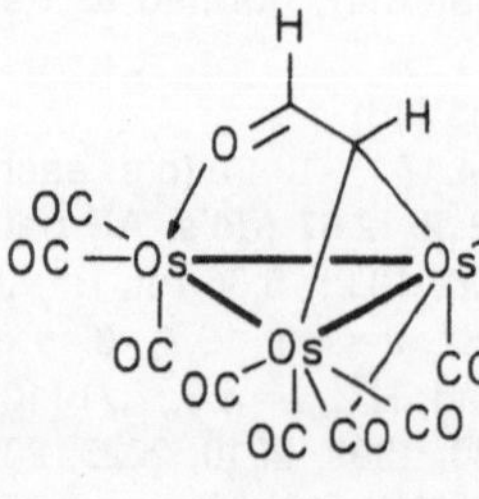

from Os$_3$(CO)$_{10}$(NCCH$_3$)$_2$ and vinylene carbonate in refluxing CH$_2$Cl$_2$ for 14 to 18 h; isolated by TLC with light petroleum as eluant (56 to 65%) [7, 9]

red crystals [9]

^{1}H NMR (CDCl$_3$): 3.43 (d, CHOs), 10.68 (d, CH=O; J(H,H) = 4.0) [7, 9]

IR (cyclohexane): 1500 (C=O), 1863 (μ-CO), 1988, 2006, 2020, 2059, 2102 (all CO) [7, 9]

thermolysis in octane at 125 °C for 1.5 h yielded (μ-H)$_2$Os$_3$(μ_3,η^2-O=CHCH)(CO)$_9$ (No. 4) in only 38%; slow conversion and low yield were associated with partial decomposition of No. 26; treatment with H$_2$ in refluxing cyclohexane gave No. 4 in 76% yield [7, 9]

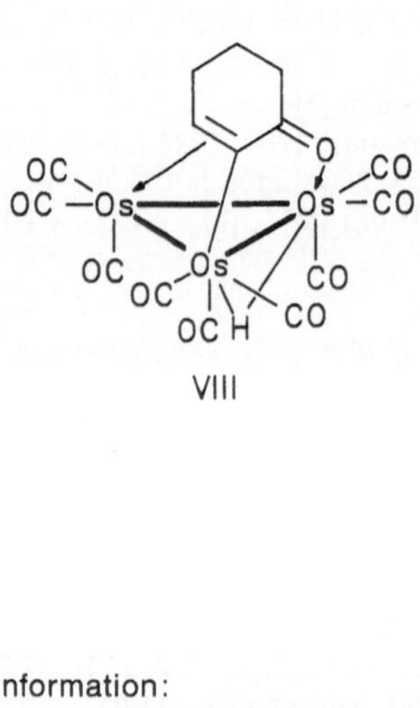

VIII

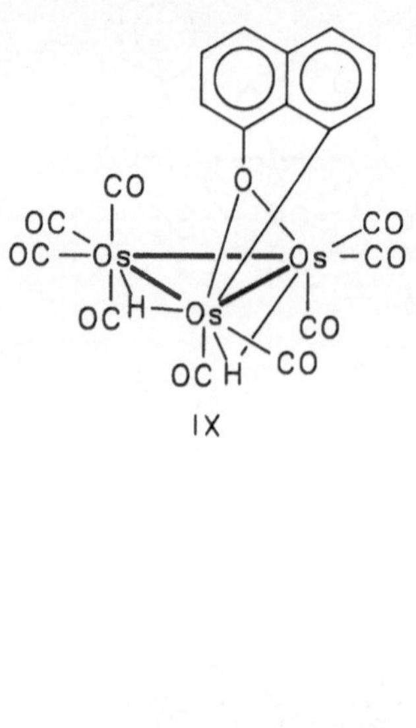

IX

*Further information:

(μ-H)Os$_3${μ_3,η^2-O=CHCP(C$_6$H$_5$)$_3$}(CO)$_9$ (Table 3, No. 1) crystallizes in the triclinic space group P$\bar{1}$ – C$_i^1$ (No. 2) with a = 10.222(2), b = 11.014(2), c = 14.454(3) Å, α = 79.05(2)°, β = 86.67(2)°, γ = 74.86(2)°; Z = 2, D$_c$ = 2.49 g/cm^3, R = 0.0481. The ^{1}L ligand is coordinated through the μ_2-C atom to Os(2) and Os(3), and through the formyl oxygen O to Os(1); the short C–O distance is consistent with a double bond. The μ_2-C–P distance clearly indicates a single bond, see **Fig. 28**. The hydride ligand, not observed crystallographically, is thought to bridge the Os(2)–Os(3) bond due to the very similar Os(1)–Os(2) and Os(1)–Os(3) bond lengths in contrast to the longer Os(2)–Os(3) distance [13].

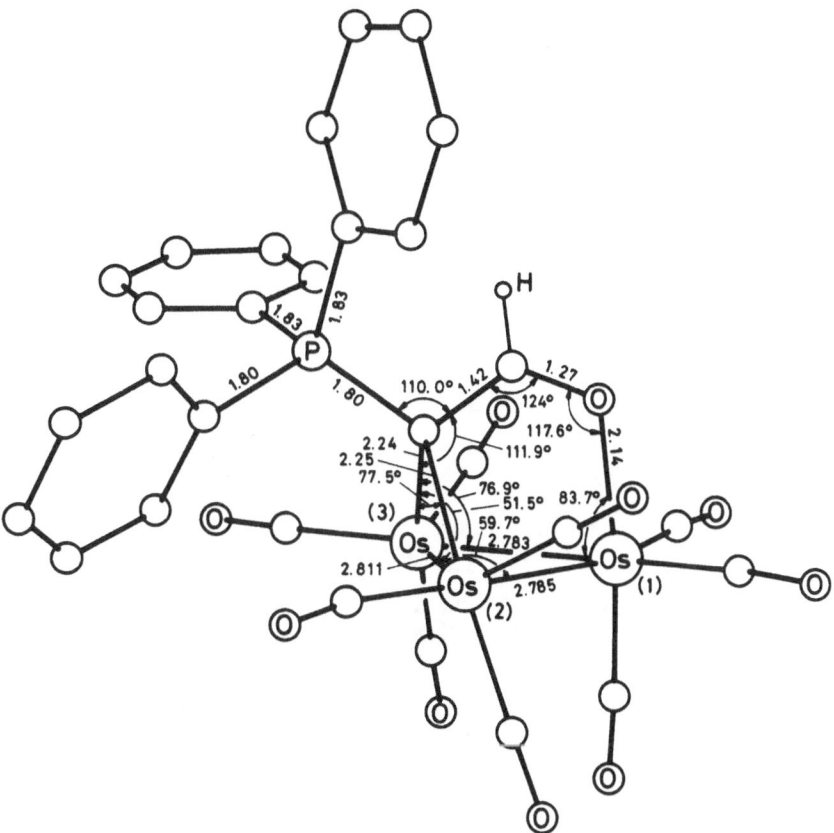

Fig. 28. Molecular structure of $(\mu\text{-H})Os_3\{\mu_3,\eta^2\text{-O=CHCP}(C_6H_5)_3\}(CO)_9$ (No. 1) with selected bond distances (in Å) and angles [13].

$(\mu\text{-H})_2Os_3(\mu_3,\eta^2\text{-O=C}_6H_3CH_2C_6H_5\text{-3})(CO)_9$ (Table 3, No. 21) crystallizes in the monoclinic space group $P2_1/c - C_{2h}^5$ (No. 14) with $a = 8.814(2)$, $b = 10.830(2)$, $c = 25.69(3)$ Å, $\beta = 92.81(2)°$; $Z = 4$, $D_c = 2.72$ g/cm³, $R = 0.0515$. The bridging hydrides were not located crystallographically, but were thought to bridge the Os(1)–Os(2) and Os(2)–Os(3) edges as consistent with their longer bond lengths and the larger Os–Os–CO angles for the equatorial CO. The difference in the bond distances of the hydride bridged Os–Os atoms is caused by the shortening effect of the bridging ${}^1L$ ligand that is counterbalanced by the lengthening effect of the bridging hydrides (see **Fig. 29**). The non-aromatic discription of the $\mu_3,\eta^2\text{-O=C}_6H_3(CH_2C_6H_5)\text{-3}$ ligand is supported by the short C=O bond length and the distribution of C–C distances in the C_6 ring [3].

References on pp. 87/8

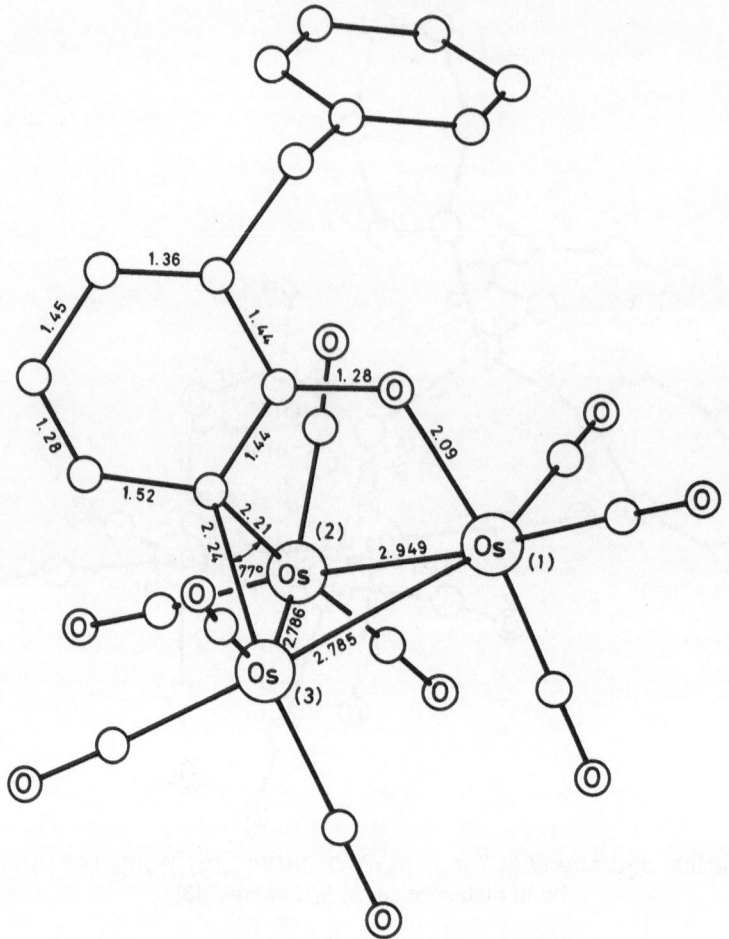

Fig. 29. Molecular structure of $(\mu-H)_2Os_3(\mu_3,\eta^2-O{=}C_6H_3CH_2C_6H_5{-}3)(CO)_9$ (No. 21) with selected bond distances (in Å) and a bond angle [3].

$Os_3(\mu_3,\eta^2-O{=}CHCH)(\mu-CO)(CO)_9$ (Table 3, No. **26**) crystallizes in the orthorhombic space group Aba2 − C_{2v}^{17} (No. 41) with a = 19.714(16), b = 12.610(24), c = 13.488(11) Å; Z = 8, D_c = 3.54 g/cm^3, R = 0.046. There is a molecular but not a crystallographic mirror plane through the O=CC atoms, Os(3), and the bridging CO. The geometry of the bridging C atom is distorted tetrahedral, see **Fig. 30**. The C–O distance in the formyl group corresponds to a double bond, but the μ_3,η^2-C–C distance implies also multiple bonding, probably being a consequence of the necessarily small Os(1)-(μ-C)-Os(2) angle of 74.9° and the state of hybridization at μ-C which allows p_π-p_π bonding between this atom and the formyl group. Average bond distance of Os–C is 1.918 Å [9].

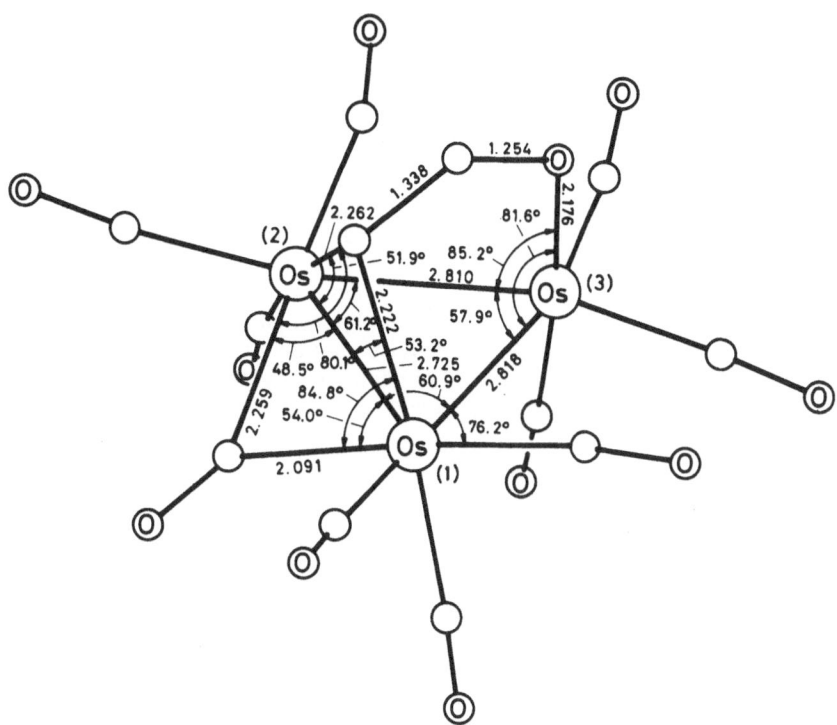

Fig. 30. Molecular structure of $Os_3(\mu_3,\eta^2\text{-}O\text{=}CHCH)(\mu\text{-}CO)(CO)_9$ (No. 26) with selected bond distances (in Å) and bond angles [9].

References:

[1] Azam, K. A.; Deeming, A. J.; Kimber, R. E.; Shukla, P. R. (J. Chem. Soc. Dalton Trans. **1976** 1853/8).

[2] Azam, K. A.; Deeming, A. J. (J. Chem. Soc. Chem. Commun. **1977** 472/3).

[3] Azam, K. A.; Deeming, A. J.; Rothwell, I. P.; Hursthouse, M. B.; New, L. (J. Chem. Soc. Chem. Commun. **1978** 1086/8).

[4] Azam, K. A.; Deeming, A. J.; Rothwell, I. P. (J. Organomet. Chem. **178** [1979] C 20/C 22).

[5] Azam, K. A.; Deeming, A. J.; Rothwell, I. P. (J. Chem. Soc. Dalton Trans. **1981** 91/8).

[6] Azam, K. A.; Deeming, A. J.; Rothwell, I. P.; Hursthouse, M. B.; Backer-Dirks, J. D. J. (J. Chem. Soc. Dalton Trans. **1981** 2039/44).

[7] Arce, A. J.; Deeming, A. J. (J. Chem. Soc. Chem. Commun. **1982** 364/5).

[8] Arce, A. J.; Deeming, A. J.; Shaunak, R. (J. Chem. Soc. Dalton Trans. **1983** 1023/5).

[9] Arce, A. J.; Deeming, A. J.; Hursthouse, M. B.; Walker, N. P. C. (J. Chem. Soc. Dalton Trans. **1987** 1861/4).

[10] Boyar, E.; Deeming, A. J.; Randle, N. P.; Bates, P. A.; Hursthouse, M. B. (J. Chem. Soc. Dalton Trans. **1987** 551/5).

[11] Deeming, A. J.; Nuel, D.; Powell, N. I.; Whittaker, C. (J. Chem. Soc. Chem. Commun. **1990** 68/70).

[12] Strickland, D. A.; Shapley, J. R. (J. Organomet. Chem. **401** [1991] 187/97).
[13] Deeming, A. J.; Nuel, D.; Powell, N. I.; Whittaker, C. (J. Chem. Soc. Dalton Trans. **1992** 757/64).

3.1.6.2.4 Compounds with Other μ_3,η^2-Bridging ^1L Ligands

The compounds covered in this section all have μ_3,η^2-bridging ^1L ligands, where a carbon atom bridges two Os atoms, and a heteroatom such as O, N, P, interacts with the third Os atom of the Os_3 triangle. Only No. 34 contains an S atom in the ^1L ligand being coordinated to two Os atoms. Most compounds can be represented by the general Formulas I to V; for the other complexes individual figures are given in Table 4. Generally, a structure that is not confirmed by X-ray analysis, can only be considered as a proposal based on spectroscopic and analytical data. For compounds of type VI, see "Organoosmium Compounds" B 5, Section 3.1.6.1.5, in preparation.

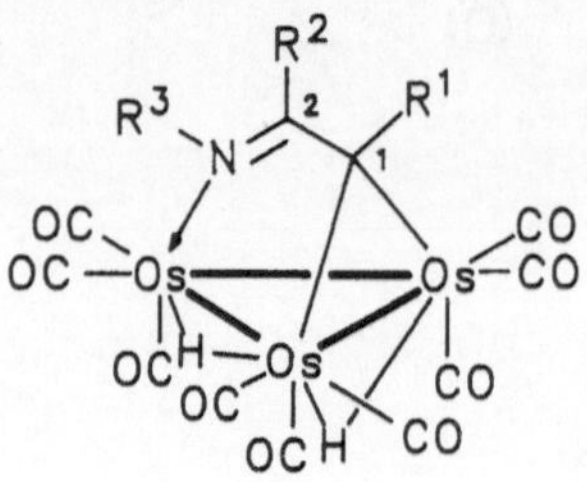

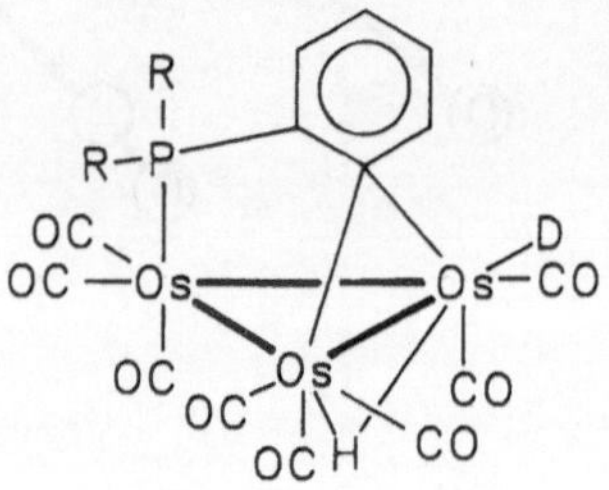

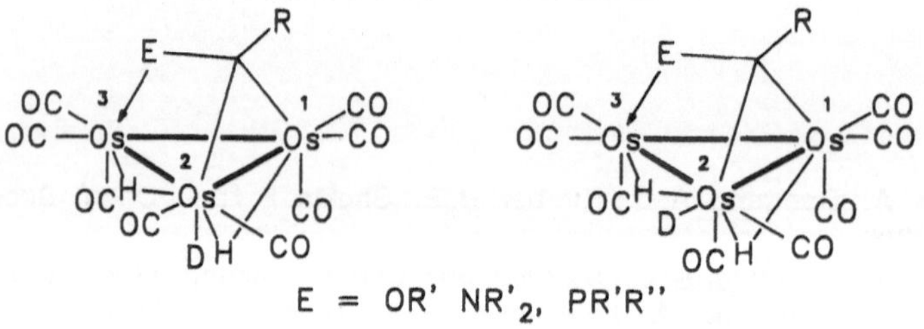

$E = OR', NR'_2, PR'R''$

Isomer 1   Isomer 2

III

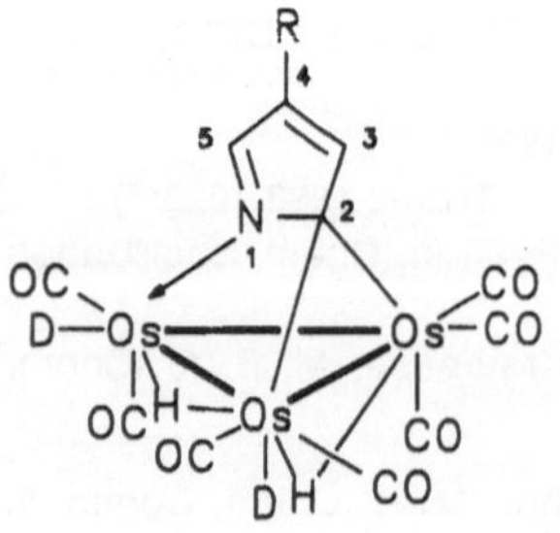

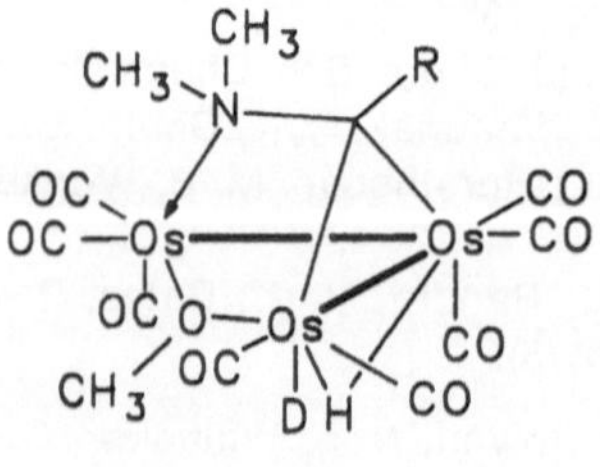

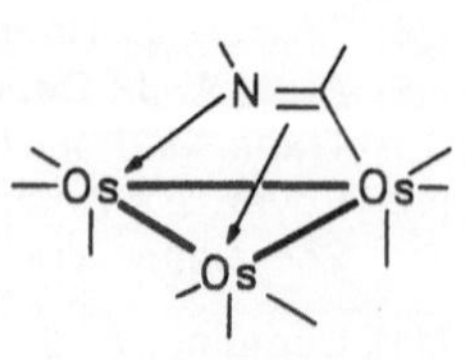

References on pp. 115/6

In compounds of Formula I, the bridging ^{1}L ligands of the type μ_3,η^2-C(1)R^{1}C(2)R^{2}NR^3 are formally four electron donor ligands. Two electrons are donated by the two fairly symmetrical Os–C σ bonds to the Os$_3$ core, the others are donated by the lone pair on nitrogen by a two–electron donor σ bond from N to Os(1) [24]. A comparison of the X-ray determinations of Nos. 3 [6], 4 [8], and 5 [24] (see pp. 104/6) revealed slight differences in the bonding mode within the ligand. The C(2)–N bond in No. 3 of 1.336 Å is slightly longer than that of a C=N double bond, whereas the C(1)–C(2) distance of 1.400 Å is shorter than a C–C single bond, indicating the presence of partial multiple bonding between the carbenoid carbon and the iminyl group. In No. 4 and No. 5 the C(2)–N bonds are 1.286 and 1.28 Å, the C(1)–C(2) distances are 1.46 and 1.41 Å, respectively.

In compounds of Formula III, the bridging ^{1}L ligands of the type μ_3,η^2-CR–E (E = OR', NR'$_2$, PR'R'') can also be considered as four-electron donors, being coordinated to the Os$_3$ core by two Os–C σ bonds and a two-electron donor interaction between E and the third Os atom [9]. $(\mu$-H)$_2$Os$_3(\mu_3,\eta^2$-CHP(CH$_3$)$_2$)(CO)$_8$P(CH$_3$)$_3$ (Nos. 16 and 17) were obtained as a separable mixture of isomers probably bearing the P(CH$_3$)$_3$ ligand either in the axial or in the equatorial position, corresponding to Isomers 1 or 2, respectively. In solution, both isomers convert at room temperature over several days into an equilibrium mixture of an Isomer 1:Isomer 2 ratio of 1.42:1.00; an intramolecular isomerization mechanism is presumed. For No. 18 only one isomer was observed with the P(C$_2$H$_5$)$_3$ ligand in the axial position (see Formula III, Isomer 1) [4].

Compounds of Formula IV. X-ray analysis of $(\mu$-H)$_2$Os$_3(\mu_3,\eta^2$-C$_4$H$_3$N)(CO)$_9$ (No. 21) revealed a bridging ligand coordinated by two Os–C σ bonds and one Os–N bond [16]. Earlier formulations with μ_3,η^2-C=CC$_2$H$_2$NH (Formula IXa) [7] or μ_3,η^2-NCC$_3$H$_3$ moieties (Formula IXb; see p. 104) [7, 10] were revised.

For $(\mu$-H)$_2$Os$_3(\mu_3,\eta^2$-C$_4$H$_3$N)(CO)$_{9-n}$D$_n$ (Nos. 25, 26, 28 to 30; D = P(C$_2$H$_5$)$_3$ or P(OCH$_3$)$_3$, n = 1 or 2) the coordination sites of the D ligands at the metal atoms were evaluated by ^{1}H NMR spectroscopy from the coupling pattern and the coupling constants of the ^{31}P nucleus with the bridging hydrides. For No. 25 and the major isomer of No. 26, the P$-\mu$-H coupling constants (12.3 and 13.3 Hz for the cis couplings, and 22.0 and 51.3 Hz for the trans couplings, respectively) are consistent with the D ligand occupying an equatorial site at the Os center (as in Formula III, Isomer 2), whereas the coupling constants of 10.2 and 12.2 Hz for the minor isomer of No. 26 (as in Formula III, Isomer 1) are in agreement with two cis couplings. For Nos. 28 and 29, which have two D ligands, the observed cis coupling constants suggested structures in which both D ligands are in axial positions, or one is in an axial, and one is in an equatorial site, resulting in one doublet and one triplet in the hydride region with cis P$-\mu$-H coupling. The latter structure proposal is consistent with the solid state structure for No. 29 evaluated by X-ray determinations [18].

Some compounds listed in Table 4 were prepared by the following methods:

Method I: $(\mu$-H)$_2$Os$_3(\mu_3,\eta^2$-C$_4$H$_3$N)(CO)$_{9-n}$D$_n$ (Formula IV; R = H, D = PR'$_3$; n = 1 or 2) was prepared by initial conversion of $(\mu$-H)$_2$Os$_3(\mu_3,\eta^2$-C$_4$H$_3$N)(CO)$_9$ (No. 21) into Nos. 24 or 27 with (CH$_3$)$_3$NO (1:1 or 1:2, respectively) in CH$_3$CN, followed by treatment with PR'$_3$ (R' = C$_2$H$_5$, OCH$_3$) in CH$_2$Cl$_2$ at room temperature for 1 to 3 h; work-up was performed by TLC with CH$_2$Cl$_2$/hexane (1:1) as eluant [18].

Method II: $(\mu$-H)$_2$Os$_3(\mu_3,\eta^2$-C$_4$H$_2$RN)(CO)$_9$ (Formula IV; R = t-C$_4$H$_9$, COCH$_3$) was prepared by dropping a CS$_2$ solution of $(\mu$-H)$_2$Os$_3(\mu_3,\eta^2$-C$_4$H$_3$N)(CO)$_9$ (No. 21) into a mixture of RCl/AlCl$_3$ in CS$_2$ at 0 °C over a period of 30 to 45 min, followed by stirring at room temperature for 30 min and purification as before [14].

Method III: $(\mu\text{-H})_2Os_3(\mu_3,\eta^2\text{-CRPR'R''})(CO)_8D$ (Formula III; R = H, CH_3, D = CO, PR'_3) [4, 5] and $(\mu\text{-H})Os_3(\mu_3,\eta^2\text{-}C_6H_4PR_2)(CO)_8D$ (Formula II; R = CH_3, C_2H_5, C_6H_5; D = CO, PR'_3) [1, 3, 5] were prepared by thermal rearrangement/decarbonylation of $Os_3(CO)_{12-n}(PR'_3)_n$ or $Os_3(CO)_{12-n}(C_6H_5PR_2)_n$ (n = 1 or 2) in nonane (for 5 to 30 h) [4], o-xylene (for 30 h) [5], or xylene (for 15 min) [1, 3] at reflux temperature. The crude products were purified by TLC with light petroleum [4, 5], column chromatography on alumina [4], or by crystallization from acetone/light petroleum [1, 3, 4, 5].

In the preparation of No. 10 from $Os_3(CO)_{10}(P(C_6H_5)_3)_2$ in refluxing xylene an additional mixture of five other Os_3 complexes was obtained; all these products were characterized by elemental and spectroscopic analysis, and in part by X-ray analysis. The yield of No. 10 decreased with increasing reaction time [1, 3].

Table 4
Compounds with Other μ_3,η^2-Bridging ^1L Ligands.
An asterisk preceding the compound number indicates further information at the end of the table, pp. 104/15.
Explanations, abbreviations, and units on p. X.

No. compound	method of preparation (yield) properties and remarks

compounds of the type $(\mu\text{-H})_2Os_3(\mu_3,\eta^2\text{-}CR^1CR^2\text{=}NR^3)(CO)_9$ (Formula I) and other compounds with μ_3,η^2-bridging CC=N units

1 $(\mu\text{-H})_2Os_3(\mu_3,\eta^2\text{-CHCH=NCH}_3)(CO)_9$

from $(\mu\text{-H})Os_3(\mu_2,\eta^2\text{-CH}_2NHC_2H_5)(CO)_{10}$ or $(\mu\text{-H})Os_3(\mu_2,\eta^2\text{-CH}_2CH=NCH_3)(CO)_{10}$ (both "Organoosmium Compounds" B 5, Section 3.1.6.1.2.7, in preparation) in refluxing heptane for 6 and 2 h, respectively; purified by TLC with CH_2Cl_2/hexane (1:5) as eluant (40 or 67%) [19]
solid [19]
^1H NMR ($CDCl_3$): −14.31, −14.14 (s's, both μ-H), 3.18 (s, CH_3), 4.15 (d, H^1; J(H,H) = 4.9), 10.31 (d, H^2; J(H,H) = 4.9) [19]
IR (hexane): 1972, 1989, 2000, 2015, 2044, 2071, 2099 (all CO) [19]

2 $(\mu\text{-H})_2Os_3(\mu_3,\eta^2\text{-CHC(C}_2H_5)\text{=}NC_3H_7\text{-}n)(CO)_9$

from $(\mu\text{-H})Os_3(\mu_3,\eta^2\text{-C(C}_2H_5)\text{=}NC_3H_7\text{-}n)(CO)_9$ ("Organoosmium Compounds" B 5, Section 3.1.6.1.5, in preparation) and an excess of CH_2N_2 (dissolved in ether) in CH_2Cl_2 at room temperature for 24 h; purified by TLC with CH_2Cl_2/hexane (1:4) as eluant (54%) [24]
yellow crystals [24]

Table 4 (continued)

No. compound	method of preparation (yield) properties and remarks

^1H NMR (CDCl$_3$): -14.59 (d, μ-H; J(H,H) = 1.5), -14.04 (partially overlapping dd, μ-H; ^{2}J(H,H) = 1.6, ^{3}J(H,H) = 0.7), 0.88 (t, CH$_3$ of C$_3$H$_7$-n?; J(H,H) = 7.5), 1.19 (t, CH$_3$ of C$_2$H$_5$?; J(H,H) = 7.6), 1.41 (m, CH$_2$), 2.02, 2.12, 3.22, 3.34 (m's, each 1 H of CH$_2$), 4.66 (s, H^1)[24]
IR (hexane): 1967, 1984, 1996, 2011, 2042, 2068, 2096 (all CO) [24]

*3 (μ-H)$_2$Os$_3$(μ_3,η^2-C(C$_3$H$_7$-i)C(CH$_3$)=NC$_6$H$_5$)(CO)$_9$

from (μ-H)$_2$Os$_3$(CO)$_{10}$ and (CH$_3$)$_2$C=CHC(CH$_3$)=NC$_6$H$_5$ in hexane at reflux temperature for 70 h; purified by chromatography on neutral alumina with hexane as eluant (27.6%, along with Os$_3$(CO)$_{12}$) [6]
yellow crystals from hexane, m. p. 169 to 174(?) °C [6]
^1H NMR (CDCl$_3$): -13.54, -12.98 (s's, both μ-H), 1.32, 1.34 (d's, C(CH$_3$)$_2$; J(H,H) = 6.1), 1.42 (s, CH$_3^2$), 2.24 (sept, CH; J(H,H) = 6.1), 6.44, 6.63, 7.12 (d's, each 1 H of C$_6$H$_5$), 7.38 (t, 2 H of C$_6$H$_5$; J(H,H) = 7); the isopropyl methyl resonances are unequivalent indicating an unsymmetric distribution of the bridging hydrides, and that hydride exchange around the cluster occurs [6]
IR (hexane): 1960, 1965, 1990, 1995, 2005, 2010, 2045, 2070, 2100 (all CO) [6]

*4

from Os$_3$(CO)$_{11}$(η^1-NH=C$_6$H$_{10}$-c) in refluxing octane for 8 h or n-heptane for 20 h; separated by TLC with cyclohexane/CH$_2$Cl$_2$ (3:2) as eluant (11%, along with (μ-H)Os$_3$(CO)$_{10}$-(μ_2,η^1-N=C$_6$H$_{10}$-c) as the main-product) [8]
pale yellow, air-stable crystals from pentane, m. p. 147 to 148 °C (dec.) [8]
^1H NMR (CDCl$_3$): -13.61, -13.56 (d, μ-H; J(H,H) = 1.7), 1.56, 1.72, 2.24 (m's, all CH$_2$), 1.92 (t, CH$_2$), 6.24 (s, NH) [8]
IR (pentane): 1967, 1974, 1987, 1997, 2012, 2041, 2070, 2095 (all CO) [8]; (KBr): 2845, 2885, 2895, 2920, 2955 (all CH$_2$), 3375, 3385 (both NH) [8]
mass spectrum: [M]$^+$, [M$-$x CO]$^+$, x = 1 to 9 [8]

References on pp. 115/6

Table 4 (continued)

No. compound	method of preparation (yield) properties and remarks

*5

from the pyrroline complex $(\mu-H)Os_3(\mu_3,\eta^2-C=N(CH_2)_3-c)(CO)_9$ ("Organoosmium Compounds" B 5, Section 3.1.6.1.5, in preparation) and CH_2N_2 (dissolved in ether) in CH_2Cl_2 for 4 h; purified by TLC with CH_2Cl_2/hexane (3:10) as eluant (75%) [19, 24]

by thermolysis of $(\mu-H)Os_3(\mu_2,\eta^2-C=N(CH_2)_3-c)(CO)_{10}$ ("Organoosmium Compounds" B 5, Section 3.1.6.1.5, in preparation) in refluxing heptane for 23 h (quantitative) [24]

orange crystals from CH_2Cl_2/hexane at $-20\,°C$ [24]

^{1}H NMR (C_6D_6): -14.56 (dd, μ-H; $^2J(H,H) = 1.7$, $^3J(H,H) = 0.8$), -14.30 (d, μ-H; $J(H,H) = 1.7$), 0.96, 1.10, 1.29, 1.52, 2.41, 2.64 (m's, each 1 H of CH_2), 3.84 (s, μ_3,η^2-CH) [24]; see also [19]

^{13}C NMR $(CDCl_3)$: 161.44, 166.49, 168.45, 171.68, 172.00, 175.73, 177.56, 179.24, 183.33 (each 1 CO), 218.07 (s, C=N) [24]

IR (hexane): 1970, 1998, 1997, 2011, 2043, 2056, 2069, 2097 (all CO) [24]

*6 C_6H_5

from $(\mu-H)_2Os_3(CO)_{10}$ and an excess of $(C_6H_5)_3P=C=C=NC_6H_5$ in THF at 20 °C for 5 min, followed by TLC with cyclohexane/CH_2Cl_2 (7:3) as eluant (32%) [20]

deep red crystals from CH_2Cl_2/pentane at $-30\,°C$ [20]

^{1}H NMR $(CDCl_3)$: -12.66 (d, μ-H; $J(H,P) = 11.27$), 6.72 to 7.79 (m, C_6H_5), 9.81 (s, CH) [20]

^{31}P NMR (CD_2Cl_2): 43.9 (s, PC_6H_5) [20]

IR (CH_2Cl_2): 1930, 1949, 1979, 2008, 2031, 2066 (all CO) [20]

compounds with μ_3,η^2-bridging CCC=N units

*7

from $Os_3(CO)_{10}(NCCH_3)_2$ and an excess of $2-C_4H_3NHCH=NCH_3$ in CH_2Cl_2 at ca. 25 °C for 2 h; purification by TLC light petroleum/ether (1:1) as eluant (60%) [23]

deep green crystals from hexane [23]

^{1}H NMR (CD_2Cl_2): -11.92 (s, μ-H), 3.95 (d, CH_3; $J(H^6,CH_3) = 1.1$), 7.09 (d, H^5), 7.43 (d, H^4; $J(H^4,H^5) = 2.2$), 8.16 (m, H^6), 9.65 (br s, NH) [23]

References on pp. 115/6

Table 4 (continued)

No. compound	method of preparation (yield) properties and remarks

IR (CH_2Cl_2): 1950, 1982, 2015, 2042, 2073 (all CO); the zwitterionic character is supported by the lower ν(CO) frequencies as compared with those for the isomer (μ-H)$Os_3(CO)_9$-(μ_3,η^2-$NC_4H_3CH{=}NCH_3$), formed at 81 °C [23]

8

from $Os_3(CO)_{10}(NCCH_3)_2$ and an excess of 2-$C_4H_3NHC_4H_6N$ as described for No. 7 (56%) [23]
green crystals [23]
^{1}H NMR (CD_2Cl_2): −11.92 (s, μ-H), 2.15 (m, H^8; J(H^8,H^9) = 1.7, 8.3), 2.88 (tt, H^9), 3.91 (tt, H^7; J(H^7,H^8) = 7.5), 7.07 (dd, H^5; J(H^5,NH) = 0.3), 7.36 (dd, H^4; J(H^4,H^5) = 2.6, J(H^4,NH) = 0.5), 9.34 (br, NH) [23]
IR (CH_2Cl_2): 1949, 1980, 2014, 2040, 2071 [23]

compounds of the type (μ-H)$Os_3(\mu_3,\eta^2$-$C_6H_4PR_2$)(CO)$_{9-n}$D$_n$ (Formula II)

9 (μ-H)$Os_3(\mu_3,\eta^2$-$C_6H_4P(C_2H_5)_2$)(CO)$_9$

III (ca. 4.2%, along with compounds VII and VIII, p. 104, as the main-products) [5]
yellow crystals [5]
^{1}H NMR ($CDCl_3$, 27 °C): −18.01 (d, μ-H; J(H,P) = 14.5), 0.2 to 3.2 (C_2H_5), 6.71, 6.96, 7.33, 8.26 (m, C_6H_4) [5]
IR (cyclohexane): 1958, 1973, 1980, 1985, 1989, 2011, 2031, 2054, 2085 (all CO) [5]
mass spectrum: [M]$^+$, [M − x CO]$^+$, x = 1 to 9 [5]

*10 (μ-H)$Os_3(\mu_3,\eta^2$-$C_6H_4P(C_6H_5)_2$)(CO)$_8$P(C_6H_5)$_3$

III (ca. 21%) [1, 3]
from $Os_3(CO)_{12}$ and P(C_6H_5)$_3$ (1:2) in refluxing xylene for 1 h; separated by leaching the residue with acetone (ca. 11%, along with eight other acetone-soluble Os_3 complexes, which were fully characterized) [1, 3]
orange-red, m.p. 198 °C [1, 3] from $CHCl_3$/light petroleum; insoluble in acetone [3]
^{1}H NMR (no medium given): −17.31 (q, μ-H; J(P, H) = 10, 17); the splitting of the signal probably resulted from coupling of the hydride ligand with two unequivalent ^{31}P nuclei [1, 3]
IR (CCl_4): 1938, 1947, 1960, 1974, 1988, 1998, 2010, 2027, 2070, 2086, 2120 (all CO) [3]

References on pp. 115/6

Table 4 (continued)

No. compound	method of preparation (yield) properties and remarks

*10 (continued)

mass spectrum: $[M - x\, CO - P(C_6H_5)_3]^+$, $x = 1$
to 8; heavier ions were also observed in
small amounts [3]
thermolysis in refluxing xylene gave
$Os_3(\mu_3,\eta^2\text{-}C_6H_4)(CO)_7(P(C_6H_5)_2)_2$,
"$Os_3(CO)_8(P(C_6H_5)_3)P(C_6H_5)_2$", and
$(\mu\text{-}H)Os_3(\mu_3,\eta^3\text{-}C_6H_3C_6H_4P(C_6H_5)_2)$-
$(CO)_7P(C_6H_5)_2$ [3]

compounds of the type $(\mu\text{-}H)_2Os_3(\mu_3,\eta^2\text{-}CRE)(CO)_{9-n}D_n$ (Formula III)

11

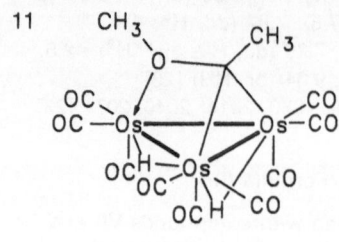

by decarbonylation of $(\mu\text{-}H)Os_3(\mu_2,\eta^2\text{-}$
$CH(CH_3)OCH_3)(CO)_{10}$ ("Organoosmium Com-
pounds" B 5, Section 3.1.6.1.2.3, in preparation) in refluxing heptane for 1 h (8%, along
with five other Os_3 clusters of which three
were characterized) [9]
^1H NMR (CDCl$_3$): -14.32 (d, μ-H; J(H,H) = 1.5),
-13.14 (d, μ-H), 2.17 (s, CCH$_3$), 3.60 (s,
OCH$_3$) [9]
IR (cyclohexane): 1977, 1996, 2006, 2018, 2026,
2049, 2055, 2077, 2102 (all CO) [9]

*12 $(\mu\text{-}H)_2Os_3(\mu_3,\eta^2\text{-}CHN(CH_3)_2)(CO)_9$

by decarbonylation of $(\mu\text{-}H)Os_3(\mu_2,\eta^2\text{-}$
$CH_2N(CH_3)_2)(CO)_{10}$ ("Organoosmium Com-
pounds" B 5, Section 3.1.6.1.2.7, in preparation) in refluxing heptane for ca. 20 min, followed by separation by TLC with
CH$_2$Cl$_2$/hexane (1:9) as eluant (28%, along
with 60% of $(\mu\text{-}H)Os_3(\mu_2,\eta^1\text{-}C=N(CH_3)_2)$-
$(CO)_{10}$, Section 3.1.6.2.2) [13]
colorless crystals from hexane at $-25\,°C$ [13]
^1H NMR (CDCl$_3$): 3.05 (s, CH$_3$), 6.64 (s, CH)
[13]; (toluene-d$_8$, $-50\,°C$): -16.34, -15.62
(s's, each μ-H), 2.12, 2.29 (s's, each CH$_3$),
5.94 (s, CH) [13]
IR (hexane): 1965, 1980, 1991, 2004, 2024, 2047,
2074, 2100 (all CO) [13]
thermolysis in octane at 125 °C for 5 h under
CO gave $(\mu\text{-}H)Os_3(\mu_2,\eta^1\text{-}C=N(CH_3)_2)(CO)_{10}$
(Section 3.1.6.2.2); the conversion proceeded
via the intermediate $(\mu\text{-}H)_3Os_3(\mu_3,\eta^1\text{-}$
$CN(CH_3)_2(CO)_9$ resulting from the cleavage
of the Os–N bond; thermolysis in heptane at
98 °C gave no reaction [13]

References on pp. 115/6

Table 4 (continued)

No. compound	method of preparation (yield) properties and remarks
13 $(\mu\text{-H})_2Os_3(\mu_3,\eta^2\text{-CHP(CH}_3)_2)(CO)_9$	III (60%) [4] pale yellow crystals, m. p. 114 °C (dec.) [4] ^{1}H NMR (CDCl$_3$, -71 °C): -20.08 (d, μ-H; J(H,P) = 6.2), -15.75 (d, μ-H; J(H,P) = 10.8), 1.82 (d, CH$_3$; J(H,P) = 9.0), 2.05 (d, CH$_3$; J(H,P) = 10.0), 4.41 (t, CH) [4]; (CDCl$_3$, 35 °C): -17.79 (br μ-H), 1.91 (d, CH$_3$; J(H,P) = 9.2), 4.42 (quint, CH; J(H,H) = 1.4, J(H,P) = 2.8) [4] at room temperature the two CH$_3$ doublets coalesced to a single doublet, and the μ-H signals broadened to a single resonance which sharpens at higher temperature demonstrating the fluxional behaviour of the complex at room temperature; the CH resonance appeared at 35 °C as a quintet (intensity ratio of 1:2:2:2:1) and was attributed to a doublet of triplets resulting from coupling to ^{31}P and to the hydrides, caused by the rapid exchange; below -50 °C only a triplet pattern was observed [4] IR (cyclohexane): 1969, 1980, 1989, 2003, 2010, 2021, 2046, 2072, 2100 (all CO) [4] mass spectrum: [M]$^+$ [4]
14 $(\mu\text{-H})_2Os_3(\mu_3,\eta^2\text{-C(CH}_3)P(C_2H_5)_2)(CO)_9$	III (32%) [4], III (ca. 19%, along with $(\mu\text{-H})Os_3(\mu_3,\eta^3\text{-CH}_2{=}CP(C_2H_5)_2)(CO)_9$ and other by-products) [5] formed by hydrogenation of $(\mu\text{-H})Os_3(\mu_3,\eta^3\text{-}CH_2{=}CP(C_2H_5)_2)(CO)_9$ in refluxing decane for 1 h (main-product as well as various decomposition products, based on IR spectra) [5] pale yellow crystals [4, 5], m. p. 177 °C (dec.) [4] ^{1}H NMR (CDCl$_3$, -59 °C): -20.05 (d, μ-H; J(H,P) = 8.0), -15.19 (d, μ-H; J(H,P) = 9.4), 1.17 (m, CH$_3$ of C$_2$H$_5$), 1.8 to 2.9 (2 m's, CH$_2$ due to the unequivalence of the C$_2$H$_5$ groups), 2.48 (d, μ-CCH$_3$; J(H,P) = 18) [4, 5]; (CDCl$_3$, 70 °C): -17.6 (br, μ-H), other signals similar as at -59 °C [4] IR (cyclohexane): 1967, 1979, 1988, 2001, 2007, 2020, 2045, 2070, 2097 (all CO) [4, 5] mass spectrum: [M]$^+$, [M $-$ C$_2$H$_5$]$^+$, then CO fragmentation [4, 5] dehydrogenation with [(C$_6$H$_5$)$_3$C]BF$_4$ in liquid SO$_2$ at room temperature for 3 weeks gave

References on pp. 115/6

Table 4 (continued)

No. compound	method of preparation (yield) properties and remarks

14 (continued)

$(\mu-H)Os_3(\mu_3,\eta^3-CH_2=CP(C_2H_5)_2)(CO)_9$; the reaction proceeded via the intermediate $[(\mu-H)_2Os_3(\mu_3,\eta^3-CH_2=CP(C_2H_5)_2)(CO)_9]^+$ [5]

15 $(\mu-H)_2Os_3\{\mu_3,\eta^2-CHP(C_2H_5)C_5H_4FeC_5H_5\}(CO)_9$

as a by-product in the preparation of $(\mu-H)_2Os_3\{\mu_3,\eta^4-C_5H_4FeC_5H_3P(C_2H_5)_2\}(CO)_8$ by thermolysis of $Os_3(CO)_{11}\{P(C_2H_5)_2C_5H_4-FeC_5H_5\}$ in refluxing octane for 8 h, followed by column chromatography with CH_2Cl_2/petroleum ether (1:3) as eluant (15%) [25]

^1H NMR (no medium given): -19.6, -15.1 (d's, each 1 μ-H; J(H,P) = 12.4, 11.1), 1.50, 2.24 (td, d, each CH_3; J(H,P) = 24.0, 24.2), 2.52, 2.88 (m's, each 1 H of CH_2), 4.16 (s, C_5H_5), 4.20, 4.38, 4.50 (m's, 1 H, 2 H, 1 H of C_5H_4); the two sharp and at room temperature well separated μ-H signals indicate that no hydride exchange processes occur as observed for No. 14; four possible positions for the μ-H ligands are considered and discussed in detail; for No. 15 both μ-H ligands are believed to span the Os(2)-Os(3) bond [25]

^{31}P $\{^1H\}$ NMR (no medium given): -50.9 [25]

FAB mass spectrum: $[M]^+$, $[M-x\ CO]^+$, x = 1 to 3 [25]

16 $(\mu-H)_2Os_3(\mu_3,\eta^2-CHP(CH_3)_2)(CO)_8P(CH_3)_3$

Isomer 1 in Formula III

III (44%), along with 31% of No. 17, Isomer 2 in Formula III; the isomers were separated by TLC, but spectroscopic investigations revealed an isomeric purity of only 80 to 90% for each compound; each solution enriched in either isomer gave the same equilibrium mixture (Isomer 1:Isomer 2 = 1.42:1) within a few days at room temperature; the isomerism is attributed the $P(CH_3)_3$ ligand occupying different coordination sites on the same Os atom [4]

pale yellow crystals, m. p. 113 to 117 °C [4]

^1H NMR ($CDCl_3$): -20.44 (dd, μ-H; J(H,P) = 8.0, 12.0), -15.68 (t, μ-H; J(H,P) = 11.2), 1.80 (d, 1 CH_3; J(H,P) = 9.3), 1.88 (d, $P(CH_3)_3$; J(H,P) = 9.5), 2.00 (d, 1 CH_3; J(H,P) = 9.4),

Table 4 (continued)

No. compound	method of preparation (yield) properties and remarks

4.48 (m, CH; J(H,P) ≤ 3, assigned to coupling with $^{31}P(CH_3)_{3,}$); μ-H and CH resonances were measured at 27 °C [4]
IR (cyclohexane): 1956, 1968, 1982, 1991, 2023, 2038, 2072 (all CO) [4]
mass spectrum: [M]$^+$ [4]

17 $(\mu-H)_2Os_3(\mu_3,\eta^2-CHP(CH_3)_2)(CO)_8P(CH_3)_3$
Isomer 2 in Formula III
III (31%); see also No. 16 [4]
pale yellow crystals, m. p. 113 to 117 °C [4]
^1H NMR (CDCl$_3$): −19.25 (dd, μ-H; J(H,P) = 8.9, 24.7), −16.06 (t, μ-H; J(H,P) = 11.2), 1.68 (d, P(CH$_3$)$_3$; J(H,P) = 10.6), 1.74 (d, 1 CH$_3$; J(H,P) = 9.1), 1.99 (d, 1 CH$_3$; J(H,P) = 9.5), 3.76 (dm, CH; J(H,P) = 12.0 assigned to coupling with $^{31}P(CH_3)_{3,}$); μ-H and CH resonances were measured at 27 °C [4]
IR (cyclohexane): 1949, 1955, 1979, 1998, 2018, 2035, 2074 (all CO) [4]
mass spectrum: [M]$^+$ [4]

18 $(\mu-H)_2Os_3(\mu_3,\eta^2-C(CH_3)P(C_2H_5)_2)(CO)_8P(C_2H_5)_3$
Isomer 1 in Formula III
III (58%; Isomer 2 not observed) [4]
yellow crystals, m. p. 98 to 102 °C [4]
^1H NMR (CDCl$_3$, 35 °C): −20.75 (dd, μ-H; J(H,P) = 9.0, 11.0), −15.30 (t, μ-H; J(H,P) = 10.0), 1.16 (m, CH$_3$ of C$_2$H$_5$), 1.94 (m, CH$_2$), 2.54 (dd, μ-CCH$_3$; J(H,P) = 4.0 assigned to coupling with $^{31}P(C_2H_5)_3$, J(H,P) = 20.4 assigned to coupling with $^{31}P(C_2H_5)_2$) [4]
IR (cyclohexane): 1951, 1960, 1965, 1979, 1989, 2018, 2036, 2068 [4]
mass spectrum: [M]$^+$ [4]

19 $(\mu-H)_2Os_3\{\mu_3,\eta^2-C(CH_3)P(C_2H_5)C_5H_4FeC_5H_5\}(CO)_8P(C_2H_5)_2C_5H_4FeC_5H_5$
Isomer 2 in Formula III
from Os$_3$(CO)$_{12}$ and C$_5$H$_5$FeC$_5$H$_4$P(C$_2$H$_5$)$_2$ (ca. 1:2) in refluxing 1,4-C$_6$H$_4$(CH$_3$)$_2$ for 10 h; work-up as described for No. 15 (as a 20:1 mixture with the isomeric complex No. 20); the isomers could be separated by repeated column chromatography; the diastereomers No. 19 and No. 20 differ in ligand orientation at the P atom in the $\mu_3,\eta^2-C(CH_3)PR'R''$ bridging moiety [25]

References on pp. 115/6

Table 4 (continued)

No. compound	method of preparation (yield) properties and remarks

19 (continued)

^{1}H NMR (no medium given): -20.42, -14.82 (pt's, each 1 μ-H; $J_1 = J_2 = 11.0$ or 9.9), 1.25 (br m, 6 H), 1.57 (m, 3 H), 2.20 (br m, 4 H), 2.42 (dd, 3 H; $J_1 = 18$, $J_2 = 5$), 2.72, 2.97 (m's, each 1 H), 4.17, 4.23 (s's, each C_5H_5), 4.28 (m, C_5H_4), 4.34, 4.49 (m's, 1 H, 3 H of C_5H_4); a structure with both μ-H ligands spanning the Os(2)-Os(3) bond is proposed for both diastereomers [25]

^{31}P $\{^1$H$\}$ NMR (no medium given): -62.7, -24.6 [25]

FAB mass spectrum: $[M]^+$, $M - x$ CO$]^+$, $x = 1$ to 5 [25]

20 (μ-H)$_2$Os$_3$\{μ$_3$,η^2-C(CH$_3$)P(C$_2$H$_5$)C$_5$H$_4$FeC$_5$H$_5$\}(CO)$_8$P(C$_2$H$_5$)$_2$C$_5$H$_4$FeC$_5$H$_5$

Isomer 2 in Formula III

for formation, see No. 19

^{1}H NMR (no medium given): -20.60 (dd, μ-H; $J_1 = 11.0$, $J_2 = 8.8$), -15.13 (pt's, μ-H; $J_1 = J_2 = 10.0$), 1.25 (br m, 6 H), 1.57 (m, 3 H), 2.20 (br m, 4 H), 2.55 (dd, 3 H; $J_1 = 18$, $J_2 = 5$), 2.72, 2.83 (m's, each 1 H), 4.22, 4.28 (s's, each C_5H_5), 4.28 (m, C_5H_4), 4.3 to 4.5 (br m, C_5H_4) [25]; see also No. 19

^{31}P $\{^1$H$\}$ NMR (no medium given): -62.1, -25.8 [25]

compounds of the type (μ-H)$_2$Os$_3$(μ$_3$,η^2-C$_4$H$_2$RN)(CO)$_{9-n}$D$_n$ (Formula IV)

*21 (μ-H)$_2$Os$_3$(μ$_3$,η^2-C$_4$H$_3$N)(CO)$_9$

from Os$_3$(CO)$_{12}$ and pyrrole in decahydronaphthalene [7] or n-decane [10] at reflux temperature for 4 h, followed by work-up by TLC with light petroleum as eluant (54%) [7] or (40%, along with compounds IXa and X, p. 104, in 25 and 20% yield, respectively) [10]

by decarbonylation of (μ-H)Os$_3$(μ$_2$,η^2-O=C(C$_4$H$_3$NH)-c)(CO)$_{10}$ ("Organoosmium Compounds" B 5, Section 3.1.6.1.2.2, in preparation) in refluxing decane for 15 min; work-up as before (40%, along with compound IXa, p. 104, in 30% yield) [10]

by slow conversion of compound IXa in D$_2$O/CDCl$_3$; conversion of N-deuterated IXa giving No. 21 with one μ-D ligand indicated an H transfer from Os to the initially coordinated β-C in the pyrrol ligand [10]

References on pp. 115/6

Table 4 (continued)

No. compound	method of preparation (yield) properties and remarks
	yellow crystals, m. p. 163 to 165 °C from CHCl$_3$/pentane [7] or from CH$_2$Cl$_2$/hexane [10]

^{1}H NMR (CDCl$_3$, ca. 30 °C): -16.3 (s, μ-H), 6.62 (dd, H^4; J(H^4,H^3) = 1.2), 7.40 (dd, H^5; J(H^5,H^4) = 4.3), 7.50 (dd, H^3; J(H^5,H^3) = 0.9); (CDCl$_3$, -40 °C): -17.1, -15.6 (d's, each μ-H); the hydride signals coalesce at room temperature into a broad singlet which sharpened at 70 °C due to a hydride exchange at higher temperatures [7, 10, 16]

^{13}C NMR (no medium given): 117.2 (C^3), 149.3 (C^4), 167.8 (C^5), 175.9 (C^2) [10]

IR (cyclohexane): 1978, 1997, 2005, 2025, 2027, 2050, 2080, 2105 (all CO) [10]; see also [7]

mass spectrum: [M]$^+$ [7]

22 (μ-H)$_2$Os$_3$(μ$_3$,η^2-C$_4$H$_2$(C$_4$H$_9$-t)N)(CO)$_9$

II (low yield) [14]

IR (hexane): 1958, 1980, 1998, 2008, 2026, 2053, 2081, 2107 (all CO) [14]

mass spectrum: [M]$^+$ [14]

23 (μ-H)$_2$Os$_3$(μ$_3$,η^2-C$_4$H$_2$(COCH$_3$)N)(CO)$_9$

II (25%) [14]

^{1}H NMR (CD$_2$Cl$_2$): -16.72, -15.61 (s's, μ-H), 2.38 (s, CH$_3$), 7.88 (s, H^3), 7.90 (s, H^5) [14]

IR (hexane): 1683 (C=O), 1985, 2005, 2013, 2033, 2058, 2086, 2112 (all CO) [14]

mass spectrum: [M]$^+$ [14]

24 (μ-H)$_2$Os$_3$(μ$_3$,η^2-C$_4$H$_3$N)(CO)$_8$NCCH$_3$

from (μ-H)$_2$Os$_3$(μ$_3$,η^2-C$_4$H$_3$N)(CO)$_9$ (No. 21) by treatment with (CH$_3$)$_3$NO (1:1) in CH$_3$CN at room temperature for 0.5 h, followed by removal of the solvent (not isolated) [18]

highly reactive and air-sensitive [18]

IR (CH$_3$CN): 1971, 1996, 2036, 2046, 2081 (all CO) [18]

reaction of in situ generated No. 24 with PR$_3$ (R = C$_2$H$_5$, OCH$_3$) gave Nos. 25 and 26, respectively, see also Preparation Method I [18]

25 (μ-H)$_2$Os$_3$(μ$_3$,η^2-C$_4$H$_3$N)(CO)$_8$P(C$_2$H$_5$)$_3$

I (ca. 50%) [18]

yellow solid [18]

^{1}H NMR (CDCl$_3$): -16.45 (d, μ-H; J(H,P) =

References on pp. 115/6

Table 4 (continued)

No. compound	method of preparation (yield) properties and remarks
25 (continued)	12.3), -14.68 (d, μ–H; J(H,P) = 22.0), 1.00 (dt, CH$_3$; J(H,P) = 16.5), 1.59 (m, CH$_2$; J(H,P) = 30.0), 6.52 (dd, 1 H of C$_4$H$_3$N; J (H,H) = 4.0, 0.8), 7.30 (d, 1 H of C$_4$H$_3$N; J (H,H) = 4.0), 7.39 (s, 1 H of C$_4$H$_3$N); data are consistent with the P(C$_2$H$_5$)$_3$ ligand occupying an equatorial site being cis to one μ–H (J(H,P) = 12.3 Hz) and trans to the other one (J(H,P) = 22.0 Hz) [18]
	IR (hexane): 1940, 1974, 1986, 2002, 2029, 2047, 2082 (all CO) [18]
	mass spectrum: [M]$^+$ [18]
26 (μ–H)$_2$Os$_3$(μ_3,η^2-C$_4$H$_3$N)(CO)$_8$P(OCH$_3$)$_3$	mixture of two isomers with the P(OCH$_3$)$_3$ ligand occupying an axial site (minor isomer) or an equatorial site (major isomer) [18]; compare also Isomers 1 and 2 in Formula III
	I (ca. 50%) [18]
	yellow solid [18]
	^{1}H NMR (CD$_2$Cl$_2$):
	minor isomer: -17.89 (d, μ–H; J(H,P) = 12.0), -16.21 (d, μ–H; J(H,P) = 10.2), 2.84 (d, CH$_3$; J(H,P) = 11.8), 5.93 (d, 1 H of C$_4$H$_3$N; J(H,H) = 4.2), 6.67 (d, 1 H of C$_4$H$_3$N; J(H,H) = 3.6), 6.77 (d, 1 H of C$_4$H$_3$N; J(H,H) = 4.3)
	major isomer: -17.46 (d, μ–H; J(H,P) = 13.3), -16.05 (d, μ–H; J(H,P) = 51.3), 3.03 (d, CH$_3$; J(H,P) = 12.0), 5.87 (d, 1 H of C$_4$H$_3$N; J(H,H) = 4.0), 6.76 (s, 1 H of C$_4$H$_3$N), 6.81 (d, 1 H of C$_4$H$_3$N; J(H,H) = 3.9) [18]
	IR (hexane): 1955, 1970, 1990, 2000, 2024, 2046, 2082, 2098 (all CO) [18]
	mass spectrum: [M]$^+$ [18]
27 (μ–H)$_2$Os$_3$(μ_3,η^2-C$_4$H$_3$N)(CO)$_7$(NCCH$_3$)$_2$	from (μ–H)$_2$Os$_3$(μ_3,η^2-C$_4$H$_3$N)(CO)$_9$ (No. 21) by treatment with (CH$_3$)$_3$NO (1:2) in CH$_3$CN at room temperature for 0.5 h, followed by removal of the solvent (not isolated) [18]
	highly reactive and air-sensitive [18]
	IR (CH$_3$CN): 1919, 1972, 1987, 2025, 2056 (all CO) [18]
	reaction of in situ generated No. 27 with PR$_3$ (R = C$_2$H$_5$, OCH$_3$) gave Nos. 28 to 30; see also Preparation Method I [18]

References on pp. 115/6

Table 4 (continued)

No. compound	method of preparation (yield) properties and remarks

28 $(\mu\text{-H})_2Os_3(\mu_3,\eta^2\text{-}C_4H_3N)(CO)_7(P(C_2H_5)_3)_2$

I (40%) [18]

yellow solid [18]

^{1}H NMR (CD_2Cl_2): -16.70 (dt, μ-H; J(H,H) = 1.0, J(H,P) = 9.9), -15.63 (d, μ-H; J(H,P) = 11.6), 1.13 (m, CH_3), 2.00 (m, CH_2), 6.50 (dd, 1 H of C_4H_3N; J(H,H) = 4.1, 1.1), 7.41 (s, 1 H of C_4H_3N), 7.42 (d, 1 H of C_4H_3N; J(H,H) = 4.1); one of the $P(C_2H_5)_3$ ligands is assumed to occupy an equatorial site, the other one an axial site to give one doublet and one triplet in the μ-H region with cis P–H coupling [18]

IR (hexane): 1928, 1954, 1962, 1976, 1993, 2022, 2054 (all CO) [18]

mass spectrum: [M]$^+$ [18]

*29 $(\mu\text{-H})_2Os_3(\mu_3,\eta^2\text{-}C_4H_3N)(CO)_7(P(OCH_3)_3)_2$

I (60%, along with 10% of No. 30) [18]

yellow platelets from CH_2Cl_2/hexane at $-5\,°C$ [18]

^{1}H NMR (CD_2Cl_2): -16.94 (d, μ-H; J(H,P) = 11.4), -15.73 (d, μ-H; J(H,P) = 10.5), 3.67 (d, 3 CH_3; J(H,P) = 11.9), 3.69 (d, 3 CH_3; J(H,P) = 12.0), 6.53 (d, 1 H of C_4H_3N; J(H,H) = 4.0), 7.42 (d, 1 H of C_4H_3N; J(H,H) = 3.9), 7.44 (s, 1 H of C_4H_3N) [18]

IR (hexane): 1948, 1968, 1982, 2006, 2044, 2060 (all CO) [18]

mass spectrum: [M]$^+$ [18]

30 $(\mu\text{-H})_2Os_3(\mu_3,\eta^2\text{-}C_4H_3N)(CO)_7(NCCH_3)P(OCH_3)_3$

I (10%, in addition to 60% of No. 29) [18]

yellow solid [18]

^{1}H NMR (CD_2Cl_2): -14.34 (d, μ-H; ^{2}J(H,P) = 12.0), -14.27 (d, μ-H; ^{3}J(H,P) = 3.0), 2.52 (s, $NCCH_3$), 3.67 (d, 3 OCH_3; J(H,P) = 12.0), 6.38 (d, 1 H of C_4H_3N; J(H,H) = 4.2), 7.25 (d, 1 H of C_4H_3N; J(H,H) = 3.9), 7.40 (s, 1 H of C_4H_3N); the $P(OCH_3)_3$ ligand has a two-bond cis coupling of 12.0 Hz with one μ-H and a three-bond coupling of 3 Hz with the other one; the structure is unclear because the $P(OCH_3)_3$ and $NCCH_3$ ligands could have different positions with respect to each other [18]

References on pp. 115/6

Table 4 (continued)

No. compound	method of preparation (yield) properties and remarks
30 (continued)	IR (hexane): 1945, 1959, 1968, 1980, 2004, 2029, 2059 (all CO) [18] mass spectrum: no [M]$^+$ observed [18]

compounds of the type $(\mu-H)Os_3(\mu_3,\eta^2-CRN(CH_3)_2)(CO)_8(\mu-OCH_3)D$ (Formula V)

*31 $(\mu-H)Os_3(\mu_3,\eta^2-CHN(CH_3)_2)(CO)_9(\mu-OCH_3)$

from $(\mu-H)Os_3(CO)_{10}(\mu-OCH_3)$ and an excess of $CH_2(N(CH_3)_2)_2$ in heptane at reflux temperature for 6 h; work-up by TLC with hexane/CH_2Cl_2 (3:2) as eluant (25%, along with $(\mu-H)Os_3(\eta^1-CHN(CH_3)_2)(CO)_9(\mu-OCH_3)$, see Formula XII, p. 112 and "Organoosmium Compounds" B 5, Section 3.1.5.1, in preparation, in 52% yield) [11, 12]

yellow air-stable crystals from CH_2Cl_2/hexane at $-25\,°C$ [11, 12]

^{1}H NMR (CDCl$_3$): -11.37 (d, μ-H; J(H,H) = 1.38), 3.03 (s, OCH$_3$?), 3.65, 4.01 (s's, both NCH$_3$?), 7.62 (d, CH; J(H,H) = 1.38) [11, 12]

IR (hexane): 1935, 1959, 1970, 1978, 1995, 2011, 2038, 2071, 2095 (all CO) [11, 12]

*32 $(\mu-H)Os_3(\mu_3,\eta^2-C(C_2H_5)N(CH_3)_2)(CO)_9(\mu-OCH_3)$

by treatment of No. 33 with CO gas (1 atm) in refluxing hexane for 30 min; purified by TLC with hexane/CH_2Cl_2 (4:1) as eluant (96%) [17]

by hydrogenation of $Os_3(\mu_3,\eta^3-CH(CH_3)CN(CH_3)_2)(CO)_9(\mu-OCH_3)$ in cyclohexane at reflux temperature for 12.5 h; work-up as before (15%, along with $(\mu-H)(H)Os_3(\mu_3,\eta^3-CH(CH_3)CN(CH_3)_2)-(CO)_8(\mu-OCH_3)$, $(\mu-H)Os_3(CO)_{10}(\mu-OCH_3)$, and $(\mu-H)_2Os_3(CO)_{10}$) [17]

yellow crystals from CH_2Cl_2/hexane at $-10\,°C$ [17]

^{1}H NMR (CDCl$_3$): -10.85 (s, μ-H), 1.41 (t, CH$_3$ of C$_2$H$_5$; ^{3}J(H,H) = 7.5), 2.54 (dq, 1 H, CH$_2$; ^{2}J(H,H) = 16.1, ^{3}J(H,H) = 7.7), 3.27, 3.64 (s's, both NCH$_3$), 3.72 (dq, 1 H, CH$_2$; ^{2}J(H,H) = 16.1, ^{3}J(H,H) = 7.1), 3.97 (s, OCH$_3$) [17]

IR (hexane): 1956, 1968, 1974, 1993, 2008, 2036, 2069, 2090 [17]

References on pp. 115/6

Table 4 (continued)

No. compound	method of preparation (yield) properties and remarks

*33 $(\mu-H)Os_3(\mu_3,\eta^2-C(C_2H_5)N(CH_3)_2)(CO)_8NC_5H_5(\mu-OCH_3)$

from $(\mu-H)(H)Os_3(\mu_3,\eta^3-CH(CH_3)CN(CH_3)_2)$-$(CO)_8(\mu-OCH_3)$ and an excess of pyridine in hexane at reflux temperature for 5 h; purified by TLC with CH_2Cl_2/hexane (3:7) as eluant (17%) [17]

pale yellow crystals from CH_2Cl_2/hexane at -10 °C [17]

^{1}H NMR $(CDCl_3)$: -9.77 (s, μ-H), 1.32 (t, CH_3 of C_2H_5; $^3J(H,H)=7.3$), 2.41 (dq, 1 H, CH_2; $^2J(H,H)=16.0$, $^3J(H,H)=7.8$), 3.31, 3.50 (s's, both NCH_3), 3.58 (s, OCH_3), 3.72 (dq, 1 H, CH_2; $^2J(H,H)=16.0$, $^3J(H,H)=6.7$), 7.4 to 7.6, 7.8 to 8.0, 9.0 to 9.1 (m's, corresponding to 2 H, 1 H, and 2 H of NC_5H_5) [17]

IR (hexane): 1931, 1953, 1959, 1985, 2002, 2030, 2070 (all CO) [17]

thermolysis in refluxing hexane under a CO atmosphere yielded No. 32 [17]

compound of the type $(\mu-H)Os_3(\mu_3,\eta^2-CHC(CH_3)_2CH_2S)(CO)_9$

*34

by decarbonylation of the 2,2-dimethylthietane complexes $Os_3(CO)_{11}SC_3H_4(CH_3)_2$ [21] or $Os_3(CO)_{10}(\mu-SC_3H_4(CH_3)_2)$ [15, 21] in refluxing heptane for 1 h, followed by separation by TLC with hexane/CH_2Cl_2 (9:1) as eluant (18 or 21%, along with $(\mu-H)Os_2(\mu,\eta^3-$SCH_2CCH_3(CH_2)_2)(CO)_6$, Formula XIV, p. 114, $Os_4(\mu-CO)(CO)_{12}(\mu-SC_3H_4(CH_3)_2)$, and small amounts of $Os_3(CO)_{12}$ as by-products) [15, 21]

orange crystals [15, 21]

^{1}H NMR $(CDCl_3)$: -17.47 (d, μ-H; $J(H,H)=1.7$), 1.31, 1.50 (s's, both CH_3), 2.11 (d, 1 H, CH_2; $J(H,H)=11.3$), 3.02 (dd, 1 H, CH_2; $J(H,H)=1.7$, $J(H,H)=11.3$), 8.55 (s, CH) [15, 21]

^{13}C NMR $(CDCl_3)$: 33.35, 33.76 (s's, both CH_3), 50.75 (d, C^2; $J(C,H)=2.3$), 62.89 (s, C^3), 132.91 (d, C^1; $J(C,H)=2.3$) [22]

IR (hexane): 1975, 1986, 2003, 2013, 2017, 2050, 2070, 2099 (all CO) [15, 21]

mass spectrum: $[M]^+$, $[M-x\ CO]^+$, $x=1$ to 9 [21]

VII

VIII

a

IX

b

X

*Further information:

$(\mu-H)_2Os_3(\mu_3,\eta^2-C(C_3H_7-i)C(CH_3)=NC_6H_5)(CO)_9$ (Table **4**, No. **3**) crystallizes in the monoclinic space group $P2_1/c - C_{2h}^5$ (No. 14) with a = 14.818(5), b = 9.051(5), c = 20.098(6) Å, β =

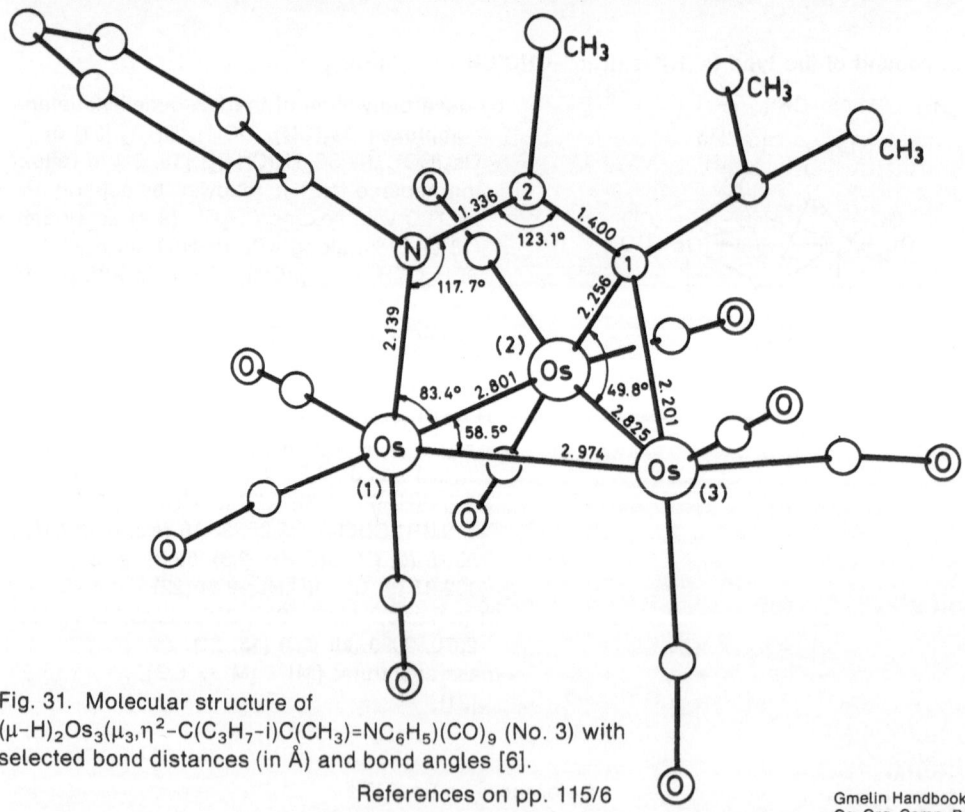

Fig. 31. Molecular structure of
$(\mu-H)_2Os_3(\mu_3,\eta^2-C(C_3H_7-i)C(CH_3)=NC_6H_5)(CO)_9$ (No. 3) with
selected bond distances (in Å) and bond angles [6].

References on pp. 115/6

108.45(3)°; Z = 4, D_c = 2.59 g/cm^3, R = 0.058. The hydride ligands were not observed crystallo-graphically but were thought to bridge the Os(2)-Os(3) and the Os(1)-Os(3) bonds. The C(2)-N as well as the C(1)-C(2) distance indicate some double bond character; see **Fig. 31**. Generally, the structure is very similar to that of $(\mu\text{-}H)_2Os_3(\mu_3,\eta^2\text{-}O\text{=}C_6H_3CH_2C_6H_5)(CO)_9$ (Section 3.1.6.2.3). Average bond distances of Os-CO and C≡O are 1.88 and 1.18 Å [6].

Thermolysis in octane at reflux temperature gave $(\mu\text{-}H)_2Os_3(\mu_3,\eta^4\text{-}CH_2C(CH_3)\text{-}CC(CH_3)\text{=}NC_6H_5)(CO)_8$ (Formula XI); thermolysis in hexane gave no reaction [6].

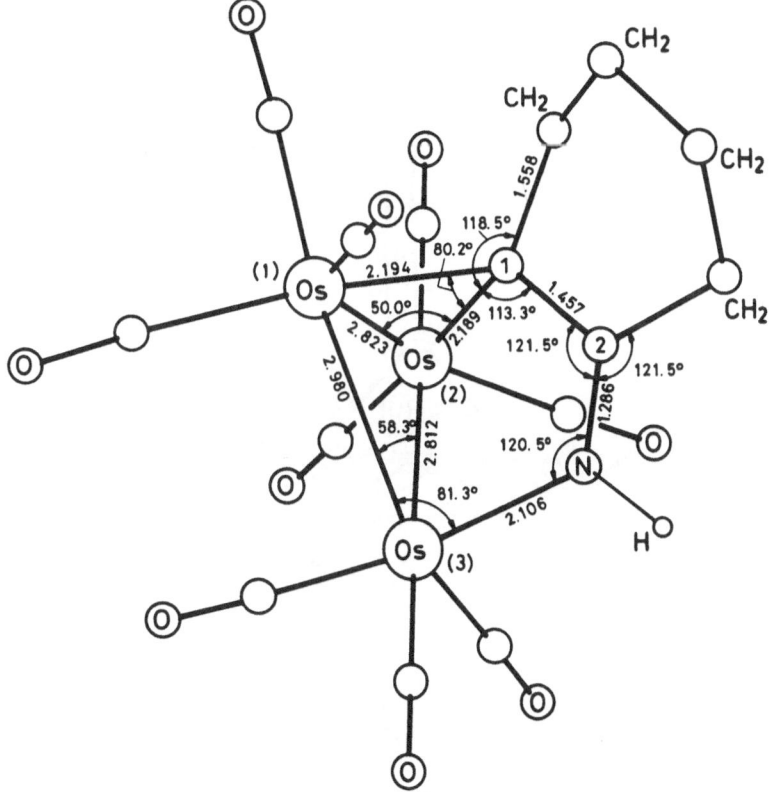

XI

$(\mu\text{-}H)_2Os_3(\mu_3,\eta^2\text{-}C_6H_8\text{=}NH)(CO)_9$ (Table 4, No. **4**) crystallizes in the monoclinic space group P2$_1$/c − C$_{2h}^5$ (No. 14) with a = 9.479(4), b = 13.498(5), c = 15.943(7) Å, β = 92.55(3)°; Z = 4, D_c = 2.99 g/cm^3, R = 0.042. The structure is shown in **Fig. 32**. One hydride bridges the

Fig. 32. Molecular structure of $(\mu\text{-}H)_2Os_3(\mu_3,\eta^2\text{-}C_6H_8\text{=}NH)(CO)_9$ (No. 4) with selected bond distances (in Å) and bond angles; μ-H ligands not shown [8].

Os(1)–Os(3) edge as confirmed by the longer bond distance and by the distribution of the carbonyls which bend away from the edge. The other hydride spans the Os(1)–Os(2) bond and is similar to the nonbridged Os–Os bond, which is consistent with the shortening effect of the μ-C moiety, and with the distribution of the CO groups. The relatively short Os(2)–Os(3) bond is probably caused by the overall capping effect of the μ_3,η^2-C_6H_8=NH ligand. Average bond distances of Os–CO and C≡O are 1.92 and 1.13 Å [8].

(μ-H)$_2$Os$_3$(μ$_3$,η^2-CHC$_4$H$_6$N)(CO)$_9$ (Table 4, No. **5**) crystallizes in the triclinic space group $P\bar{1}$–C_i^1 (No. 2) with a = 13.394(8), b = 15.601(8), c = 9.069(7) Å, α = 88.51(5)°, β = 80.61(6)°, γ = 86.96(5)°; Z = 4, D_c = 3.22 g/cm^3, R = 0.07. The asymmetric unit contains two nearly identical molecules. Molecule **A** is shown in **Fig. 33**. In Molecule **B** the Os(1)–Os(2), Os(2)–Os(3), and Os(1)–Os(3) bond lenghts are 2.999, 2.802, and 2.808 Å, respectively. The positions of the μ-H ligands were calculated. In Molecule **A** the Os(1)–Os(3) bridging hydride is located almost in the plane of the metal atoms, whereas the other hydride is positioned below the Os(2)–Os(3) plane by 1.16 Å. In Molecule **B** the μ-H ligands are located at the Os(1)–Os(2) and Os(2)–Os(3) edges. In both Molecule **A** and Molecule **B** the pyrrolidine ring tilts to

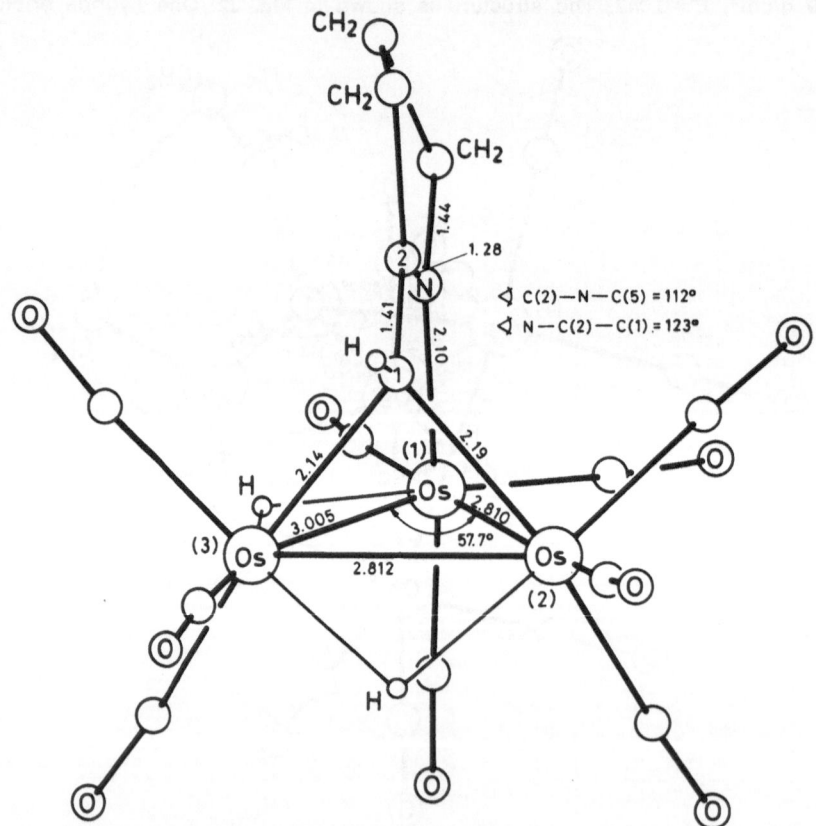

Fig. 33. Molecular structure of Molecule **A** of (μ-H)$_2$Os$_3$(μ$_3$,η^2-CHC$_4$H$_6$N)(CO)$_9$ (No. **5**) with selected bond distances (in Å) and bond angles [23].

References on pp. 115/6

the side toward the Os(1)–Os(3) edge by 8.7° and 6.4°, respectively, relative to the Os_3 plane. Average bond distances of Os–CO and C≡O are 1.90 and 1.15 Å [23].

(μ-H)Os₃(μ₃,η²-C(P(C₆H₅)₃)CH=NC₆H₅)(CO)₉ (Table **4**, No. **6**) crystallizes in the triclinic space group P1̄–C_i^1 (No. 2) with a = 9.568(1), b = 17.881(2), c = 20.583(7) Å, α = 91.47(2)°, β = 90.82(2)°, γ = 90.31(1)°; Z = 4, D_c = 2.272 g/cm³, R = 0.025. The unit cell contains two independent molecules, which differ slightly in the position of the bridging carbon resulting in small distortions in the ¹L ligand and in the Os_3 triangle. The significant shortening of the Os–Os edges is caused by the triply-bridging ¹L ligand with the cramping effect of the P(C₆H₅)₃ ligand. In the molecule shown in **Fig. 34**, the hydride was crystallographically observed; it bridges the Os(1)–Os(2) nearly symmetrically. The averaged Os–CO bond distances are given as 1.863 and 1.927 Å for the molecules [20].

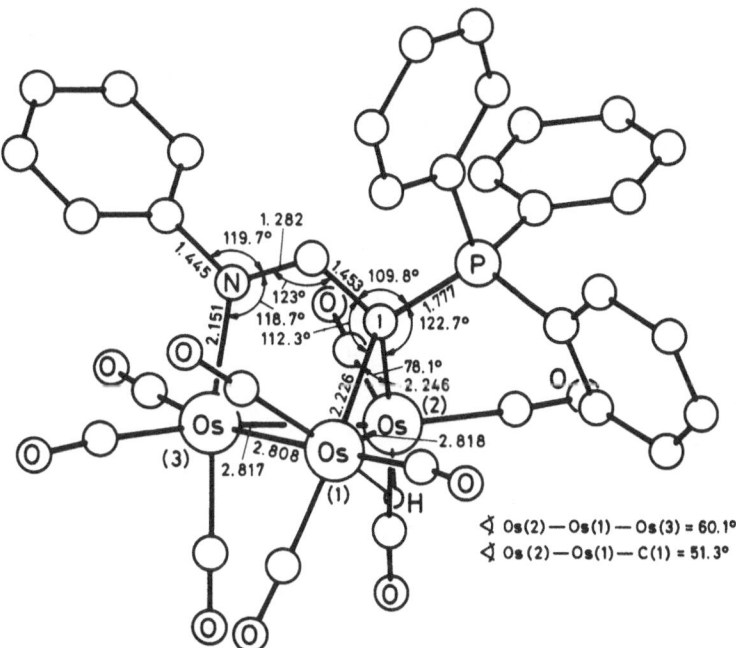

Fig. 34. Molecular structure of (μ-H)Os₃(μ₃,η²-C(P(C₆H₅)₃)CH=NC₆H₅)(CO)₉ (No. 6) with selected bond distances (in Å) and bond angles [20].

(μ-H)Os₃(μ₃,η²-C₄H₂NHCH=NCH₃)(CO)₉ (Table **4**, No. **7**) crystallizes in the monoclinic space group P2₁/n–C_{2h}^5 (No. 14) with a = 14.004(3), b = 19.943(5), c = 14.028(4) Å, β = 97.59(2)°; Z = 8, D_c = 3.18 g/cm³, R = 0.0777. The unit cell contains two independent but very similar molecules; one is shown in **Fig. 35**. The four-atom bridge across the Os_3 triangle is unusual in a trimetallic cluster. The μ₃,η²-C₄H₂NHCH=NCH₃ ligand is essentially planar with an approximate mirror plane passing through the organic ligand and through Os(1). The hydride was not observed crystallographically but it is believed to bridge the Os(2)–Os(3) bond which is consistent with the positions of the carbonyl ligands. However, the quality of the X-ray structure study does not allow a decision between N and C atoms; the arrangement with a bridging carbon rather than a nitrogen was indicated by ¹H NMR spectroscopy. The structure of the other molecule of the unit cell with selected bond lenghts and angles is given [23].

References on pp. 115/6

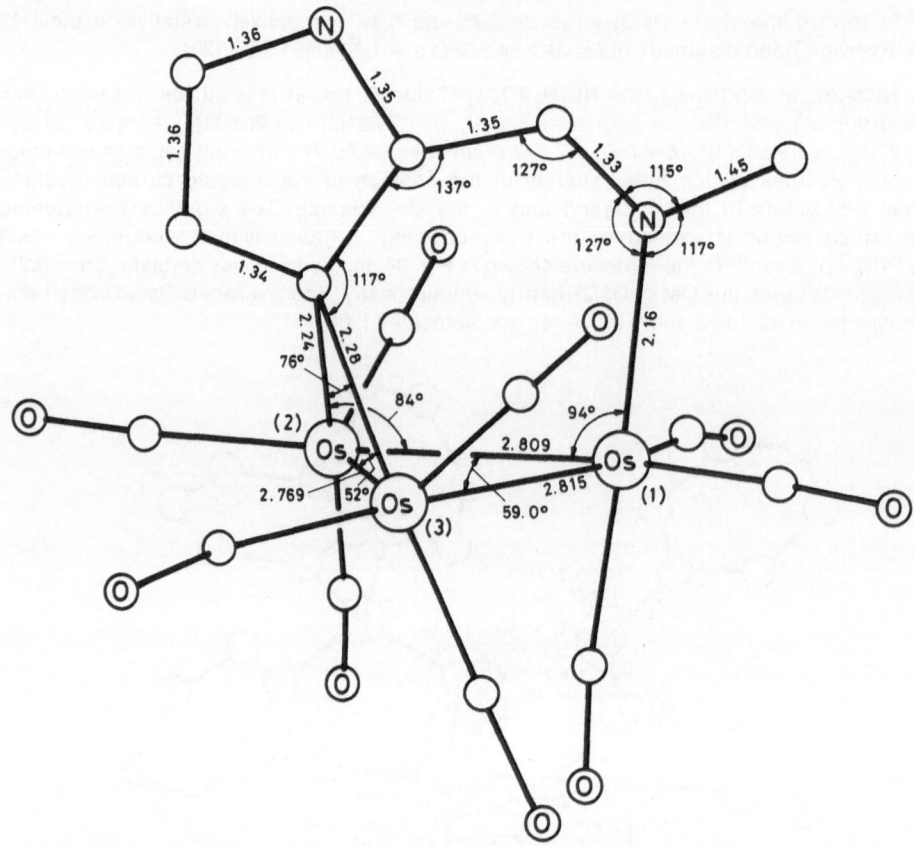

Fig. 35. Molecular structure of one of the two independent molecules of $(\mu-H)Os_3(\mu_3,\eta^2-C_4H_2NHCH{=}NCH_3)(CO)_9$ (No. 7) with selected bond distances (in Å) and bond angles [23].

$(\mu-H)Os_3(\mu_3,\eta^2-C_6H_4P(C_6H_5)_2)(CO)_8P(C_6H_5)_3$ (Table **4,** No. **10**) crystallizes in the monoclinic space group C2/c $-$ C$_{2h}^6$ (No. 15) with a = 20.981, b = 13.179, c = 31.422 Å, β = 112.25°; Z = 8, R = 0.08. The structure is shown in **Fig. 36**. The hydride ligand probably bridges the Os(2)-Os(3) bond [2].

$(\mu-H)_2Os_3(\mu_3,\eta^2-CHN(CH_3)_2)(CO)_9$ (Table **4,** No. **12**) crystallizes in the orthorhombic space group P2$_1$2$_1$2$_1$ $-$ D$_2^4$ (No. 19) with a = 13.769(6), b = 14.841(8), c = 9.058(3) Å; Z = 4, D$_c$ = 3.16 g/cm^3, R = 0.0544. The structure is shown in **Fig. 37**. The variation of the Os-Os distances between 2.775 and 2.977 Å is attributed to the presence of the two not crystallographically observed hydride ligands, that probably span the two longer Os(1)-Os(3) and Os(1)-Os(2) edges; the elongation of the Os(1)-Os(2) bond is counteracted by the bridging carbene ligand. Average bond distances of Os-CO and C≡O are 1.85 and 1.18 Å [13].

Low-temperature ^{1}H NMR spectra clearly established the existence of two unequivalent bridging hydride ligands which undergo dynamic exchange processes. The μ-H resonances coalesce at ca. 281 K to a broad singlet at $\delta = -15.89$ ppm (in toluene-d$_8$?; ΔG = 15.2 kcal/mol at 284 K). Similarly, the two sharp CH$_3$ resonances coalesce at ca. 263 K (ΔG = 14.5 kcal/mol at 263 K); at 298 K broad resonances at δ = 2.38 (in toluene-d$_8$?) or 3.05 (in CDCl$_3$?)

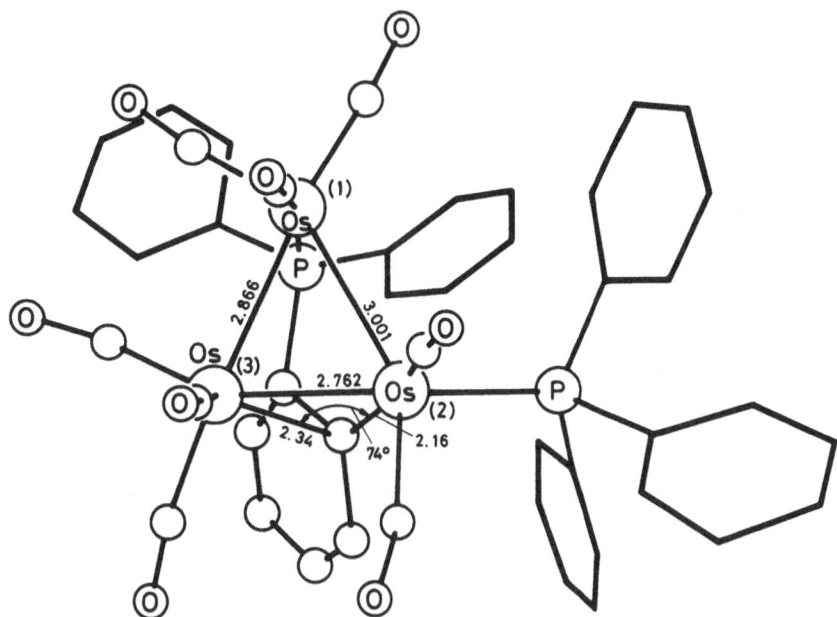

Fig. 36. Molecular structure of $(\mu\text{-H})Os_3(\mu_3,\eta^2\text{-}C_6H_4P(C_6H_5)_2)(CO)_8P(C_6H_5)_3$ (No. 10) with selected bond distances (in Å) and a bond angle [2].

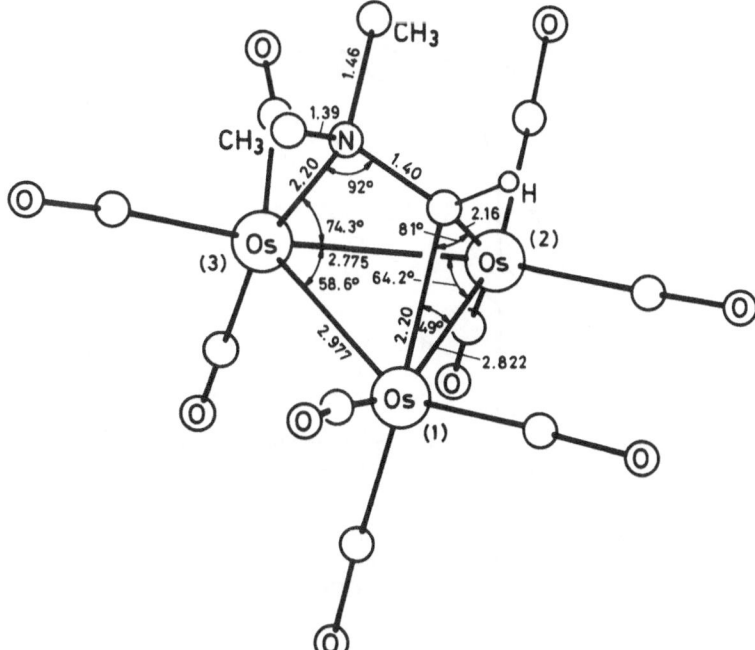

Fig. 37. Molecular structure of $(\mu\text{-H})_2Os_3(\mu_3,\eta^2\text{-}CHN(CH_3)_2)(CO)_9$ (No. 12) with selected bond distances (in Å) and bond angles [13].

References on pp. 115/6

ppm were observed. The analysis of the line shapes in the broadened regions indicated two exchange processes: fast shifts of μ-H(1) between the Os(1)-Os(3) and Os(2)-Os(3) bonds averaging the CH₃, but not the hydride environments, and approximately three times slower shifts of μ-H(2) between the Os(1)-Os(2) and Os(2)-Os(3) bonds which, along with the first process, also averaged the hydride resonances [13].

(μ-H)₂Os₃(μ₃,η²-C₄H₃N)(CO)₉ (Table **4**, No. **21**) crystallizes in the monoclinic space group P2₁/c−C$_{2h}^5$ (No. 14) with a = 11.308(2), b = 12.457(2), c = 15.384(3) Å, β = 123.59(2)°; Z = 4, D$_c$ = 3.27 g/cm³, R = 0.041. The structure is shown in **Fig. 38**. The C₄H₃N ring ligand is nearly planar and the plane is almost perpendicular to the Os₃ plane with a dihedral angle of 93.0(4)°. All interatomic distances in the ring ligand are fairly similar indicating that the aromatic character of the ring is not totally quenched in this mode of bonding. The hydride ligands were not observed crystallographically but probably bridge the Os(1)-Os(2) and Os(2)-Os(3) bonds due to the lenghtening of the metal-metal bond distances and CO positions. Bond distances for Os-CO are found in the range of 1.85 to 1.91 Å [16].

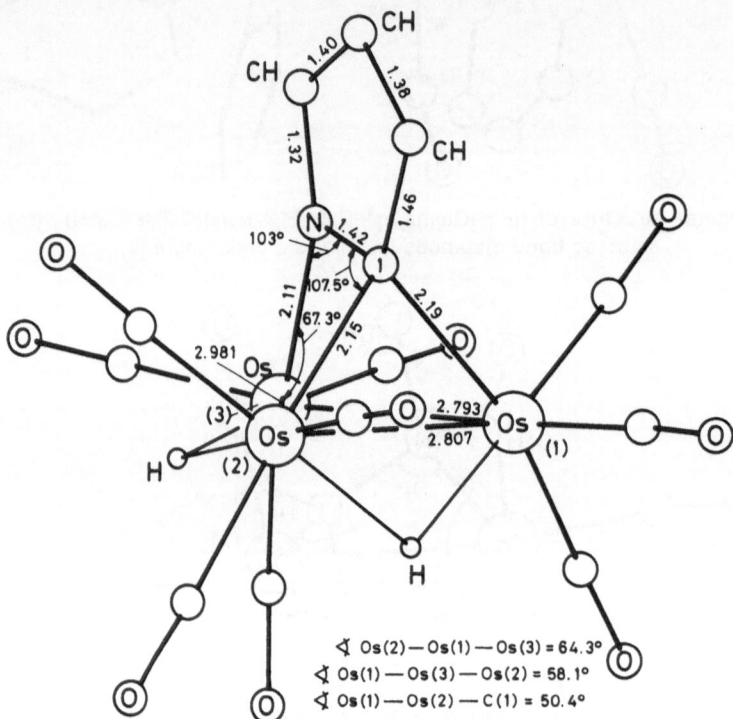

Fig. 38. Molecular structure of (μ-H)₂Os₃(μ₃,η²-C₄H₃N)(CO)₉ (No. 21) with selected bond distances (in Å) and bond angles [16].

Reaction with RCl/AlCl₃ (R = t-C₄H₉, CH₃CO) in CS₂ yielded Nos. 22 and 23 [14]. Treatment with (CH₃)₃NO (1:1 or 1:2) in CH₃CN at ca. 25 °C resulted in highly reactive and air-sensitive (μ-H)₂Os₃(μ₃,η²-C₄H₃N)(CO)₉₋ₙ(NCCH₃)ₙ (n = 1 or 2, Nos. 24 and 27, respectively); see also Preparation Method I [18].

(μ-H)₂Os₃(μ₃,η²-C₄H₃N)(CO)₇(P(OCH₃)₃)₂ (Table **4**, No. **29**) crystallizes in the triclinic space group P$\bar{1}$−C$_i^1$ (No. 2) with a = 9.519(3), b = 10.165(3), c = 14.942(4) Å, α = 81.12(2)°, β = 80.73(2)°, γ = 80.50(3)°; Z = 2, D$_c$ = 2.575 g/cm³, R = 0.051. The structure is shown in **Fig. 39**.

References on pp. 115/6

The μ_3,η^2-C_4H_3N ligand is planar and caps the Os_3 triangle perpendicularly; the dihedral angle is 90.5(5)°. One $P(OCH_3)_3$ group occupies an axial site on Os(1) pseudo-trans to μ-C (P-Os(1)-(μ-C) = 171.1(4)°), whereas the other one occupies an equatorial site on Os(3). The two hydride ligands were not observed crystallographically but potential energy calculations revealed that they bridge the Os(1)-Os(2) and Os(1)-Os(3) edges. Average bond distances of Os-CO and C≡O are 1.90 and 1.13 Å [18].

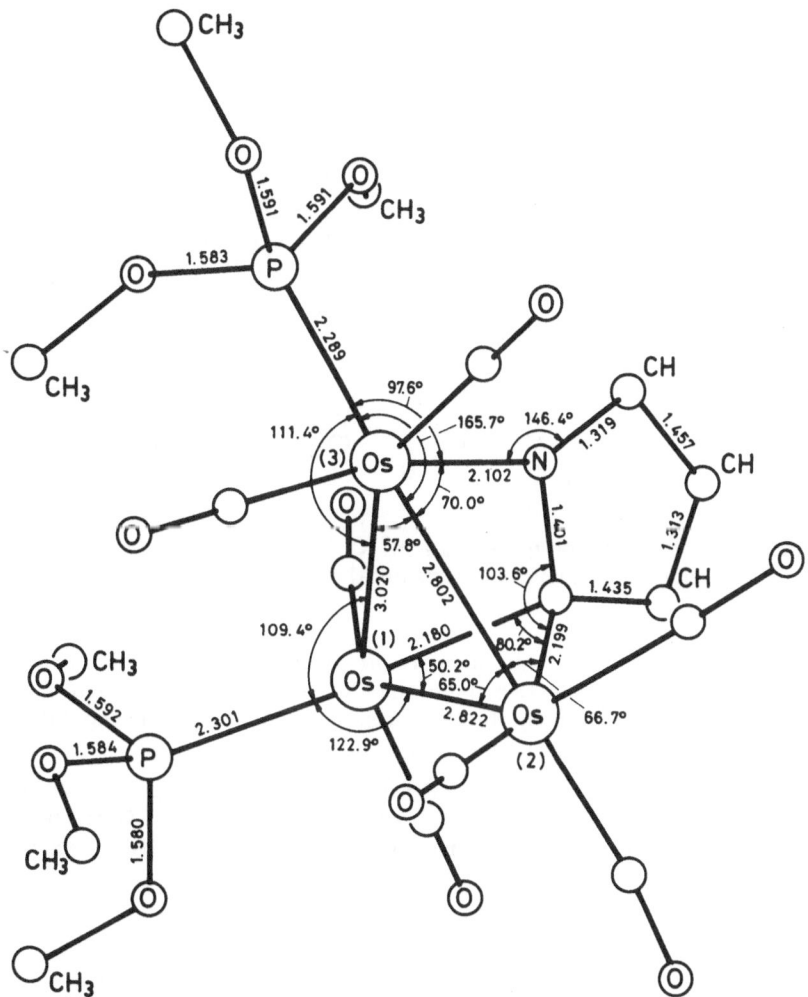

Fig. 39. Molecular structure of $(\mu$-H)$_2$Os$_3(\mu_3,\eta^2$-$C_4H_3N)(CO)_7(P(OCH_3)_3)_2$ (No. 29) with selected bond distances (in Å) and bond angles [18].

(μ-H)Os$_3(\mu_3,\eta^2$-CHN(CH$_3$)$_2$)(CO)$_9(\mu$-OCH$_3$) (Table 4, No. 31) crystallizes in the monoclinic space group P2$_1$/n − C$_{2h}^5$ (No. 14) with a = 8.878(2), b = 26.287(8), c = 9.243(2) Å, β = 115.19(1)°; Z = 4, D$_c$ = 3.09 g/cm³, R = 0.0381. The molecule consists of an open Os$_3$ triangle with a μ_3,η^2-bridging CHN(CH$_3$)$_2$ carbene moiety and an OCH$_3$ group bridging the nonbonded pair of metal atoms; see **Fig. 40**. The very long C–N distance is attributed to the lack of C–N multiple bonding, probably due to the coordination of the nitrogen lone pair of electrons to an Os atom. Average bond distances of Os–CO and C≡O are 1.90 and 1.14 Å [11, 12].

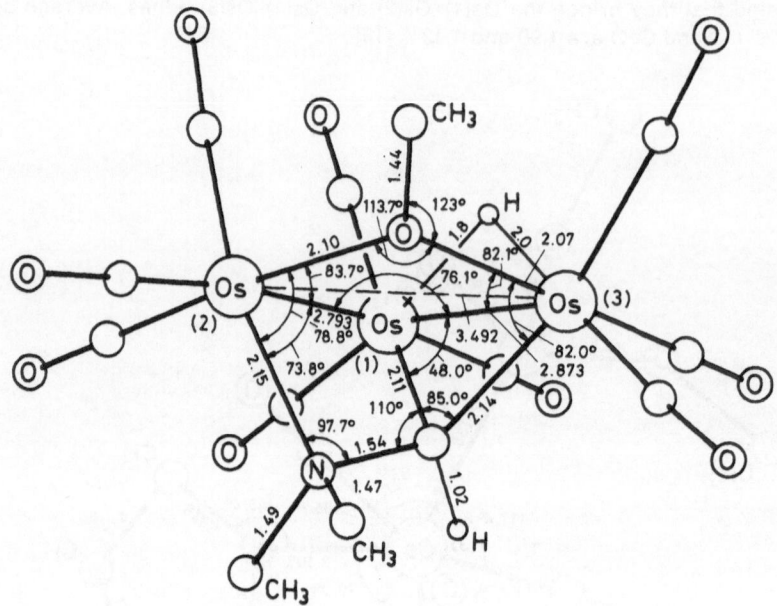

Fig. 40. Molecular structure of (μ-H)Os₃(μ₃,η²-CHN(CH₃)₂)(CO)₉(μ-OCH₃) (No. 31) with selected interatomic distances (in Å) and bond angles [11, 12].

Thermolysis in heptane at reflux temperature under a CO atmosphere for 4 h gave (μ-H)Os₃(η¹-CHN(CH₃)₂)(CO)₉(μ-OCH₃) (Formula XII, R = H; "Organoosmium Compounds" B 5, Section 3.1.5.1, in preparation); the Os(2)–Os(3) bond formation closing the Os₃ cluster is believed to be the driving force for the thermal isomerization [11, 12].

Reaction with an excess of CH₃O₂CC≡CH in refluxing cyclohexane for 40 min yielded the addition product Os₃(μ₃,η³-C(CO₂CH₃)=CHC=N(CH₃)₂)(CO)₉(μ₂,η¹-OCH₃) (Formula XIII) by formation of a C–C bond and the elimination of 1 equivalent of H₂; a mechanism was proposed [12].

XII

XIII

(μ-H)Os$_3$(μ$_3$,η^2-C(C$_2$H$_5$)N(CH$_3$)$_2$)(CO)$_9$(μ-OCH$_3$) (Table **4**, No. **32**) crystallizes in the mono-clinic space group P2$_1$/n − C$_{2h}^5$ (No. 14) with a = 12.354(2), b = 13.017(2), c = 13.371(2) Å, β = 96.10(1)°; Z = 4, D$_c$ = 2.92 g/cm^3, R = 0.025. The structure, shown in **Fig. 41**, is similar to that of No. 31 consisting of an open Os$_3$ cluster and having a very long C–N distance in the μ$_3$,η^2-C(C$_2$H$_5$)N(CH$_3$)$_2$ unit. Average bond distances of Os–CO and C≡O are 1.92 and 1.14 Å [17].

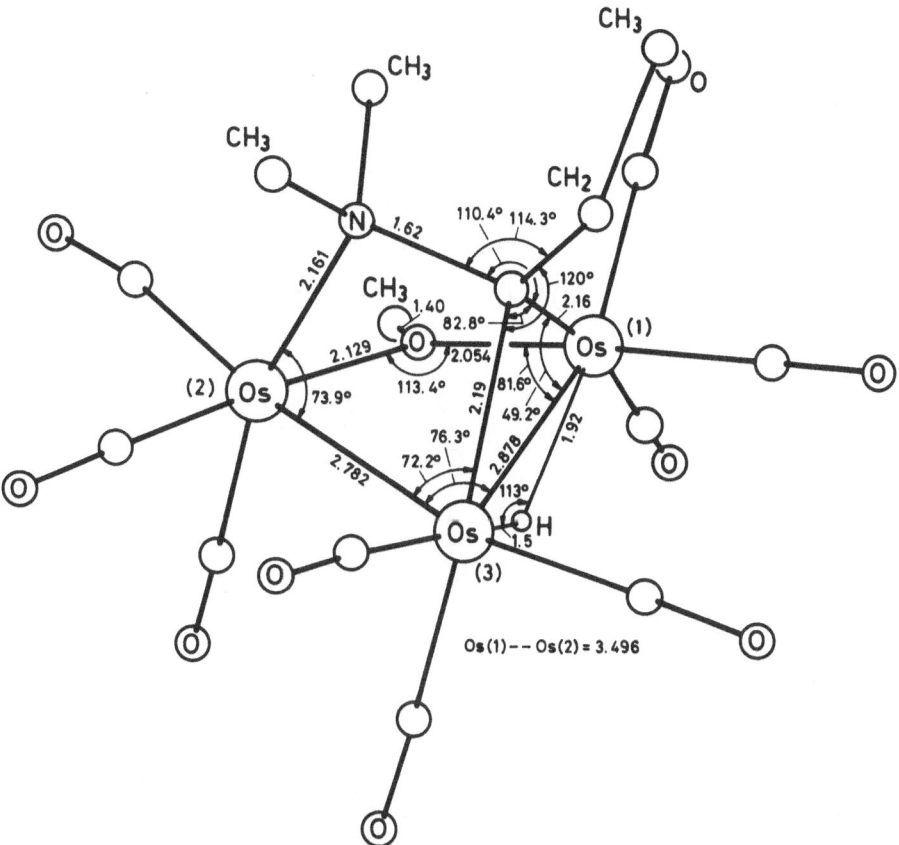

Fig. 41. Molecular structure of (μ-H)Os$_3$(μ$_3$,η^2-C(C$_2$H$_5$)N(CH$_3$)$_2$)(CO)$_9$(μ-OCH$_3$) (No. 32) with selected interatomic distances (in Å) and bond angles [17].

Isomerization in octane at 125 °C for 5 h under a CO atmosphere resulted in (μ-H)Os$_3$(η^1-C(C$_2$H$_5$)N(CH$_3$)$_2$)(CO)$_9$(μ-OCH$_3$) (Formula XII, R = C$_2$H$_5$; "Organoosmium Compounds" B 5, Section 3.1.5.1, in preparation) [17].

(μ-H)Os$_3$(μ$_3$,η^2-C(C$_2$H$_5$)N(CH$_3$)$_2$)(CO)$_8$(NC$_5$H$_5$)(μ-OCH$_3$) (Table **4**, No. **33**) crystallizes in the triclinic space group P1̄ − C$_i^1$ (No. 2) with a = 9.381(1), b = 16.261(3), c = 8.828(1) Å, α = 97.72(2)°, β = 115.82(3)°, γ = 78.61(2)°; Z = 2, D$_c$ = 2.74 g/cm^3, R = 0.022. The structure, shown in **Fig. 42**, is similar to those of Nos. 31 and 32. Average bond distances of Os–CO and C≡O are 1.90 and 1.15 Å, respectively [17].

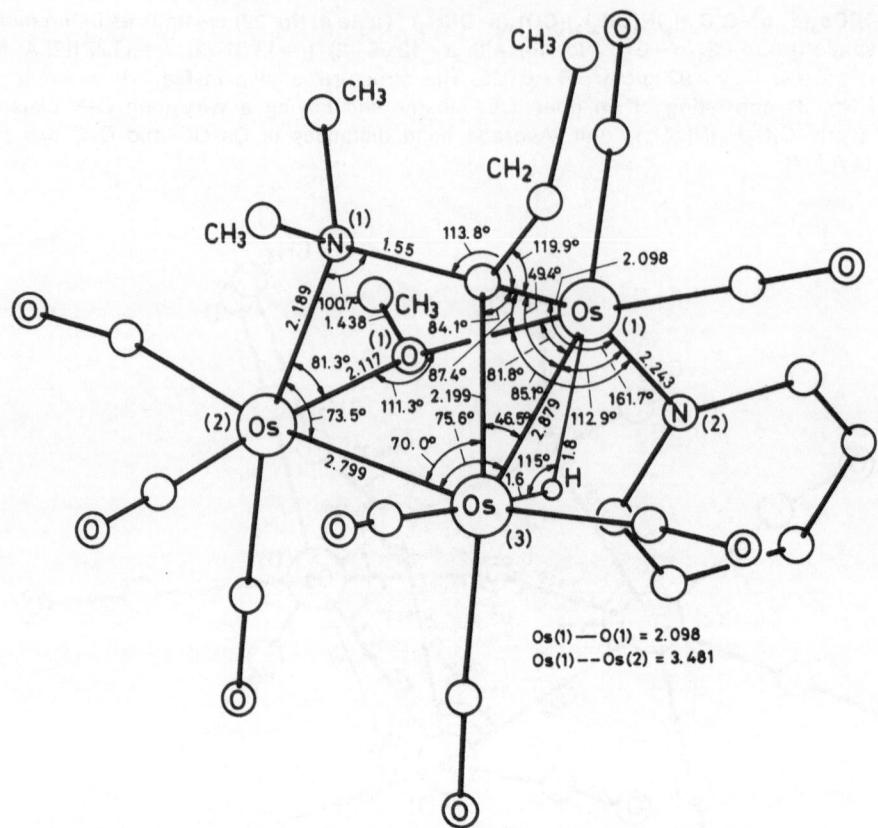

Fig. 42. Molecular structure of $(\mu\text{-H})Os_3(\mu_3,\eta^2\text{-C}(C_2H_5)N(CH_3)_2)(CO)_8(NC_5H_5)(\mu\text{-OCH}_3)$ (No. 33) with selected interatomic distances (in Å) and bond angles [17].

$(\mu\text{-H})Os_3(\mu_3,\eta^2\text{-CHC}(CH_3)_2CH_2S)(CO)_9$ (Table 4, No. 34) crystallizes in the triclinic space group $P\bar{1}\text{-C}_i^1$ (No. 2) with a = 9.330(2), b = 14.352(2), c = 8.656(2) Å, α = 93.73(2)°, β = 117.58(1)°, γ = 101.09(2)°; Z = 2, D_c = 3.10 g/cm³, R = 0.023. The $\mu_3,\eta^2\text{-CHC}(CH_3)_2CH_2S$ ligand bridges one face of the closed Os_3 cluster, in which the Os(1)–Os(3) edge is spanned by the S atom and the Os(1)–Os(2) by the alkylidene carbon. The third edge of the cluster is bridged by a hydride (located crystallographically); see **Fig. 43**. Average bond distances of Os–CO and C≡O are 1.91 and 1.14 Å [15, 21].

The reaction with an H_2/CO mixture (68 and 13.6 atm, respectively) in heptane at 100 °C for 8 h yielded $(\mu\text{-H})Os_2(\mu,\eta^3\text{-SCH}_2CCH_3(CH_2)_2)(CO)_6$ (Formula XIV) and $(\mu\text{-H})Os_3(CO)_{10}$-$(\mu\text{-SCH}_2C_4H_9\text{-t})$ [15, 21].

XIV

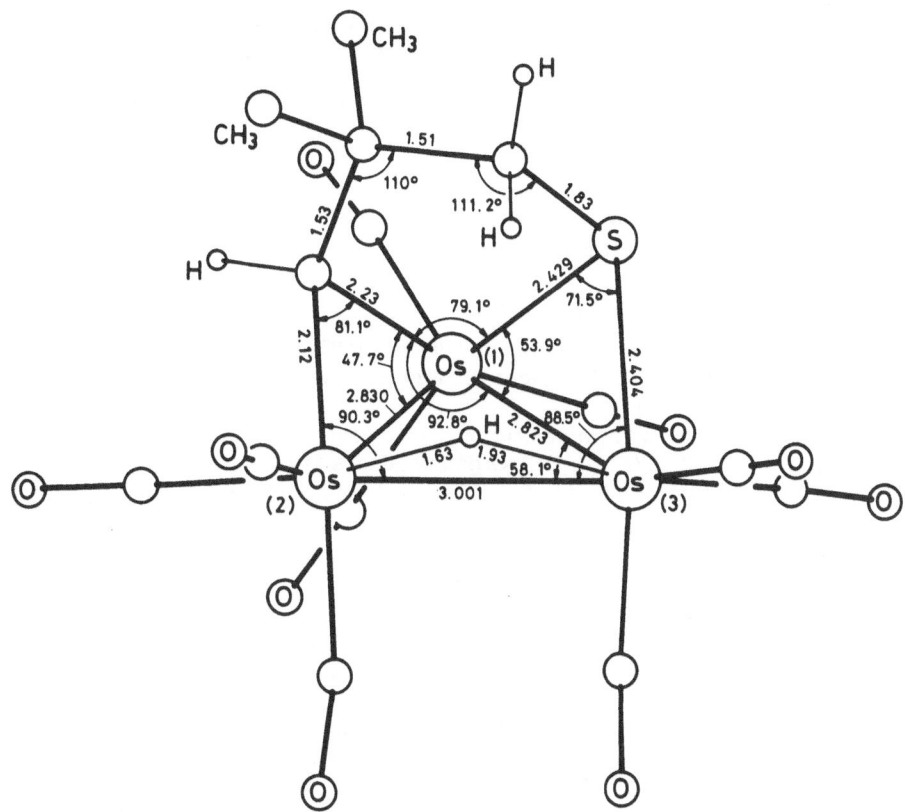

Fig. 43. Molecular structure of $(\mu\text{-H})Os_3(\mu_3,\eta^2\text{-CHC}(CH_3)_2CH_2S)(CO)_9$ (No. 34) with selected bond distances (in Å) and bond angles [15, 21].

References:

[1] Bradford, C. W.; Nyholm, R. S.; Gainsford, G. J.; Guss, J. M.; Ireland, P. R.; Mason, R. (J. Chem. Soc. Chem. Commun. **1972** 87/9).

[2] Gainsford, G. J.; Guss, J. M.; Ireland, P. R.; Mason, R.; Bradford, C. W.; Nyholm, R. S. (J. Organomet. Chem. **40** [1972] C 70/C 72).

[3] Bradford, C. W.; Nyholm, R. S. (J. Chem. Soc. Dalton Trans. **1973** 529/33).

[4] Deeming, A. J.; Underhill M. (J. Chem. Soc. Dalton Trans. **1973** 2727/30).

[5] Deeming, A. J. (J. Organomet. Chem. **128** [1977] 63/72).

[6] Adams, R. D.; Selegue, J. P. (Inorg. Chem. **19** [1980] 1795/801).

[7] Yin, C. C.; Deeming, A. J. (J. Chem. Soc. Dalton Trans. **1982** 2563/4).

[8] Süss-Fink, G.; Raithby, P. R. (Inorg. Chim. Acta **71** [1983] 109/14).

[9] Boyar, E.; Deeming, A. J.; Arce, A. J.; De Sanctis, Y. (J. Organomet. Chem. **276** [1984] C 45/C 48).

[10] Arce, A. J.; De Sanctis, Y.; Deeming, A. J. (J. Organomet. Chem. **311** [1986] 371/8).

[11] Adams, R. D.; Babin, J. E. (Organometallics **6** [1987] 1364/5).

[12] Adams, R. D.; Babin, J. E. (Organometallics **7** [1988] 2300/6).

[13] Adams, R. D.; Babin, J. E. (Organometallics **7** [1988] 963/9).

[14] Chen, H.; Johnson, B. F. G.; Lewis, J. (Organometallics **8** [1989] 2965/7).

[15] Adams, R. D.; Pompeo, M. P. (Organometallics **9** [1990] 2651/3).

[16] Day, M. W.; Hardcastle, K. I.; Deeming, A. J.; Arce, A. J.; De Sanctis, Y. (Organometallics **9** [1990] 6/12).

[17] Adams, R. D.; Pompeo, M. P.; Tanner, J. T. (Organometallics **10** [1991] 1068/78).

[18] Chen, H.; Johnson, B. F. G.; Lewis, J.; Raithby, P. R. (J. Organomet. Chem. **406** [1991] 219/36).

[19] Kabir, S. E.; Day, M. W.; Irving, M.; McPhillips, T.; Minassian, H.; Rosenberg, E.; Hardcastle, K. I. (Organometallics **10** [1991] 3997/4004).

[20] Langenbahn, M.; Stoeckli-Evans, H.; Süss-Fink, G. (J. Organomet. Chem. **402** [1991] C 12/C 15).

[21] Adams, R. D.; Belinski, J. A.; Pompeo, M. P. (Organometallics **11** [1992] 2016/24).

[22] Adams, R. D.; Pompeo, M. P. (Organometallics **11** [1992] 2281/9).

[23] Arce, A. J.; De Sanctis, Y.; Hernandez, L.; Marquez, M.; Deeming, A. J. (J. Organomet. Chem. **436** [1992] 351/65).

[24] Day, M. W.; Freeman, W.; Hardcastle, K. I.; Isomaki, M.; Kabir, S. E.; McPhillips, T.; Rosenberg, E.; Scott, L. G.; Wolf, E. (Organometallics **11** [1992] 3376/84).

[25] Cullen, W.R.; Rettig, S. J.; Zheng, T. C. (Can. J. Chem. **71** [1993] 399/409).

3.1.6.2.5 Compounds with μ_3,η^3-Bridging ^{1}L Ligands

$(\mu\text{-}H)_2Os_3(\mu_3,\eta^3\text{-}6\text{-}(CH)C_5H_3N(CH\text{=}NC_3H_7\text{-}i)\text{-}2)(CO)_8$ (see **Fig. 44**) was prepared from $(\mu\text{-}H)Os_3(\mu_2,\eta^3\text{-}6\text{-}(CH_2)C_5H_3N(CH\text{=}NC_3H_7\text{-}i)\text{-}2)(CO)_9$ by decarbonylation/hydrogen transfer in n-nonane at 130 °C for 75 min, followed by purification by column chromatography with hexane/ether (1:1) as eluant, and crystallization from the concentrated eluate at −20 °C; dark red air-stable crystals in 50% yield; sparingly soluble in aliphatic solvents but dissolves readily in polar ones [6].

^{1}H NMR (C_6D_6): −14.40, −10.98 (d's, both μ-H; J(H,H) = 2 Hz), 0.94, 1.10 (d's, both CH_3; J(H,H) = 6 Hz), 2.96 (sept, CH of i-C_3H_7; J(H,H) = 6 Hz), 5.22 (s, μ_2-CH), 5.79, 6.31 (d's, each 1 H of C_5H_3N; J(H,H) = 8 Hz), 6.58 (dd, 1 H of C_5H_3N; J(H,H) = 8 Hz), 6.93 (s, CH=N) ppm; the hydride signal appearing at −10.98 ppm is at a rather low field for a bridging hydride [6].

^{13}C NMR (CD_2Cl_2): 24.0, 25.9 (both CH_3), 45.4 (μ_2-CH), 67.8 (CH of C_3H_7-i), 120.7, 125.1, 133.8, 155.3, 165.5 (each 1 C of C_5H_3N), 163.0 (CH=N), 169.2, 171.1, 174.3, 181.3, 183.7, 183.8, 190.2, 195.6 (all CO) ppm [6].

IR (n-hexane): 1936, 1963, 1983, 1993, 2001, 2010, 2053, 2086 (all CO) cm^{-1} [6].

Mass spectrum: [M]$^+$ [6].

The complex crystallizes in the monoclinic space group $P2_1/n - C_{2h}^5$ (No. 14) with a = 16.284(1), b = 16.325(1), c = 8.605(1) Å, β = 93.80(1)°; Z = 4, D_c = 2.78 g/cm^3, R = 0.051. The structure is shown in **Fig. 44**. The hydride ligands were not observed crystallographically, but one is assumed to bridge the Os(1)–Os(3) bond, probably out of the Os_3 plane opposite to the alkylidene bridge due to the large carbonyl osmium bond angles of the axial carbonyls. The second hydride probably bridges the Os(2)–Os(3) bond at an equatorial site. Average bond distances of Os–C and C≡O are 1.91 and 1.12 Å [6].

$(\mu\text{-}H)Os_3(\mu_3,\eta^3\text{-}C_6H_4P(C_6H_5)CH_2P(C_6H_5)_2)(CO)_8$ (see **Fig. 45**) was prepared by thermolysis of $Os_3(CO)_{10}(\mu_2,\eta^2\text{-}(C_6H_5)_2PCH_2P(C_6H_5)_2)$ or $HOs_3(\mu_3,\eta^3\text{-}C_6H_4P(C_6H_5)CH_2P(C_6H_5)_2)(CO)_9$ (see below) in refluxing toluene for 4 h; dark green crystals in 77% yield [1, 2, 3].

Gmelin Handbook
Os-Org. Comp. B6

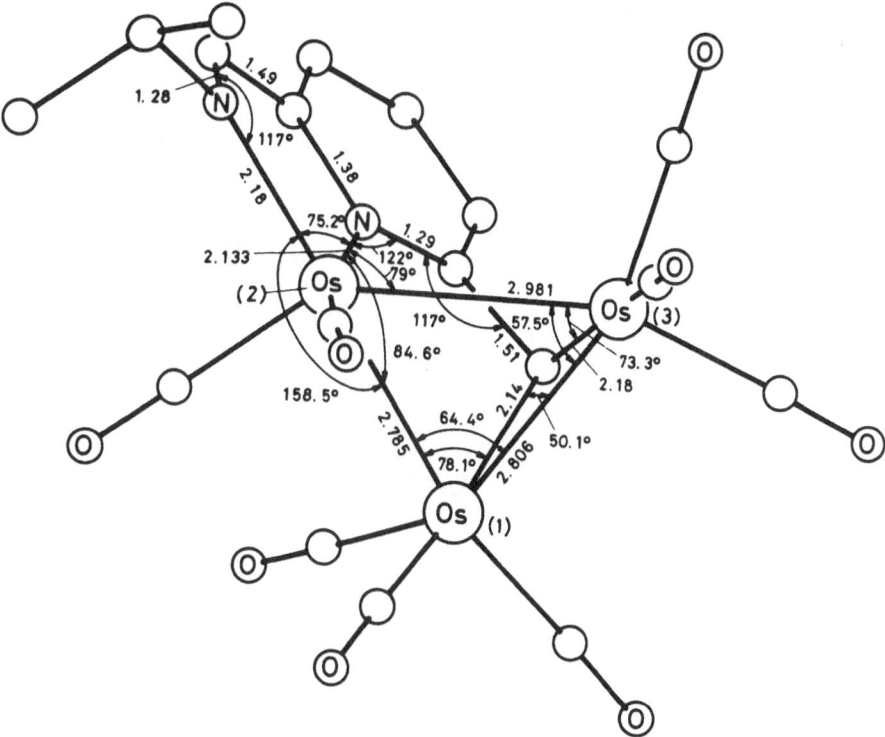

Fig. 44. Molecular structure of $(\mu\text{-H})_2Os_3(\mu_3,\eta^3\text{-}6\text{-}(CH)C_5H_3N(CH{=}NC_3H_7\text{-}i)\text{-}2)(CO)_8$ with selected bond distances (in Å) and bond angles [6].

^{1}H NMR (CD$_2$Cl$_2$): -13.43 (dd, μ-H; J(H,P) = 28.9, J(H,P') = 12.1) [1, 9], 4.20, 5.05 (ddd's, each 1 H of CH$_2$; J(H,H) = 14.2, J(H,P) = 11.4, J(H,P') = 7.4 or 9.0 Hz) ppm [1]. ^{31}P NMR (CD$_2$Cl$_2$): -20.2 (d; J(P,P) = 69.6 Hz), -18.0 (d) ppm [1, 9]. IR (CH$_2$Cl$_2$): 1928, 1991, 2020, 2066 (all CO) cm^{-1} [1, 9].

The complex crystallizes in the monoclinic space group P2$_1$/c – C$_{2h}^5$ (No. 14) with a = 19.546(6), b = 12.386(3), c = 14.022(4) Å, β = 103.28(3)°; Z = 4, D$_c$ = 2.37 g/cm^3, R = 0.043. The structure is shown in **Fig. 45**. It reveals that the Os(1)-Os(3) bond distance is significantly shorter than the other Os-Os bonds due to a formal double bond character. The hydride ligand was not observed crystallographically but the distribution of the carbonyl groups indicated it bridges the Os(1)-Os(3) bond. Average bond distance of Os-CO is 1.90 Å [1].

The coordinative unsaturation of the 46-electron compound is demonstrated by the facile CO addition in toluene or CH$_3$CN at room temperature resulting in **HOs$_3$(μ_3,η^3-C$_6$H$_4$P(C$_6$H$_5$)CH$_2$P(C$_6$H$_5$)$_2$)(CO)$_9$** of structure I [1, 2] or II [2, 9]; the compound is described in detail in "Organoosmium Compounds" B 5, Section 3.1.6.1.6, in preparation. The CO addition reaction has been followed step-by-step by cyclic voltammetry [7].

Treatment with a large excess of PR$_3$ (R = C$_2$H$_5$, n-C$_4$H$_9$, C$_6$H$_5$, OCH$_3$ [4, 9], OC$_3$H$_7$-i, OC$_4$H$_9$-n, OC$_6$H$_5$) in toluene at room temperature for 1 to 7 h resulted in mixtures of $(\mu\text{-H})Os_3(\mu_3,\eta^3\text{-}C_6H_4P(C_6H_5)CH_2P(C_6H_5)_2)(CO)_8PR_3$ (Formula II; see also "Organoosmium Compounds" B 5, Section 3.1.6.1.6, in preparation) with Os$_3$(CO)$_8$(μ_2,η^2-

References on pp. 119/20

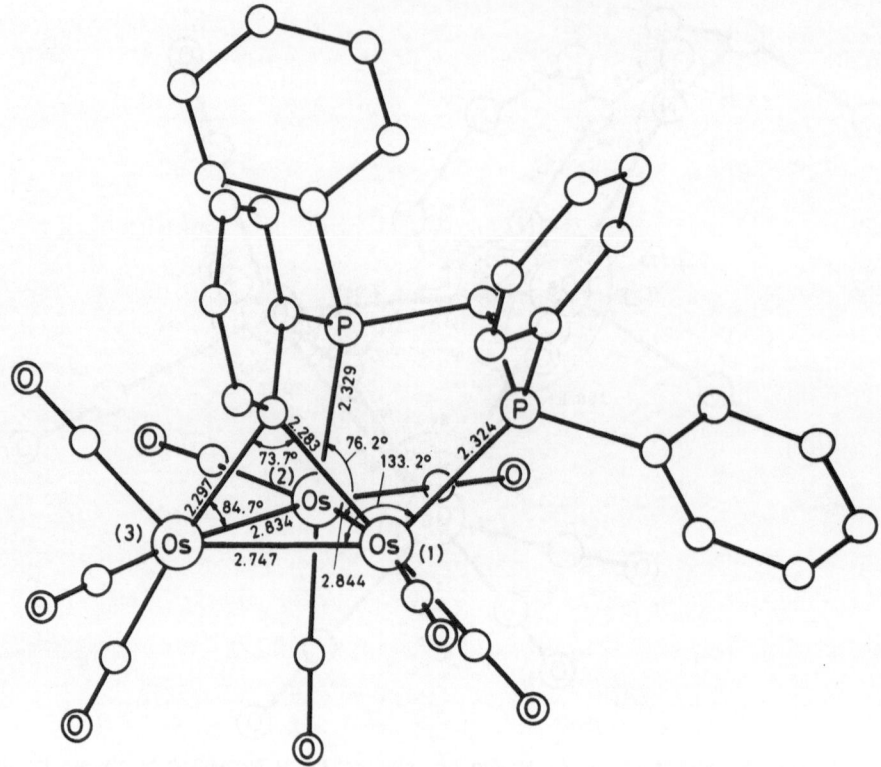

Fig. 45. Molecular structure of $(\mu\text{-}H)Os_3(\mu_3,\eta^3\text{-}C_6H_4P(C_6H_5)CH_2P(C_6H_5)_2)(CO)_8$ with selected bond distances (in Å) and bond angles [1].

$(C_6H_5)_2PCH_2P(C_6H_5)_2)(PR_3)_2$; the reaction with $P(i\text{-}C_3H_7)_3$ in $CHCl_3$ gave only the product of type II [9].

Hydrogenation in toluene at 80 °C gave $(\mu\text{-}H)_2Os_3(CO)_8(\mu_2,\eta^2\text{-}(C_6H_5)_2PCH_2P(C_6H_5)_2)$ [2, 3].

Reaction with $RC{\equiv}CR$ ($R=C_6H_5$ [4, 8], $C_6H_4CH_3\text{-}4$ [8]) in toluene at 80 °C for 5 to 8 h yielded $Os_3(\mu_3,\eta^2\text{-}RC{\equiv}CR)(CO)_7(\mu_2,\eta^2\text{-}(C_6H_5)_2PCH_2P(C_6H_5)_2)$ by rearrangement of the $\mu_3,\eta^3\text{-}C_6H_4P(C_6H_5)CH_2P(C_6H_5)_2$ unit; the $RC{\equiv}CR$ ligand is coordinated in a $\mu_3,\eta^2\text{-}\perp$ mode [4, 8]. Treatment with $RC{\equiv}CR$ ($R=CH_3$, CF_3) gave compounds of the type $Os_3(\mu_3,\eta^2\text{-}RC{\equiv}CR)\text{-}$

References on pp. 119/20

$(\mu\text{-CO})(CO)_7(\mu_2,\eta^2\text{-}(C_6H_5)_2PCH_2P(C_6H_5)_2)$ with the $RC\equiv CR$ unit being bonded in a $\mu_3,\eta^2\text{-}\|$ mode [8].

Treatment with an excess of $Sn\{CH(Si(CH_3)_3)_2\}_2$ in refluxing hexane yielded $Os_3(CO)_8(\mu_2,\eta^2\text{-}(C_6H_5)_2PCH_2P(C_6H_5)_2)\{\mu\text{-}Sn[CH(Si(CH_3)_3)_2]_2\}_2$ [5].

Cyclovoltammetric reduction in CH_3CN ($c = 10^{-3}$ M, 0.1 M $[N(C_4H_9)_4][BF_4]$, Hg-electrode, scan rate 400 mV/s) occurred at -1.18 V, and reoxidation at -1.11 V vs. SCE; an additional oxidation peak was attributed to a conversion of the reduced complex into a radical anion oxidizable at -0.96 V vs. SCE. Controlled potential coulometry at -1.25 V vs. SCE and polarography indicated a quasi-reversible one-electron reduction at ca. -1.14 V vs. SCE, followed by a slow chemical change; assuming a first-order reaction following the electron transfer, a half-life of ca. 10 s could be estimated for the radical anion. Electrochemical reduction occurred in acetone at -1.08 and in CH_2Cl_2 at -1.23 V vs. SCE [7].

The X-band EPR spectrum of electrochemically reduced $(\mu\text{-H})Os_3(\mu_3,\eta^3\text{-}C_6H_4P(C_6H_5)CH_2P(C_6H_5)_2)(CO)_8$ in CH_3CN showed at ca. -196 °C anisotropic g values with $g_\perp = 2.011$ and $g_\| = 2.095$, and a line width of $\Delta H_{pp} = 145$ G. Upon increasing the temperature to the glass-liquid transition, the signal became isotropic with $g_{iso} = 2.04$ and $\Delta H_{pp} = 11$ G. The absence of hyperfine couplings with H, P and magnetically active Os isotopes indicates that the added electron is highly delocalized [7].

$(\mu\text{-H})Os_3(\mu_3,\eta^3\text{-}C_6H_4P(C_6H_5)C(=CH_2)P(C_6H_5)_2)(CO)_8$ (Formula III) was prepared by thermolysis of $Os_3(CO)_{10}(\mu_2,\eta^2\text{-}(C_6H_5)_2PC(=CH_2)P(C_6H_5)_2)$ in refluxing toluene for 5 h, followed by TLC separation with CH_2Cl_2/heptane (1:1) as eluant; dark green crystals from acetone/C_2H_5OH in 45% yield [9].

^{1}H NMR (no medium given): -13.3 (dd, μ-H; J(H,P) = 30.0, J(H,P') = 10.0). ^{31}P NMR (no medium given): -1.2 (d), 8.6 (d, J(P,P) = 132) ppm. IR (CH_2Cl_2): 1930, 1990, 2018, 2060 (all CO) cm^{-1} [9].

Reaction with CO in toluene at room temperature led to **$HOs_3(\mu_3,\eta^3\text{-}C_6H_4P(C_6H_5)C(=CH_2)P(C_6H_5)_2)(CO)_9$** (compare Formula I and II) which is described in detail in "Organoosmium Compounds" B 5, Section 3.1.6.1.6, in preparation [9].

III

References:

[1] Clucas, J. A.; Foster, D. F.; Harding, M. M.; Smith, A. K. (J. Chem. Soc. Chem. Commun. **1984** 949/50).

[2] Clucas, J. A.; Harding, M. M.; Smith, A. K. (J. Chem. Soc. Chem. Commun. **1985** 1280/1).

[3] Bruce, M. I.; Horn, E.; Shawkataly, O. B.; Snow, M. R.; Tiekink, E. R. T.; Williams, M. L. (J. Organomet. Chem. **316** [1986] 187/211).

[4] Clucas, J. A.; Dolby, P. A.; Harding, M. M.; Smith, A. K. (J. Chem. Soc. Chem. Commun. **1987** 1829/31).

[5] Bartlett, R. A.; Cardin, C. J.; Cardin, D. J.; Lawless, G. A.; Power, J. M.; Power, P. P. (J. Chem. Soc. Chem. Commun. **1988** 312/3).

[6] Zoet, R.; van Koten, G.; Vrieze, K.; Jansen, J.; Goubitz, K.; Stam, C. H. (Organometallics **7** [1988] 1565/72).

[7] Osella, D.; Ravera, M.; Smith, A. K.; Mathews, A. J.; Zanello, P. (J. Organomet. Chem. **423** [1992] 255/61).

[8] Brown, M. P.; Dolby, P. A.; Harding, M. M.; Mathews, A. J.; Smith, A. K.; Osella, D.; Arbrun, M.; Gobetto, R.; Raithby, P. R.; Zanello, P. (J. Chem. Soc. Dalton Trans. **1993** 827/34).

[9] Brown, M. P.; Dolby, P. A.; Harding, M. M.; Mathews, A. J.; Smith, A. K. (J. Chem. Soc. Dalton Trans. **1993** 1671/9).

3.1.6.3 Compounds with ^{1}L Ligands Bonded by μ_3-Bridging C Atoms

The compounds dealt with in this section consist of an $Os_3(\mu_3,\eta^1-{}^1L)(CO)_n$ skeleton bearing additional heteroatom-bonded ligands D and μ-hydrido and/or other bridging units. The structures can be represented by the general Formulas I to III. The compounds of type IV have two $\mu_3,\eta^1-{}^1L$ ligands. The structures that are not confirmed by X-ray analysis, can only be considered as a proposal based on spectroscopic and analytical data. For compounds of the types $(\mu-H)Os_3(C(OCH_3)_2)(\mu_3,\eta^1-CC_6H_5)(CO)_9$ and $Os_3(C(OCH_3)R)$-$(\mu_3,\eta^1-CR')(\mu_3,\eta^1-CR'')(CO)_8$, see "Organoosmium Compounds" B 5, Section 3.1.5.2, in preparation.

For the compounds of type II (Nos. 39 to 42) the structural disposition of the bridging μ_3,η^1-CCO ligand and reactivity vary considerably depending on the charge of the complexes. For neutral $(\mu-H)_2Os_3(\mu_3,\eta^1-CCO)(CO)_9$ (No. 41) an upright orientation of the

μ_3,η^1-CCO moiety nearly perpendicular to the Os_3 plane was observed suggesting a valence bond formation such as $Os_3\equiv C^--C\equiv O^+$ for the complexed carbonyl methylidyne ligand [25]. The earlier proposed tilted configuration of the C=C=O type [20] was refuted by an X-ray analysis [25]. On the contrary, the structure of the dianion $[Os_3(\mu_3,\eta^1\text{-CCO})(CO)_9]^{2-}$ (No. 39) is significantly different from that of its protonated analog. Here, the μ_3,η^1-CCO unit is better formulated as a ketenylidene ligand which is approximately linear (the C=C=O angle is 175.5(5)°) and tilted over an Os center by an angle of 26° between the least-squares line through C-C-O and the perpendicular to the metal triangle plane. The three Os-C distances are not equal, with the μ-C being furthest away from that Os center toward which the CCO ligand is tilted. The μ_3,η^1-CCO is considered as a six-electron donor [51]. Furthermore, the μ_3,η^1-CCO ligand acts as either an electrophile, a nucleophile, or a carbide precursor [36, 51].

The structurally interesting features of the compounds No. 43 to 48, represented by Formula III, are the only semi-triply bridging μ_3,η^1-^{1}L ligands exhibiting two stronger symmetrical interactions with the two $Os(CO)_3$ centers, and one weaker interaction with the $Os(CO)_4$ group. The $(\mu_3\text{-C})\cdots Os(CO)_4$ interatomic distances range from 2.35 to 2.63 Å, and the dihedral angles $(CO)_3Os\cdots(\mu_3\text{-C})\cdots Os(CO)_3$ are between 69.7° and 82.1° [24, 28, 38, 63]. Thus, these bondings can be regarded as two-electron donations from the formally saturated (18 electron) $Os(CO)_4$ center to the nominally sp^2-hybridized bridging carbon atom [24]. In contrast to $(\mu\text{-H})Os_3(\mu_2,\eta^1\text{-}^1L)(CO)_{10}$ complexes, the semi-triply bridging μ_3,η^1-^{1}L ligands lie over the Os_3 plane rather than external to it. The R substituents are bent out of the $(CO)_3Os\cdots(\mu_3\text{-C})\cdots Os(CO)_3$ plane toward the $Os(CO)_4$ moiety leading to close contacts between R and an axial CO group which cause significant distortions in the intraligand angles at the $Os(CO)_4$ center [24, 38].

Variable-temperature ^{13}C NMR of Nos. 43 [24], 45 [51], 46 [63], and 47 [38] showed single resonances for the $Os(CO)_4$ groups (CO-d,e,f; see Formula III) which broaden between -50 and -90 °C [24, 38, 51, 63], and split into 2:1:1 patterns below -100 °C. The splitting pattern revealed that CO-d and one of the axial CO groups remained unchanged, whereas the other axial CO, probably CO-f, was displaced to higher field. This high-field shift was attributed to the distortions in the $Os(CO)_4$ moiety due to localized three-fold exchanges. The activation energies ΔG^+ for the axial-equatorial carbonyl exchange at the coalescence point at ca. -70 °C were estimated as ca. 9 kcal/mol for Nos. 43 and 47. The low activation energy values indicated the involvement of the pseudo-seven-coordinated Os centers interacting with the carbyne carbon [24, 38]. The resonances of the μ_3,η^1-carbyne carbons of Nos. 45 to 47 are temperature dependent and are shifted to lower field by 6 to 10 ppm if the temperature is raised from ca. -90 to ca. $+30$ °C. The low-field shift was assigned either to unspecified conformational equilibria in solution [28, 38, 51] or to an equilibrium between the μ_3-C and the μ_2-C coordination of the bridging ligand. At low temperature the μ_3-C conformation is thermodynamically favoured, in which the electron deficiency at the alkylidyne carbon is stabilized by its electrophilic interaction with the $Os(CO)_4$ group [63].

The molecular orbital patterns of Nos. 43 and 45 to 47, evaluated by extended Hückel calculations, revealed only a μ_2-bridging arrangement for the carbyne in these compounds. The bridging mode of the carbyne carbon and the $Os(1)\cdots(\mu\text{-C})$ antibonding interactions are influenced by the R substituents as indicated by the variation of the dihedral angle θ between the Os_3 plane and the $Os(2)\cdots(\mu\text{-C})\cdots Os(3)$ plane [77].

The two capping alkylidyne moieties in the compounds of Formula IV donate a total of six electrons to the cluster, resulting in a 48-electron cluster [51].

References on pp. 169/71

Most compounds listed in Table 5 were prepared by the following methods:

Method I: Starting from $(\mu-H)_3Os_3(\mu_3,\eta^1-CBr)(CO)_9$ (No. 5).

    a. $(\mu-H)_3Os_3(\mu_3,\eta^1-CR)(CO)_9$ and $(\mu-H)_3Os_3(\mu_3,\eta^1-CX)(CO)_9$ (Formula I) were prepared from No. 5 by nucleophilic substitution of Br by an R or X ligand. Most of the reactions were performed with an excess of $AgSbF_6$ as the Lewis acid and a suitable nucleophile in CH_2Cl_2 between 20 and 40 °C for 5 min to 12 h [65]; the nucleophile precursors are given in Table 5 at the individual compound. Similarly, in the presence of CO, the reaction of No. 5 with $AgSbF_6$ [65] or $AlCl_3$ [20, 78], followed by quenching with the nucleophilic agent after 30 min, led to compounds of the type $(\mu-H)_3Os_3(\mu_3,\eta^1-CCOR')(CO)_9$ (Nos. 28 and 30 to 33) [20, 65, 78], and also to No. 18 (R = CH_3) and No. 27 (R = CO_2H; probably by hydrolyzation of the initial reaction products formed from No. 5 and $AgSbF_6/CO/SnR_4$); the reactions in the presence of CO probably proceeded via the intermediate $[(\mu-H)_3Os_3(\mu_3,\eta^1-CCO)(CO)_9]^+$ (No. 42) [65]. The crude products were purified by extraction of the residue with ether followed by TLC with CH_2Cl_2/hexane (1:3) as eluant [65].

    $(\mu-H)_3Os_3(\mu_3,\eta^1-CF)(CO)_9$ (No. 3) was obtained by treatment of No. 5 with $AgBF_4/[N(C_2H_5)_4]BF_4$ at 0 °C for 50 min [57].

    b. $(\mu-H)_3Os_3(\mu_3,\eta^1-CR)(CO)_9$ and $(\mu-H)_3Os_3(\mu_3,\eta^1-CX)(CO)_9$ of Formula I were obtained by photolysis of a solution of No. 5 in the presence of $Re_2(CO)_{10}$ (ca. 1:1) at room temperature in a sealed NMR tube under a CO atmosphere (1 or 2 atm) for 5 to 10 h. The product formation depended on the solvent: With cyclohexane or toluene No. 1 (R = H) was formed, whereas benzene as solvent mainly led to No. 34 (R = C_6H_5) with No. 1 as a by-product; the products were not isolated [58]. Photolysis in C_6F_6 mainly gave No. 38 (R = C_6F_5), whereas in C_6H_5F a mixture of No. 34 (R = C_6H_5), No. 37 (R = C_6H_4F-4), No. 3 (X = F), and No. 1 (R = H) in 40%, 9%, 5%, and 10%, respectively, was obtained. The product mixtures were separated by TLC with petroleum ether or CH_2Cl_2/petroleum ether as eluant [76]. The reactions probably proceeded via the $(\mu-H)_3Os_3(\mu_3-C^{\cdot})(CO)_9$ radical initially generated by $^{\cdot}Re(CO)_5$ as was concluded from the additional formation of $Re(CO)_5Br$ [58].

Method II: Starting from $(\mu-H)_3Os_3(\mu_3,\eta^1-CBX_2)(CO)_8D$ (Nos. 9, 11, and 12; X = Cl, Br, D = CO, $P(C_6H_5)_3$).

    a. $(\mu-H)_3Os_3(\mu_3,\eta^1-CH)(CO)_8D$ were prepared by hydrolysis from Nos. 9, 11, or 12 upon condensing excess H_2O into a CH_2Cl_2 solution at −78 °C and stirring for 24 h at room temperature? [48, 74], or by treatment with glacial acetic acid [48]; purification by extraction of the crude products with CH_2Cl_2 [74].

    b. $(\mu-H)_3Os_3(\mu_3,\eta^1-CBCl_2R)(CO)_9$ (R = $N(CH_3)_3$, $P(CH_3)_3$, $P(C_6H_5)_3$) were obtained by condensing an equimolar amount of the Lewis base R at −196 °C into an NMR tube containing No. 9 in CD_2Cl_2. The products were only characterized by ^1H and ^{11}B NMR spectra, but not isolated [74].

Method III: $(\mu-H)_3Os_3(\mu_3,\eta^1-CR)(CO)_8D$ (Formula I) were prepared by hydrogenation of the appropriate precursors in n-heptane, n-octane, n-decane, or toluene between 98 and 120 °C for 15 min to 12 h under a H_2 pressure of usually 1 atm [1, 10, 11, 19, 22, 25, 27, 38, 42, 44, 47, 62], or at 90 °C for 3 d under 50 atm of H_2 [33]; the precursors are given in Table 5 at the individual compound. Work-up of the crude products was performed by TLC [10, 19, 22, 25, 42] with hexane/ether (5:1) [19], cyclohexane [10, 22], or pentane/CH_2Cl_2 [42]. The hydrogenation of No. 23 to give No. 21 was catalyzed by Pd/C in CH_3OH as solvent [65]. Most of the hydrogenation reactions proceeded by the elimination of a carbonyl group or another ligand.

Method IV: $(\mu\text{-H})_3Os_3(\mu_3,\eta^1\text{-CC}_6H_5)(CO)_9$ (No. 34) and $(\mu\text{-H})_3Os_3(\mu_3,\eta^1\text{-CX})(CO)_9$ (X = Cl, Br, B_5H_8, 1,2-$C_2B_{10}H_{11}$; Formula I, Nos. 4, 5, 16, and 17) were prepared from $\{(\mu\text{-H})_3Os_3(\mu_3\text{-C})(CO)_9\}_3O_3B_3O_3$ (in situ prepared from $(\mu\text{-H})_2Os_3(CO)_{10}$ and B_2H_6 [59]) by treatment with BX_3 (X = Cl, Br) [31, 81], BF_3/HX (X = B_5H_8, 1,2-$C_2B_{10}H_{11}$) [59], or with BF_3/C_6H_6 (C_6H_6 also as the solvent) [31]. The reactions with BX_3 were done in CH_2Cl_2 at room temperature for 5 min, followed by hydrolyzation of excess BX_3 with CH_3OH and washing the residue with pentane [81]. No. 5 (X = Br) could also be prepared from $(\mu\text{-H})_3Os_3(\mu_3,\eta^1\text{-COCH}_3)(CO)_9$ (No. 6) as before but purification by TLC with CH_2Cl_2/hexane (1:5) [12], or from $(\mu\text{-H})_3Os_3(\mu_3,\eta^1\text{-COBO}_2C_6H_4)(CO)_9$ (No. 8) in toluene onto which an excess of BBr_3 was condensed at $-196\,°C$, followed by stirring at ambient temperature for 1 h [81]. The preparations of Nos. 16 and 17 were performed by condensing an excess of BF_3/HX into a CH_2Cl_2 solution of the starting material at $-196\,°C$, followed by stirring at room temperature for 40 h. No. 17 was purified in the air by column chromatography on silica gel with acetone/CH_2Cl_2. No reaction was observed with BF_3/1,7-$C_2B_{10}H_{12}$ [59].

Method V: $(\mu\text{-H})_3Os_3(\mu_3,\eta^1\text{-CCOR})(CO)_9$ (R = OCH_3, $N(C_2H_5)_2$, $N(CH_3)C_6H_5$; Formula I, Nos. 28, 30, and 31) were prepared from $[(\mu\text{-H})_3Os_3(\mu_3,\eta^1\text{-CCO})(CO)_9]BF_4$ (No. 42) and RH in CH_2Cl_2 at $-80\,°C$ [20, 78] or at room temperature, followed by extraction into toluene at $-78\,°C$ [20, 78]. For all preparations the extremely moisture-sensitive No. 42 was prepared in situ by protonation of No. 41 with $HBF_4\cdot O(C_2H_5)_2$ in ether at $-80\,°C$ [20] or in CH_2Cl_2 at room temperature [78].

Method VI: a. $(\mu\text{-H})_3Os_3(\mu_3,\eta^1\text{-CBX}_2)(CO)_8D$ (D = CO, $P(C_6H_5)_3$; X = Cl, Br) were prepared by condensing an excess of BX_3 at low temperature into a solution of $(\mu\text{-H})_3Os_3(CO)_8D(\mu_3,\eta^1\text{-BCO})$ in CH_2Cl_2. After warming to room temperature the solution was stirred for ca. 30 min followed by separation from B_2O_3 by extraction with CH_2Cl_2 and washing of the residue with hexane [48, 74].

b. $(\mu\text{-H})_3Os_3(\mu_3,\eta^1\text{-CBClX})(CO)_9$ (X = Cl, Br) were obtained from $(\mu\text{-H})_3Os_3(\mu_3,\eta^2\text{-C(OR)BCl})(CO)_9$ (R = 9-borabicyclo[3.3.1]nonane-9-yl, $BClC_6H_5$; both "Organoosmium Compounds" B 5, Section 3.1.6.1.5, in preparation) and BX_3 as described before, but with a reaction time of 2 h; the products were only identified by ^{11}B NMR spectra but not isolated [73, 74].

Method VII: $(\mu\text{-H})_3Os_3(\mu_3,\eta^1\text{-CR})(CO)_9$ (R = H, CH_3, C_6H_5; Formula I, Nos. 1, 18, and 34) and $(\mu\text{-H})_2Os_3(\mu_3,\eta^1\text{-CCO})(CO)_9$ (Formula II, No. 41) were prepared by thermal decarbonylation and/or rearrangement of the bridging unit of a suitable Os_3-cluster precursor upon heating in a solvent such as cyclohexane, hexane, heptane, nonane, n-tridecane, benzene, toluene, or xylene between 65 and 220 °C for 30 min to 72 h; the precursors are given in Table 5 at the individual compound [5, 6, 7, 13, 17, 18, 20, 25, 26, 32, 60]. The crude product can be purified by TLC with n-pentane/CH_2Cl_2 (9:1) as eluant [60].

    References on pp. 169/71

Method VIII: $[(\mu-H)_n Os_3(\mu_3,\eta^1-CCO)(CO)_9]^{n-2}$ ($n = 1$, 2, or 3; Nos. 40, 41, and 42) were prepared by successive protonation of $[Os_3(\mu_3,\eta^1-CCO)(CO)_9][N(P(C_6H_5)_3)_2]_2$ (No. 39) with equivalent amounts of CF_3SO_3H or with successive equivalents of HCl in CH_2Cl_2 at room temperature for 5 min followed by addition of ether to remove $[N(P(C_6H_5)_3)_2][CF_3SO_3]$ or $[N(P(C_6H_5)_3)_2]Cl$, respectively. The products were isolated by precipitation with pentane from ether solution. The formation of No. 42 proceeded via No. 40 and No. 41.

Methylation of No. 39 with $CF_3CO_3CH_3$ over a period of 12 h led to $[Os_3(\mu_3,\eta^1-CCH_3)(\mu-CO)(CO)_9][N(P(C_6H_5)_3)_2]$ (No. 50) [51].

A silica gel-anchored variant of $(\mu-H)_3Os_3(\mu_3,\eta^1-CCH_3)(CO)_8P(C_6H_5)_2C_2H_5$ (No. 19), $(\mu-H)_3Os_3(\mu_3,\eta^1-CCH_3)(CO)_8P(C_6H_5)_2C_2H_4SIL$ (SIL = silica gel) was obtained as a by-product in ethylene hydrogenation reactions upon treatment of $(\mu-H)_2Os_3(CO)_9$-$P(C_6H_5)_2C_2H_4SIL$ or $Os_3(CO)_{11}P(C_6H_5)_2C_2H_4SIL$, suspended in isooctane or n-heptane with C_2H_4/H_2 (4 or 16 atm, respectively) at reflux temperature for 20 h [11, 14].

Table 5
Compounds with ^1L Ligands Bonded by μ_3-Bridging C Atoms.
An asterisk preceding the compound number indicates further information at the end of the table, pp. 150/69.
Explanations, abbreviations, and units on p. X.

No. compound	method of preparation (yield in %) properties and remarks

compounds of the type $(\mu-H)_3Os_3(\mu_3,\eta^1-CR)(CO)_8D$, $(\mu-H)_3Os_3(\mu_3,\eta^1-CX)(CO)_8D$ (Formula I) and $[(\mu-H)_2Os_3(\mu_3,\eta^1-CBCl_2)(CO)_9]M$

*1 $(\mu-H)_3Os_3(\mu_3,\eta^1-CH)(CO)_9$	Ia (99%, $HSi(C_2H_5)_3$ as the nucleophile precursor); Ia (69%, $Sn(C_2H_5)_4$ as the nucleophile precursor); Ia (18%; main-product in the preparation of No. 23 by treatment of No. 5 with $AgSbF_6$ at $-78\,°C$ and immediate quenching with allylsilane; the reaction probably proceeded via the carbocationic species $[(\mu-H)_3Os_3(\mu_3-C)(CO)_9]^+)$ [65] Ib (10%, in C_6H_5F, along with 40% of No. 34, and Nos. 3, and 37) [76]; Ib (not isolated, more than 80% in toluene or cyclohexane, based on ^1H NMR); Ib (not isolated, in benzene along with No. 34 as the main-product, based on ^1H NMR) [58] IIa (95%) [48, 74], III (54%; from $(\mu-H)_2Os_3$-$(\mu_3,\eta^1-CCO)(CO)_9$, No. 41) [25] VII (from $(\mu-H)_2Os_3(\mu_2,\eta^1-CH_2)(CO)_{10}$, Section 3.1.6.2.1) [6]; VII (26 to 39%, from $(\mu-H)Os_3(CO)_{10}(\mu-OCH=CH_2)$, or from $(\mu-H)Os_3(\mu_2,\eta^2-O=CCH_3)(CO)_{10}$, see "Organoosmium Compounds" B 5, Section 3.1.6.1.2.2, in preparation) [5, 7, 13], along with 3.5% of $(\mu-H)_2Os_3(\mu_3,\eta^2-O=CHCH)(CO)_9$, Section 3.1.6.2.3 [7, 13]

Table 5 (continued)

No. compound	method of preparation (yield in %) properties and remarks

from No. 5 and $HSn(C_4H_9-n)_3$ in heptane at 60 °C under a CO atmosphere (85%) [58]

by hydrolysis of $(\mu-H)_3Os_3(\mu_3,\eta^1-CB_5H_8)(CO)_9$ (No. 16) with glacial acetic acid by condensation at -78 °C and then at 80 °C for 24 h; recrystallization of the crude product from CH_2Cl_2/hexane (90%); hydrolysis of No. 16 and conversion into No. 1 occurs also with the moisture in air [59]

by hydrolysis of $[(\mu-H)_2Os_3(\mu_3,\eta^2-C\equiv CH)-(CO)_9]^+$ between 0 and ca. 20 °C, followed by subsequent TLC separation of the CH_2Cl_2 extract with pentane as eluant (36%, along with 7% of No. 26) [41], or (16%, along with 20% of No. 26) [61]; the cation was generated in situ either from $(\mu-H)Os_3(\mu_3,\eta^2-C\equiv CH)(CO)_9$ with conc. H_2SO_4, or from $(\mu-H)Os_3-(\mu_3,\eta^2-C=CHOC_2H_5)(CO)_9$ (or the isomer $(\mu-H)Os_3(\mu_3,\eta^2-CH=COC_2H_5)(CO)_9)$ with $HBF_4 \cdot O(C_2H_5)_2$, $[(C_6H_5)_3C]BF_4$ [41], or with CF_3CO_2H [61]

by protonation of $(\mu-H)_3Os_3(\mu_3,\eta^2-C(OBC_8H_{14})BCl)(CO)_9$ or $(\mu-H)_3Os_3(\mu_3,\eta^2-C(OBClC_6H_5)BCl)(CO)_9$ (both "Organoosmium Compounds" B 5, Section 3.1.6.1.5, in preparation) with anhydrous HCl (1:1, condensed at -196 °C) in CD_2Cl_2 at room temperature for 2 h (quantitative or ca. 25%, respectively, based on spectroscopic data but not isolated) [74]; $BC_8H_{14} = 9$-borabicyclo[3.3.1]nonane-9-yl

as a by-product in the preparation of No. 28 from No. 6 and $AlCl_3/CO/CH_3OH$ in CH_2Cl_2 at room temperature (not isolated, only small amounts, based on ^1H NMR) [78]

from $Os_3(CO)_{12}$ and $C_6H_5N(CH_3)_2$ in decahydronaphthalene at reflux temperature for 6 h under CO atmosphere; separated by TLC with $CHCl_3$/pentane (3:7) as eluant (4%) [7, 9]

colorless crystals from pentane [6, 7, 13]

^1H NMR (CD_2Cl_2): -20.02 (d, $\mu-H$), 9.36 (q, CH; $J(H,H) = 1.1$) [6]; ($CDCl_3$, ca. 27 °C): -19.43 (d, $\mu-H$), 9.27 (q, CH; $J(H,H) = 1.2$) [5, 7, 13]; ($CDCl_3$, 17 °C): -19.58 ($\mu-H$; $J(H^a,^{187}Os) = 27.5$, $J(H^a,H^b) = 1.5$); species

References on pp. 169/71

Table 5 (continued)

No. compound	method of preparation (yield in %) properties and remarks

*1 (continued)

containing one ^{187}Os have a lower cluster symmetry and exhibit an approximate A_2BX spin system; the two hydrides adjacent to ^{187}Os (H^a) became equivalent to each other but inequivalent to the third hydride (H^b) [35]

^{13}C NMR (acetone-d_6): 118.4 (CH); the resonance at $\delta = 68.2$ ppm originally reported [6] was due to a fold-over peak [8]

IR (cyclohexane): 2010, 2021, 2083 (all CO) [7, 13]

FT-IR (KBr, $-195\,°C$; calculated values in { }): 600 to 750 ($\gamma(OsHOs)$ and $\nu(Os-\mu_3,\eta^1-C)_{sym}$; {625(5)}), 870 to 910 (total 4 bands, C–H bending mode; band at 887 ($\delta(CH)$, e-mode in C_{3v}; {867(20)}), 1330 to 1400 (4 bands, $\nu(OsHOs)_{asym/sym}$; {1332(45), 1395(30)}), 3020 ($\nu(CH)_{sym}$, a_1-mode in C_{3v}; {2870(150)}), 3031 ($\nu(CH)_{asym}$); the calculated values for the vibration modes resulted from the mean-square amplitudes of vibration of a low temperature neutron diffraction; the C–H bending mode would be degenerated for strict C_{3v} symmetry, but the actual solid-state symmetry is C_1 [30]

mass spectrum: $[M]^+$, $[M-x\,CO]^+$, $x=1$ to 9 [6, 7]

2 $(\mu-H)_3Os_3(\mu_3,\eta^1-CH)(CO)_8P(C_6H_5)_3$

IIa (94%) [74]
yellow solid [74]
^1H NMR (CD_2Cl_2?): -19.58 (s, 1 $\mu-H$), -18.42 (d, 2 $\mu-H$; J(H,P) = 10.0), 7.33 (m, C_6H_5), 9.82 (s, CH); the H,P-coupling indicates that the phosphane ligand is in an axial position cis to the bridging hydrogens [74]
mass spectrum: $[M]^+$ [74]

3 $(\mu-H)_3Os_3(\mu_3,\eta^1-CF)(CO)_9$

Ia (63%) [57], Ib (15 or 5%, as a by-product in the preparation of Nos. 38 or 34) [76]
white solid [57]
^1H NMR ($CDCl_3$): -18.80 (s, $\mu-H$) [57]
^{13}C NMR ($CDCl_3$): 165.5 (6 equatorial CO), 165.9 (3 axial CO), 212.45 (μ_3,η^1-C; J(C,F) = 370.9) [57]
^{19}F NMR ($CDCl_3/CFCl_3$): -77.73 (CF) [57]
IR (cyclohexane): 1067 (CF), 1986, 2015, 2027, 2076, 2083 (all CO) [57]
mass spectrum: $[M]^+$, $[M-x\,CO]^+$, $x=1$ to 9 [57]

Table 5 (continued)

No. compound	method of preparation (yield in %) properties and remarks
*4 $(\mu\text{-H})_3Os_3(\mu_3,\eta^1\text{-CCl})(CO)_9$	IV (44%) [31, 81] by heating of No. 1 at 45 °C in CO-saturated CCl_4 with an excess of t-C_4H_9OCl; purified by TLC with pentane as eluant (ca. 90%) [58] as a by-product in the preparation of No. 27 from No. 41 and HCl gas in undried CD_2Cl_2 (identified from X-ray analysis) [49] as a by-product in the preparation of No. 28 from No. 6 upon treatment with $AlCl_3/CO/CH_3OH$ in CH_2Cl_2 at room temperature (not isolated, only small amounts) [78] white crystalline solid from CH_2Cl_2/hexane at -15 °C [31, 81] ^1H NMR $(CDCl_3)$: -18.9 (μ-H) [31, 81] IR (CH_2Cl_2): 2004, 2038, 2070, 2092 (all CO) [31, 81] He-I photoelectron spectrum (91 °C): 7.87 sh, 8.28, 9.50, 10.67, 10.86 sh, 12.29, 15.57 (all vertical ionization potentials), 7.50 (adiabatic ionization potential) eV; the spectroscopic results in addition to molecular orbital calculations of the Fenske-Hall type indicate that apparently the orbitals of apical carbon character are more stable than those of apical boron character in $(\mu\text{-H})_3Os_3(\mu_3\text{-BD})(CO)_9$ complexes (D=CO, $P(CH_3)_3$) [39] mass spectrum: $[M]^+$, $[M-x\ CO-y\ H]^+$, $x=1$ to 9, $y=1$ to 3 [31, 81] reaction with B_5H_9 in the presence of $AlCl_3$ yielded $(\mu\text{-H})_3Os_3(\mu_3,\eta^1\text{-CB}_5H_8)(CO)_9$ (No. 16) [59]
*5 $(\mu\text{-H})_3Os_3(\mu_3,\eta^1\text{-CBr})(CO)_9$	IV (98%) [12], IV (85%) [81], IV (60%) [31, 81] from No. 1 at 45 °C in CO-saturated $CBrCl_3$ with an excess of t-C_4H_9OCl (ca. 90%) [58] white to pale yellow, air-stable crystalline solid from CH_2Cl_2/hexane at -15 °C [12, 31, 81] ^1H NMR $(CDCl_3)$: -19.2 (μ-H) [12]; $(CDCl_3,$ 17 °C): -18.9 (μ-H) [31, 35, 81]; species containing one ^{187}Os have a lower cluster symmetry and exhibit an approximate A_2BX spin system; the two hydrides adjacent to ^{187}Os (H^a) became equivalent to each other but inequivalent to the third hydride (H^b); $J(H^a,^{187}Os)=28.9$, $J(H^a,H^b)=1.5$ [35] IR (cyclohexane): 1979, 2022, 2035, 2095 (all CO) [31, 81]; (cyclohexane): 1990, 2021, 2032, 2069, 2087 [12]

References on pp. 169/71

Table 5 (continued)

No. compound	method of preparation (yield in %) properties and remarks
*5 (continued)	He-I photoelectron spectrum (94 °C): 7.87 sh, 8.32, 9.43, 10.34, 10.70 sh, 11.51, 12.09, 15.58 (all vertical ionization potentials), 7.63 (adiabatic ionization potential) eV; the spectroscopic results in addition to molecular orbital calculations of the Fenske-Hall type indicate that apparently the orbitals of apical carbon character are more stable than those of apical boron character in $(\mu\text{-H})_3Os_3(\mu_3\text{-BD})$-$(CO)_9$ complexes (D = CO, P(CH$_3$)$_3$) [39] mass spectrum: $[M]^+$ [12, 31, 81], $[M - x\ CO - y\ H]^+$, x = 1 to 9, y = 1 to 3 [31, 81]
*6 $(\mu\text{-H})_3Os_3(\mu_3,\eta^1\text{-COCH}_3)(CO)_9$	III (75 to 79%; from $(\mu\text{-H})Os_3(\mu_2,\eta^1\text{-COCH}_3)(CO)_{10}$, Section 3.1.6.2.2) [10, 22, 33], along with 11% of $(\mu\text{-H})_2Os_3(CO)_{10}$ [33], III (54%, along with $(\mu\text{-H})_2Os_3(\mu_3,\eta^2\text{-C(C}_6H_5)\text{=COCH}_3)(CO)_9$; from $Os_3(\mu_3,\eta^1\text{-CC}_6H_5)(\mu_3,\eta^1\text{-COCH}_3)(CO)_9$, No. 53, under 4 atm of H$_2$; only 12% under 1 atm of H$_2$) [44] yellow solid from CH$_3$OH [22] ^{1}H NMR (CDCl$_3$): -18.53 (s, μ-H), 3.8 (s, CH$_3$) [22]; (CDCl$_3$, 17 °C): -18.58 (μ-H; J(H^a,^{187}Os) = 28.4, J(H^a,H^b) = 1.6); species containing one ^{187}Os have a lower cluster symmetry and exhibit an approximate A$_2$BX spin system; the two hydrides adjacent to ^{187}Os (H^a) became equivalent to each other but inequivalent to the third hydride (H^b) [35] ^{13}C NMR (CDCl$_3$): 69.3 (CH$_3$), 166.4 (3 CO-axial), 167.0 (6 CO-equatorial), 205.2 (μ_3,η^1-C) [10, 22] IR (no medium given): 1140, 1195 (μ_3,η^1-CO) [22]; (cyclohexane): 1995, 2008, 2013, 2022, 2074, 2077, 2107 (all CO) [12, 22] mass spectrum: $[M]^+$, $[M - x\ CO - COCH_3]^+$, x = 1 to 9; intense doubly charged trinuclear ions were also observed [22]
7 $(\mu\text{-H})_3Os_3(\mu_3,\eta^1\text{-COCH}_3)(CO)_8P(C_6H_5)_3$	no preparation and data reported oxidation with Ag$^+$ (probably derived from AgSO$_3$CF$_3$) yielded $[(\mu\text{-H})_3Os_3(CO)_9P\text{-}(C_6H_5)_3]^+$, probably via intermediate $[(\mu\text{-H})_3Os_3(\mu_3,\eta^1\text{-COCH}_3)(CO)_8P(C_6H_5)_3]^+$ as concluded from the initially dark green color of the solution [70]

References on pp. 169/71

Table 5 (continued)

No. compound	method of preparation (yield in %) properties and remarks
8 $(\mu\text{-H})_3Os_3(\mu_3,\eta^1\text{-COBO}_2C_6H_4)(CO)_9$	by treatment of $(\mu\text{-H})_2Os_3(CO)_{10}$ with $C_6H_4O_2BH$ (catechol borane; ca. 1:1) in THF at $-78\,°C$, followed by stirring at $25\,°C$ for 2 h; purified by washing of the residue with ether (82%) [81]

cream-colored solid [81]

^1H NMR (THF-d_8, 30 °C): -18.45 (s, μ-H), 6.96 (m, 2 H, C_6H_4), 7.29 (d, 2 H, C_6H_4; $J(H,C)=3$) [81]

^{13}C NMR (THF-d_8, $-85\,°C$): 112.37 (dd, 2 C, C_6H_4; $^1J(H,C)=166.4$, $^2J(C,H)=4.8$), 121.53 (dd, 2 C, C_6H_4; $^1J(C,H)=161.6$, $^2J(C,H)=7.5$), 150.49 (s, 2 C, C_6H_4), 168.37 (d, 6 equatorial CO; $J(C,H)=10.5$), 168.70 (s, 3 axial CO), 191.12 (q, μ_3,η^1-C; $J(C,H)=3.5$); at room temperature the CO groups are fluxional exhibiting one singlet at $\delta=167.5$ ppm; the favored exchange process is a localized axial-equatorial exchange [81]

^{11}B NMR (THF-d_8, 30 °C): 19.1 (s, $BO_2C_6H_4$) [81]

IR (THF): 1981, 2023, 2045, 2059, 2071, 2108 (all CO) [81]

mass spectrum: $[M]^+$ [81]

reaction with BBr_3 in toluene yielded No. 5 [81]

| *9 $(\mu\text{-H})_3Os_3(\mu_3,\eta^1\text{-CBCl}_2)(CO)_9$ | VIa (93 to 95%) [48, 74], VIb [73, 74] obtained by slow conversion of $(\mu\text{-H})_3Os_3(\mu_3,\eta^2\text{-C(OBClC}_6H_5)BCl)(CO)_9$ ("Organoosmium Compounds" B 5, Section 3.1.6.1.5, in preparation) at room temperature [74] |

white moisture-sensitive crystals; stable at room temperature under vacuum [48, 74]

^1H NMR (CD$_2$Cl$_2$): -19.43 (s, μ-H) [48, 74]

^{13}C NMR (CD$_2$Cl$_2$): 132.5 (s, μ_3,η^1-C), 165.26 (d, 6 CO-equatorial; $J(C,H)=11.3$) 167.41 (s, 3 CO-axial); the 6 equatorial CO groups are trans to the μ-H moieties [48, 74]

^{11}B NMR (CD$_2$Cl$_2$): 57.4 (br s, BCl_2); experiments with $^{10}BCl_3$ indicated that the formation of No. 9 from the μ_3,η^1-BCO precursor complex (see Preparation Method VIa) occurred intramolecularly by movement of the B atom from a vertex site to a terminal site; no incorporation of a ^{10}B atom from $^{10}BCl_3$

References on pp. 169/71

Table 5 (continued)

No. compound	method of preparation (yield in %) properties and remarks
*9 (continued)	was observed; the isotope exchange pro- ceeded extremely slow (within 2 d) [48, 74] IR (CH$_2$Cl$_2$): 2020, 2089 (both CO) [48, 74]
10 [(μ–H)$_2$Os$_3$(μ$_3$,η^1-CBCl$_2$)(CO)$_9$]M M = NH(CH$_3$)$_3$	by condensation of N(CH$_3$)$_3$ into a CH$_2$Cl$_2$ solu- tion of No. 9 at −196 °C, followed by stirring of the mixture at ca. 25 °C for 1 h; purified by washing of the residue with ether (90%); the formation proceeded via the instable ad- duct (μ–H)$_3$Os$_3$(μ$_3$,η^1-CBCl$_2$N(CH$_3$)$_3$)(CO)$_9$ (No. 13) which converted to the salt above −10 °C [74] white solid [74] ^{1}H NMR (CD$_2$Cl$_2$, 30 °C): −19.23 (s, μ-H), 2.87 (s, CH$_3$), 11.79 (s, NH) [74] ^{11}B NMR (CD$_2$Cl$_2$, 30 °C): 45.0 (br s, BCl$_2$) [74] IR (CH$_2$Cl$_2$, 30 °C): 2013, 2083 (both CO) [74]
M = K	by treatment of No. 9 with an excess of KH in the presence of B(CH$_3$)$_3$ in THF at ca. 25 °C [74] ^{1}H NMR (CD$_2$Cl$_2$, 30 °C): −19.42 (s, μ-H) [74] ^{11}B NMR (CD$_2$Cl$_2$, 30 °C): 46.2 (br s, BCl$_2$) [74] IR (CH$_2$Cl$_2$, 30 °C): 2013, 2083 (both CO) [74]

11 (μ–H)$_3$Os$_3$(μ$_3$,η^1-CBCl$_2$)(CO)$_8$P(C$_6$H$_5$)$_3$	VIa (78%) [74]
	yellow very moisture-sensitive solid [74]
	^{1}H NMR (CD$_2$Cl$_2$, 30 °C): −19.63 (s, 1 μ-H),
	   −18.41 (d, 2 μ-H; J(H,P) = 10.5), 7.42 (m,
	   C$_6$H$_5$); H,P-coupling indicates that the phos-
	   phane ligand is in an axial position cis to the
	   bridging hydrides [74]
	^{11}B NMR (CD$_2$Cl$_2$, 30 °C): 53.4 (s, BCl$_2$) [74]
	^{31}P NMR (CD$_2$Cl$_2$, 30 °C): −10.83 (m, P(C$_6$H$_5$)$_3$)
	   [74]
	IR (CH$_2$Cl$_2$): 1968, 2013, 2034, 2082, 2101 (all
	   CO) [74]
	hydrolysis in CH$_2$Cl$_2$ solution at low tempera-
	ture yielded No. 2; see Preparation Method
	IIa [74]

12 (μ–H)$_3$Os$_3$(μ$_3$,η^1-CBBr$_2$)(CO)$_9$	VIa (87%) [74] white solid [74] ^{1}H NMR (CD$_2$Cl$_2$, 30 °C): −19.36 (s, μ-H) [74]

Table 5 (continued)

No. compound	method of preparation (yield in %) properties and remarks

^{13}C NMR (CD$_2$Cl$_2$, 30 °C): 138.45 (μ_3,η^1-C), 164.55 (d, 6 CO-equatorial; J(C,H) = 12.5), 166.83 (s, 3 CO-axial); the magnitude of the C,H coupling indicates that the 6 equatorial CO groups are in trans positions to the hydrides [74]
^{11}B NMR (CD$_2$Cl$_2$, 30 °C): 52.5 (s, BBr$_2$) [74]
IR (CH$_2$Cl$_2$): 2029, 2091 (both CO) [74]
mass spectrum: [M − 1]$^+$ [74]
hydrolysis in CH$_2$Cl$_2$ at low temperature yielded No. 1; see Preparation Method IIa [74]

12a (μ-H)$_3$Os$_3$(μ_3,η^1-CBClBr)(CO)$_9$
VIb [73, 74]
^{11}B NMR (CD$_2$Cl$_2$, 30 °C): 55.4 (br s, BClBr) [74]

13 (μ-H)$_3$Os$_3$(μ_3,η^1-CBCl$_2$N(CH$_3$)$_3$)(CO)$_9$
IIb [74]
^{1}H NMR (CD$_2$Cl$_2$, −40 °C): −18.98 (s, μ-H), 3.18 (s, CH$_3$) [74].
^{11}B NMR (CD$_2$Cl$_2$, −40 °C): 21.0 (s, BCl$_2$) [74]
stable below −10 °C but converted above this temperature to No. 10 [74]

14 (μ-H)$_3$Os$_3$(μ_3,η^1-CBCl$_2$P(CH$_3$)$_3$)(CO)$_9$
IIb [74]
^{1}H NMR (CD$_2$Cl$_2$, −40 °C): −19.07 (s, μ-H), 1.62 (d, CH$_3$; J(H,P) = 105) [74]
^{11}B NMR (CD$_2$Cl$_2$, −40 °C): 13.1 (s, BCl$_2$) [74]
stable below −10 °C [74]

15 (μ-H)$_3$Os$_3$(μ_3,η^1-CBCl$_2$P(C$_6$H$_5$)$_3$)(CO)$_9$
IIb [74]
^{1}H NMR (CD$_2$Cl$_2$, 30 °C): −19.00 (s, μ-H), 6.99 (d, 6 H, C$_6$H$_5$; J(H,H) = 5.2), 7.11 (d, 6 H, C$_6$H$_5$; J(H,H) = 5.0), 7.32 (m, 3 H, C$_6$H$_5$) [74]
^{11}B NMR (CD$_2$Cl$_2$, 30 °C): 3.98 (d, BCl$_2$; J(B,P) = 154) [74]
stable below 30 °C but above this temperature free P(C$_6$H$_5$)$_3$ was noted in the ^{31}P NMR spectrum [74]

References on pp. 169/71

Table 5 (continued)

No. compound	method of preparation (yield in %) properties and remarks

16

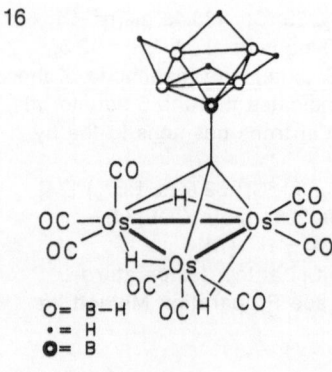

O= B–H
•= H
O= B

IV (ca. 33%) [59]
from $(\mu\text{-H})_3Os_3(\mu_3,\eta^1\text{-CCl})(CO)_9$ (No. 4) and
B_5H_9 in the presence of $AlCl_3$ [59]
light orange solid; well soluble in CH_2Cl_2 and
$CHCl_3$, slightly soluble in hexane; stable at
room temperature under vacuum, reaction
with moisture in the air resulted in
$(\mu\text{-H})_3Os_3(\mu_3,\eta^1\text{-CH})(CO)_9$ (No. 1) [59]
^{1}H NMR (no medium given): -19.23 (s, μ-H),
-0.86 (br s, B–H–B), 3.06 (q, BH; J(H,B) =
140) [59]
^{11}B NMR (no medium given): -25.3 (s, apical
B), -12.2 (d, B_4H_8; J(H,B) = 140); upon proton spin decoupling the doublet collapsed to
a singlet [59]
IR (CH_2Cl_2): 1359, 1409, 1430 (not assigned),
2005, 2013, 2019, 2074, 2079 (all CO), 2599
(BH) [59]
mass spectrum: $[M]^+$, $[M - x\, CO - B_5H_8]^+$, x = 1
to 9 [59]
reaction with glacial acetic acid at $-78\,°C$ and
then at 80 °C for 24 h yielded No. 1 [59]

17

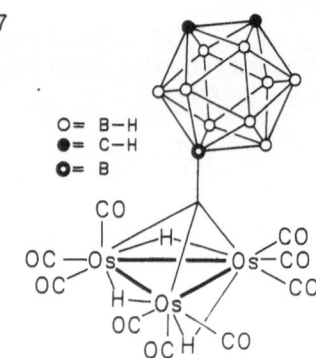

O= B–H
●= C–H
O= B

IV (32%) [59]
light yellow solid; slightly soluble in CH_2Cl_2,
$CHCl_3$ and acetone; air stable at room temperature [59]
^{1}H NMR (no medium given): -19.23 (s, μ-H);
B–H region not resolved; the proposed structure based on the fact that the favored site
for electrophilic substitution of carbaboranes
is the boron atom farthest from the carbons
[59]
^{11}B NMR (no medium given): -13.73 (J = 151),
-10.65 (J = 164), -5.71 (J = 148), 0.97 (J =
129); no assignment given [59]
IR (KBr): 602, 696, 1084, 1093, 1106, 1109, 1119
(unassigned), 1998, 2073 (both CO), 2607
(BH) [59]
mass spectrum: $[M]^+$, $[M - x\, CO]^+$, x = 1 to 9 [59]

*18 $(\mu\text{-H})_3Os_3(\mu_3,\eta^1\text{-CCH}_3)(CO)_9$

Ia (85%, $Sn(CH_3)_4$ as the nucleophile precursor) [65], Ia (61%, $CO/HSi(C_2H_5)_3$ as the nucleophile precursor) [65], III (73%, from
$(\mu\text{-H})_2Os_3(\mu_3,\eta^2\text{-CCH}_2)(CO)_9$ [1, 27], VII (2%,
from $(\mu\text{-H})Os_3(\mu_2,\eta^2\text{-CH}(CH_3)OCH_3)(CO)_{10}$,
"Organoosmium Compounds" B 5, Section

Table 5 (continued)

No. compound	method of preparation (yield in %) properties and remarks

3.1.6.1.2.3, in preparation) [26]

^{1}H NMR (CDCl$_3$): -18.58 (s, μ-H), 4.45 (s, CH$_3$) [1, 2, 27]

^{1}H NMR (nematic-phase, in Vari-Light liquid crystal VL-3268-N, 37 °C): dipolar H,H-couplings (in Hz) from Spectrum 1: 781.6 ± 3.3 (intra CH$_3$), 286.4 ± 1.6 (intra μ-H), 80.4 ± 3.1 (inter CH$_3$- μ-H); from Spectrum 2: 873.2 ± 1.7, 317 ± 3.3, 88.1 ± 3.0; assignments as in Spectrum 1 [2]

^{13}C NMR (CDCl$_3$, -50 to 100 °C [4]): 45.3 (CH$_3$), 154.7 (μ_3,η^1-C), 166.9 (s, 3 CO-axial), 167.8 (d, 6 CO-equatorial; J(C,H)=8); the spectrum is consistent with the expected C_{3v} symmetry [4, 27]; the resonances were compared to those of the related Ru- and Co-complexes, and with Os-complexes bearing other C_2 moieties [27]

IR (cyclohexane): 2004, 2017, 2069, 2076 (four of the five expected νCO bands: 2 a$_1$ +3 e modes); (CCl$_4$): 1035, 1118, 1144, 1367, 1438, 2711, 2836, 2864, 2892, 2916; (CS$_2$): 697, 734, 1034, 1116, 1143, 1366, 2707, 2831, 2866, 2889, 2911 [27]

IR (KBr, 298 K): 693, 704 (δOsH, a$_1$ mode, the splitting is probably a correlation effect), 732 (δOsH, e mode); (KBr, 77 K): 1365, 1388 (both νOsH); (CsI, 298 K): 390 to 560 given (not assigned); (CsI, 77 K): additional bands at 1366 and 1392 [27]

Raman (solid): 44, 68, 88, 109 (all β, γOs$_3$(CO)$_3$), 118 (νOsOs, e mode), 178 (νOsOs, a$_1$ mode), 1988 (νCO, formally forbidden a$_2$ mode and therefore absent in solution), 2007 (νCO), 2105 (νCO, a$_1$ mode) [27]

Raman (CH$_2$Cl$_2$): 702 (δOsH, a$_1$ mode), 2016, 2108 (both CO) [27]

protonation with HSO$_3$F or CF$_3$SO$_3$H at 22 °C yielded not isolable [(μ-H)$_3$Os$_3$(μ_2,η^1-HCCH$_3$) (CO)$_9$]$^+$ (Section 3.1.6.2.1; compare also Formula V, p. 150; R=CH$_3$) having probably an agostic bonded hydrogen; work-up of the acid solution recovered No. 18 in 87% yield [40, 66]; see also [3, 43]; neither protonation in CF$_3$CO$_2$H nor H/D exchange in CF$_3$CO$_2$D could be observed [3]

Table 5 (continued)

No. compound	method of preparation (yield in %) properties and remarks

19 $(\mu-H)_3Os_3(\mu_3,\eta^1-CCH_3)(CO)_8P(C_6H_5)_2C_2H_5$

III (quantitative; from $(\mu-H)_2Os_3(\mu_3,\eta^2-C=CH_2)(CO)_8P(C_6H_5)_2C_2H_5)$ [11, 19]
colorless crystals [19]
^1H NMR (CDCl$_3$, 31 °C): -18.76 (s, 1 μ-H), -18.03 (d, 2 μ-H; J(H,P) = 11.8), 0.96 (dt, CH$_3$; J(H,H) = 7, J(H,P) = 19), 2.60 (m, CH$_2$; J(H,H) = 7), 4.63 (d, μ_3,η^1-CCH$_3$; J(H,P) = 4), 7.43 (m, C$_6H_5$); the μ-H,P coupling indicates that the phosphane ligand is in an axial position cis to the bridging hydrides; the complex appears to possess a plane symmetry [19]
^{13}C NMR (CDCl$_3$, 31 °C): 168.4 (CO-a,a'), 168.7(d, CO-b,b'; J(C,H) = 8.0), 169.0 (d, CO-c,c'; J(C,H) = 7.0), 176.0 (d, CO-d,d'; J(C,H) = 5.0); for assignment, see Formula I [19]
IR (cyclohexane): 1954, 1966, 1979, 1991, 2006, 2022, 2026, 2069, 2091.5 (all CO) [19]; spectrum depicted [11]
mass spectrum: [M]$^+$ [19]

20 $(\mu-H)_3Os_3(\mu_3,\eta^1-CC_2H_5)(CO)_9$

from $(\mu-H)_2Os_3(CO)_{10}$ and CH$_2$=CHCH$_2$OH in cyclohexane at room temperature for 48 h and isolation by TLC (4%) [13]
^1H NMR (CDCl$_3$, 27 °C): -19.06, (s, μ-H), 1.57 (t, CH$_3$), 4.27 (q, CH$_2$) [13]
IR (cyclohexane): 2008, 2021, 2080 (all CO) [13]

21 $(\mu-H)_3Os_3(\mu_3,\eta^1-CC_3H_7-n)(CO)_9$

III (81%, from No. 23) [65]
^1H NMR (CDCl$_3$): -18.81 (s, μ-H), 1.12 (t, CH$_3$; J(H,H) = 6.9), 1.79 to 1.90 (m, CH$_2$), 4.13 to 4.21 (m, CCH$_2$) [65]
IR (cyclohexane): 2005, 2017, 2076 (all CO) [65]

22 $(\mu-H)_3Os_3(\mu_3,\eta^1-CCH_2Cl)(CO)_9$

probably formed as a by-product in the preparation of No. 27 from No. 41 (only identified by mass spectrum) [49]
mass spectrum: [M]$^+$ [49]

23 $(\mu-H)_3Os_3(\mu_3,\eta^1-CCH_2CH=CH_2)(CO)_9$

Ia (96%, CH$_2$=CHCH$_2$Si(CH$_3$)$_3$ as the nucleophile precursor) [65]
^1H NMR (CDCl$_3$): -18.86 (s, μ-H), 4.82 (d, CCH$_2$; J(H,H) = 7.4), 5.05 (dd, H$_{cis}$ of =CH$_2$; J(H,H) = 2.2, 9.8), 5.21 (dd, H$_{trans}$ of =CH$_2$; J(H,H) = 2.0, 17.7), 6.24 (ddt, =CH; J(H,H) = 7.4, 9.8, 17.7) [65]

References on pp. 169/71

Table 5 (continued)

No. compound	method of preparation (yield in %) properties and remarks
	IR (cyclohexane): 2006, 2018, 2077 (all CO) [65] reaction with H_2/Pd/C in CH_3OH yielded No. 21; see also Preparation Method III [65]
24 $(\mu\text{-}H)_3Os_3(\mu_3,\eta^1\text{-}CCH_2C(O)C_4H_9\text{-}t)(CO)_9$	Ia (55%, $CH_2\text{=}C(C_4H_9\text{-}t)OSi(CH_3)_3$ as the nucleophile precursor) [65] ^{1}H NMR ($CDCl_3$): -18.84 (s, μ-H), 1.23 (s, CH_3), 5.19 (s, CH_2) [65] IR (cyclohexane): 1715 (C=O), 2003, 2016, 2024, 2079 (all CO) [65]
25 $(\mu\text{-}H)_3Os_3(\mu_3,\eta^1\text{-}CCH_2C_6H_5)(CO)_9$	III (17%, from $Os_3(\mu_3,\eta^2\text{-}C\equiv CC_6H_5)(\mu_2,\eta^1\text{-}C\equiv CC_6H_5)(CO)_9$) [47], III (40%, from $(\mu\text{-}H)Os_3(\mu_3,\eta^2\text{-}C\text{=}CHC_6H_5)(\mu\text{-}Br)(CO)_9$) [62] solid [62] ^{1}H NMR ($CDCl_3$): -18.86 (s, μ-H), 5.23 (s, CH_2), 7.23 to 7.34 (m, C_6H_5) [62] IR (cyclohexane): 2004, 2007, 2016, 2069, 2077 (all CO) [62] mass spectrum: $[M]^+$ [62]
26 $(\mu\text{-}H)_3Os_3(\mu_3,\eta^1\text{-}CCHO)(CO)_9$	by hydrolysis of $[(\mu\text{-}H)_2Os_3(\mu_3,\eta^2\text{-}C\equiv CH)(CO)_9]^+$ between 0 and ca. 20 °C as described for No. 1 (20%, along with 16% of No. 1) [61], (7%, along with 36% of No. 1) [41] pale yellow crystals from CH_3OH [61] ^{1}H NMR ($CDCl_3$): -19.40 (s, μ-H), 10.68 (s, CHO) [41]
*27 $(\mu\text{-}H)_3Os_3(\mu_3,\eta^1\text{-}CCO_2H)(CO)_9$	Ia [65] by reaction of No. 41 with dilute HCl; similarly upon treatment with HCl gas in undried CD_2Cl_2 in an NMR tube at room temperature under vacuum for 3 weeks; work-up by successive fractional crystallization at -10 °C (47%, along with small amounts of No. 4 and No. 22) [49] also formed by slow conversion of moisture-sensitive Nos. 41 [25] and 42 [78] crystals [49] ^{1}H NMR (CD_2Cl_2, 30 °C): -19.46 (s, μ-H), 10.76 (s, CO_2H) [49] ^{13}C NMR (CD_2Cl_2, 30 °C): 165.5 (6 CO-equatorial), 167.8 (3 CO-axial), 170.1 (CO_2H) [49] IR (CH_2Cl_2): 1639, 1786 (both CO_2H), 2031, 2068, 2079, 2092, 2128 (all CO), 3497 (OH)

References on pp. 169/71

Table 5 (continued)

No. compound	method of preparation (yield in %) properties and remarks
*27 (continued)	[49]; (Nujol mull): 1634 (CO$_2$H), 1988, 2015, 2024, 2033, 2082 (all CO) [78]; (Nujol mull): 1644 (CO$_2$H) [25] mass spectrum: [M]$^+$ [49]
28 (μ-H)$_3$Os$_3$(μ_3,η^1-CCO$_2$CH$_3$)(CO)$_9$	Ia (69%, CH$_3$OH as the nucleophile precursor) [20, 78], Ia (45%, CH$_3$OH as the nucleophile precursor) [65], V [20, 78] from (μ-H)$_2$Os$_3$(μ_3,η^1-CCO)(CO)$_9$ (No. 41) upon heating in CH$_3$OH [20] from (μ-H)$_3$Os$_3$(μ_3,η^1-COCH$_3$)(CO)$_9$ (No. 6) upon treatment with AlCl$_3$/CO and CH$_3$OH under the conditions of Preparation Method Ia (48%, along with small amounts of No. 1 and No. 4) [78] colorless solid from CH$_2$Cl$_2$/CH$_3$OH [78] ^{1}H NMR (CDCl$_3$): −19.40 (s, μ-H), 3.82 (d, CH$_3$; ^{3}J(H,C)=4.1) [20, 65, 78] ^{13}C NMR (CDCl$_3$): 187.1 (CO$_2$) [65] IR (cyclohexane): 1688 (C=O), 2015, 2025, 2030, 2081, 2089, 2116 (all CO) [20, 78] mass spectrum: [M]$^+$ [20, 78] esterification by treatment with conc. H$_2$SO$_4$ followed by quenching with C$_2$H$_5$OH yielded No. 29 [78] pyrolysis in toluene at 120 to 125 °C for 21 h under 1 atm of CO resulted in (μ-H)$_2$Os$_3$(μ_3,η^2-O=C(OCH$_3$)CH)(CO)$_9$ (Section 3.1.6.2.3); the reaction is of first order with ΔG^+=30 kcal/mol at 398 K; an alkylidyne/alkylidene rearrangement mechanism is discussed that involved the reversible formation of intermediate (μ-H)$_2$Os$_3$(μ_3,η^2-HCCO$_2$CH$_3$)(CO)$_9$ having an agostic bonded hydrogen (see Formula V, p. 150; R=CO$_2$CH$_3$); the probably irreversible second step might be the rate determining one [78]
29 (μ-H)$_3$Os$_3$(μ_3,η^1-CCO$_2$C$_2$H$_5$)(CO)$_9$	by esterification of No. 28 with conc. H$_2$SO$_4$/C$_2$H$_5$OH followed by treatment with CH$_2$Cl$_2$ and subsequent hydrolysis; purified by TLC (92%); similar esterification, but in the presence of catalytic amounts of conc. H$_2$SO$_4$, occurred in refluxing C$_2$H$_5$OH within 1 h (not isolated) [78] colorless solid [78]

Table 5 (continued)

No. compound	method of preparation (yield in %) properties and remarks

^{1}H NMR (CDCl$_3$): -19.35 (s, μ-H), 1.39 (t, CH$_3$; ^{3}J(H,H) = 6.9), 4.31 (q, CH$_2$) [78]

IR (cyclohexane): 1684 (C=O), 1983, 2013, 2022, 2027, 2079, 2087, 2115 (all CO) [78]

mass spectrum: [M]$^+$ and fragment ions [78]

30 $(\mu$-H$)_3$Os$_3(\mu_3,\eta^1$-CC(O)N(C$_2$H$_5$)$_2$)(CO)$_9$

Ia (44%, NH(C$_2$H$_5$)$_2$ as the nucleophile precursor) [78], V (not isolated, but evidenced by ^{1}H NMR spectrum) [78]

colorless solid [78]

^{1}H NMR (CDCl$_3$): -19.02 (s, μ-H), 1.25 (t, CH$_3$; ^{3}J(H,H) = 7.0), 3.70 (q, CH$_2$) [78]

IR (cyclohexane): 1582 (C=O), 2012, 2019, 2026, 2033, 2075, 2085, 2113 (all CO) [78]

mass spectrum: [M]$^+$ and fragment ions [78]

pyrolysis in toluene at 120 to 125 °C under 1 atm of CO for 10 min yielded $(\mu$-H$)_2$Os$_3(\mu_3,\eta^2$-O=C(N(C$_2$H$_5$)$_2$)CH)(CO)$_9$ (Section 3.1.6.2.3); the reaction is of first order with ΔG$^+$ = 27 kcal/mol at 398 K; an alkylidyne/alkylidene rearrangement mechanism is discussed that involved the reversible formation of intermediate $(\mu$-H$)_2$Os$_3(\mu_3,\eta^2$-HCC(O)N(C$_2$H$_5$)$_2$)(CO)$_9$ having an agostic bonded hydrogen (see Formula V, p. 150; R = CO$_2$N(C$_2$H$_5$)$_2$) [78]; see also No. 28

31 $(\mu$-H$)_3$Os$_3(\mu_3,\eta^1$-CC(O)N(CH$_3$)C$_6$H$_5$)(CO)$_9$

Ia (40%, C$_6$H$_5$NHCH$_3$ as the nucleophile precursor) [78], V (not isolated, but evidenced by ^{1}H NMR spectrum) [78]

colorless solid [78]

^{1}H NMR (CDCl$_3$): -19.19 (s, μ-H), 3.37 (s, CH$_3$), 7.33 (m, C$_6$H$_5$) [78]

IR (cyclohexane): 1582 (C=O), 1977, 2005, 2018, 2025, 2030, 2072, 2086, 2113 (all CO) [78]

mass spectrum: [M]$^+$ [78]

32 $(\mu$-H$)_3$Os$_3(\mu_3,\eta^1$-CC(O)CH$_2$CH=CH$_2$)(CO)$_9$

Ia (25%, CH$_2$=CHCH$_2$Si(CH$_3$)$_3$ as the nucleophile precursor) [65]

^{1}H NMR (CDCl$_3$): -19.31 (s, μ-H), 3.71 (dt, C(O)CH$_2$; J(H,H) = 2.0, 6.9), 5.19 (dd, H$_{cis}$ of =CH$_2$; J(H,H) = 1.4, 10.4), 5.20 (dd, H$_{trans}$ of =CH$_2$; J(H,H) = 1.4, 15.9), 6.18 (ddt, =CH; J(H,H) = 6.9, 10.4, 15.9) [65]

References on pp. 169/71

Table 5 (continued)

No. compound	method of preparation (yield in %) properties and remarks
32 (continued)	IR (cyclohexane): 1631, 1645 (both C=O), 2023, 2027, 2088 (all CO) [65]
33 $(\mu-H)_3Os_3(\mu_3,\eta^1-CC(O)CH_2C(O)C_4H_9-t)(CO)_9$	Ia (34%, $CH_2=C(C_4H_9-t)OSi(CH_3)_3$ as the nucleophile precursor) [65] appeared to be a 10:1 mixture of the enol and keto tautomers in $CHCl_3$ solution ^1H NMR ($CDCl_3$): Enol Tautomer: -19.23 (s, $\mu-H$), 1.19 (s, CH_3), 6.14 (s, CH), 16.40 (s, OH); Keto Tautomer: -19.33 (s, $\mu-H$), 1.20 (s, CH_3), 4.14 (s, CH_2) [65] IR (cyclohexane): 1591 (C=O), 2025, 2087 (both CO); the weak, but very broad band at 1591 cm^{-1} is characteristic for the enol form of a 1,3-diketone [65]
*34 $(\mu-H)_3Os_3(\mu_3,\eta^1-CC_6H_5)(CO)_9$	Ia (97%, C_6H_6 as the nucleophile precursor) [65], Ia (56%, $Sn(C_6H_5)_4$ as the nucleophile precursor) [65], Ib (40%, in C_6H_5F) [76], Ib (not isolated, in C_6H_6 along with No. 1 as a by-product, based on ^1H NMR) [58] III (95%, from No. 47 in decane at 150 °C (20 min) or 110 °C for 90 min) [38], III (moderate yield, from $(\mu-H)Os_3(\mu_3,\eta^3-CHC_6H_4)(CO)_9$) [38] IV (70%, in CH_2Cl_2/C_6H_6) [81], IV (45%, in C_6H_6) [31] VII (ca. 10%, from $(\mu-H)_2Os_3(\mu_3,\eta^3-C_6H_3CH_3)(CO)_9$; only in traces under an additional H_2 atmosphere [60] pale yellow air-stable crystals from CH_2Cl_2/CH_3OH [31, 38, 81] ^1H NMR (acetone-d_6, 30 °C): -18.6 (s, $\mu-H$), 7.01 (m, 1 H, C_6H_5), 7.22 (m, 2 H, C_6H_5), 7.76 (m, 2 H, C_6H_5) [81]; similar in $CDCl_3$ [31]; ($CDCl_3$): -18.7 (s, $\mu-H$), 6.8 to 7.8 (m, C_6H_5) [38] IR (cyclohexane): 1980, 2007, 2017, 2022, 2073, 2080 [31, 38, 81] He-I photoelectron spectrum (88 °C): 7.35, 8.08, 9.32, 10.64, 11.99, 14.97 (vertical ionization potentials), 7.03 (adiabatic ionization potential) eV; the spectroscopic results in addition to molecular orbital calculations of the Fenske-Hall type indicate that apparently the

Table 5 (continued)

No. compound	method of preparation (yield in %) properties and remarks
	orbitals of apical carbon character are more stable than those of apical boron character in $(\mu\text{-H})_3Os_3(\mu_3\text{-BD})(CO)_9$ complexes (D = CO, P(CH$_3$)$_3$) [39] mass spectrum: $[M]^+$, $[M-x\,CO-C_6H_5-y\,H]^+$, x = 1 to 9, y = 1 to 3 [31, 81]
35 $(\mu\text{-H})_3Os_3(\mu_3,\eta^1\text{-CC}_6H_4CH_3\text{-3})(CO)_9$	Ia (83% as 1:2.2 mixture with No. 36, C$_6$H$_5$CH$_3$ as the nucleophile precursor) [65] ^{1}H NMR (CD$_2$Cl$_2$): −18.59 (s, μ-H), 2.35 (s, CH$_3$), 6.88, 7.11, 7.47, 7.53 (d, t, d, s, each 1 H of C$_6$H$_4$; J(H,H) ≈ 7.6) [65]
36 $(\mu\text{-H})_3Os_3(\mu_3,\eta^1\text{-CC}_6H_4CH_3\text{-4})(CO)_9$	Ia (51%, Sn(C$_6$H$_4$CH$_3$-4)$_4$ as the nucleophile precursor); for formation using C$_6$H$_5$CH$_3$ see No. 35 [65], III (94%; from $(\mu\text{-H})_2Os_3(\mu_3,\eta^1\text{-}$ CC$_6$H$_4$CH$_3$-4)(CO)$_9$W(η^5-C$_5$H$_5$)(CO)$_3$, Section 3.1.6.4.1) [42] from $(\mu\text{-H})_2Os_3(\mu_3,\eta^1\text{-CC}_6H_4CH_3\text{-4})\text{-}$ (CO)$_9$W(η^5-C$_5$H$_5$)(CO)$_3$ (Section 3.1.6.4.1) and (η^5-C$_5$H$_5$)W(≡CC$_6$H$_4$CH$_3$-4)(CO)$_2$ in CH$_2$Cl$_2$ at room temperature for 40 h, or in refluxing benzene for 25 min; isolated by TLC with pentane/CH$_2$Cl$_2$ (1:1) as eluant (22 or 13%; along with 22 or 13% of $(\mu\text{-H})Os_3(\mu_3,\eta^1\text{-}$ CC$_6$H$_4$CH$_3$-4)(CO)$_{10}$, No. 48, and Os$_3$(μ_3,η^1-CC$_6$H$_4$CH$_3$-4)$_2$(CO)$_9$W(η^5-C$_5$H$_5$)H, Section 3.1.6.4.1, in traces or in 53% yield); upon preparation in CH$_2$Cl$_2$ additional $(\mu\text{-H})Os_3(\mu_3,\eta^1\text{-CC}_6H_4CH_3\text{-4})_2(\mu_3\text{-CO})\text{-}$ (CO)$_9$W(η^5-C$_5$H$_5$) (Section 3.1.6.4.1) was obtained in 26% yield [42] yellow crystalline solid from CH$_2$Cl$_2$ [42] ^{1}H NMR (CD$_2$Cl$_2$): −18.58 (s, μ-H), 2.35 (s, CH$_3$), 7.04, 7.57 (d's, each 2 H of C$_6$H$_4$; J(H,H) = 8.2) [65]; (C$_6$D$_6$): −18.83 (s, μ-H), 2.06 (s, CH$_3$), 6.88 to 7.74 (m, C$_6$H$_4$) [42] IR (cyclohexane): 2007, 2017, 2021, 2080 (all CO) [42, 65] mass spectrum: $[M]^+$ [42]
37 $(\mu\text{-H})_3Os_3(\mu_3,\eta^1\text{-CC}_6H_4F\text{-4})(CO)_9$	Ib (9%) [76] pale yellow solid [76] ^{1}H NMR (CDCl$_3$): −18.61 (s, μ-H), 7.52, 7.7 (m's, each 2 H of C$_6$H$_4$F) [76] ^{19}F NMR (CDCl$_3$): −99.6 (s, C$_6$H$_4$F) [76] IR (CH$_2$Cl$_2$): 1992, 2022, 2073, 2082, 2110 (all CO) [76]

References on pp. 169/71

Table 5 (continued)

No. compound	method of preparation (yield in %) properties and remarks
*38 $(\mu\text{-H})_3Os_3(\mu_3,\eta^1\text{-CC}_6F_5)(CO)_9$	Ib (55% in C_6F_6, along with 15% of No. 3, 15% of $(\mu\text{-H})_2Os_3(\mu_2,\eta^1\text{-CC}_6F_5)(CO)_9Re(CO)_5$, Section 3.1.6.4.3, and 3% of $[(\mu\text{-H})_3Os_3(\mu_3,\eta^1\text{-CCO})(CO)_9]_2$) [76]
	pale yellow microcrystals [76]
	^{1}H NMR (CDCl$_3$): -18.82 (s, μ-H) [76]
	^{19}F NMR (CDCl$_3$): -163.6 (m, C_6F_5) [76]
	IR (CH$_2$Cl$_2$): 1514 (C_6F_5), 1986, 2013, 2023, 2081, 2085, 2113 (all CO) [76]
	mass spectrum: [M]$^+$ [76]

compounds of the type $[(\mu\text{-H})_nOs_3(\mu_3,\eta^1\text{-CCO})(CO)_9]^{n-2}$ (Formula II)

*39 $[Os_3(\mu_3,\eta^1\text{-CCO})(CO)_9]M_2$ $\quad M = N(P(C_6H_5)_3)_2$	from $[Os_3(\mu\text{-CO})(CO)_{10}]M_2$ and CH$_3$COCl (ca. 1:1) in CH$_2$Cl$_2$, resulting in intermediate $[Os_3(\mu_3,\eta^1\text{-COCOCH}_3)(CO)_{10}]M$ (No. 44); to remove MCl, the reduced solution was treated with ether followed by filtration, removal of the solvent, and treatment with THF and Na/(C$_6$H$_5$)$_2$CO; isolated by removal of the solvent, dissolving of the residue in CH$_2$Cl$_2$ and addition of $[N(P(C_6H_5)_3)_2]$Cl (47%) [36, 51]
	orange crystals from CH$_2$Cl$_2$/ether [51]
	^{13}C NMR (CD$_2$Cl$_2$/CH$_2$Cl$_2$, 30 °C): 31.2 (s, μ_3,η^1-C), 168.3 (s, CCO), 186.0 (s, 9 CO; ^{1}J(C,^{187}Os) = 42); ^{1}J(C,C) = 81; below -90 °C the signal at 186.0 broadens and collapses into resonances at 181.1 and 190.9 (2:1), which remain down to -140 °C; these observations are probably due to a freezing out of the (CO)$_3$ turnstile motions, CO exchange, or CCO precession motion [51]
	IR (CH$_2$Cl$_2$): 1889, 1956, 1994, 2031 (all CO) [51]
$\quad M = P(C_6H_5)_4$	probably similarly prepared as described for the N(P(C$_6$H$_5$)$_3$)$_2$ salt; isolated with $[P(C_6H_5)_4]$Br [51]
40 $[(\mu\text{-H})Os_3(\mu_3,\eta^1\text{-CCO})(CO)_9][N(P(C_6H_5)_3)_2]$	VIII (13%) [51]
	yellow powder from ether/pentane [51]
	^{1}H NMR (CD$_2$Cl$_2$): -20.01 (s, μ-H) [51]
	^{13}C NMR (CD$_2$Cl$_2$/CH$_2$Cl$_2$, 30 °C): 17.2 (s, μ_3,η^1-C), 163.4 (s, CCO), 178.0 (s, 9 CO); J(C,C) = 86 [51]; (CD$_2$Cl$_2$/CH$_2$Cl$_2$, -90 °C): 18.2 (s, μ_3,η^1-C), 162.5 (s, CCO), 168.2, 172.9, 179.9, 183.3 (s's, each 2 CO), 185.6 (s, 1 CO);

Table 5 (continued)

No. compound	method of preparation (yield in %) properties and remarks
	$^1J(C,C) = 86$; the splitting pattern of the CO resonances at low temperature is consistent with the existence of a mirror plane passing through the molecule [51] IR (CH_2Cl_2): 1912, 1965, 1997, 2035, 2075 (all CO) [51] protonation with HCl yielded No. 41 [51]
*41 $(\mu-H)_2Os_3(\mu_3,\eta^1-CCO)(CO)_9$	earlier formulated as $Os_3(CH_2)(CO)_{10}$ [15, 17] VII (72 to 84%, from $Os_3(\mu_2,\eta^1-CH_2)(\mu-CO)$-$(CO)_{10}$, Section 3.1.6.2.1) [17, 18, 20, 25, 32]; only traces were formed under a H_2 atmosphere [15, 17, 20, 32]; the formation probably proceeded via No. 43 [25]; VII (ca. 75%, from No. 43), or by heating of solid No. 43 in a sublimation apparatus at 85 °C/0.1 Torr [25] VIII (not isolated) [51] yellow crystals from pentane at −15 °C [20, 25, 32] ^1H NMR $(CD_2Cl_2, -80$ to 25 °C): −19.73 (s, $\mu-H$); off resonance decoupling $(\mu_3,\eta^1-^{13}C$-enriched sample) showed a doublet for the $\mu-H$ resonance with $^2J(H,C) = 2.9$ [20]; (C_6D_6): −20.1 (s, $\mu-H$) [17, 32] ^{13}C NMR $(CDCl_3)$: 8.6 (s, μ_3,η^1-C; and a superimposed doublet, ca. 40% total intensity, $^1J(C,C) = 86$), 160.3 (d, CCO; $^1J(C,C) = 86$), 165.8 (6 CO-equatorial), 175.6 (3 CO-axial) [20]; $(CD_2Cl_2, -60$ °C): 158.3, 168.0 (each 2 CO), 169.9 (1 CO), 173.0, 179.6 (each 2 CO) [20] IR (cyclohexane): 1984, 1994, 2006, 2034, 2055, 2064, 2086, 2121 (all CO) [20, 32]; the band at 1644 cm^{-1}, earlier assigned to $\nu(CCO)$ [20], arises from the $\nu(CO_2H)$ vibrations of $(\mu-H)_3Os_3(\mu_3,\eta^1-CCO_2H)(CO)_9$ (No. 27) which is formed by conversion of No. 41 in moisture [25] He-I photoelectron spectrum (81 °C): 8.00, 8.88, 9.80, 10.83, 11.37, 12.36, 15.22 (vertical ionization potentials), 7.61 (adiabatic ionization potential) eV; the spectroscopic results in addition to molecular orbital calculations of the Fenske-Hall type indicate that apparently the orbitals of apical carbon character are

References on pp. 169/71

Table 5 (continued)

No. compound	method of preparation (yield in %) properties and remarks
*41 (continued)	more stable than those of apical boron character in $(\mu-H)_3Os_3(\mu_3-BD)(CO)_9$ complexes $(D=CO, P(CH_3)_3)$ [39]
	mass spectrum: $[M]^+$, $[M-x\,CO-y\,H]^+$, $x=1$ to 10, $y=1, 2$ [17, 20, 32]
42 $[(\mu-H)_3Os_3(\mu_3,\eta^1-CCO)(CO)_9]X$	VIII (not isolated) [51]
	observed as an intermediate in the reaction of $(\mu-H)_3Os_3(\mu_3,\eta^1-CBr)(CO)_9$ (No. 5) with $AgSbF_6/CO$/nucleophile; see Preparation Method Ia [65]
X=BF_4	by protonation of No. 41 [25, 78] with $HBF_4 \cdot O(C_2H_5)_2$ in CH_2Cl_2 at room temperature, followed by washing of the precipitate with small amounts of ether or benzene [78]
	intermediate in the preparation of No. 28 from No. 41; see Preparation Method V [20]
	white solid [20, 25, 78]; very moisture-sensitive with conversion into No. 27, based on IR data [78]; apparently partial deprotonation in an excess of ether [78]
	^1H NMR (CH_2Cl_2, 35 °C [25]): -19.35 (s, $\mu-H$) [25, 78]
	IR (Nujol): 2039, 2069, 2125, 2155 (all CO) [25, 78]
	the μ_3,η^1-CCO ligand is believed to be perpendicular to the Os_3 plane as concluded from molecular orbital analysis [25]
	reaction of the in situ generated compound with dilute HCl or RH ($R=OCH_3$, $N(C_2H_5)_2$, $N(CH_3)C_6H_5$) in CH_2Cl_2 yielded Nos. 27, 28, 30, or 31 [20, 78]; see also Preparation Method V

compounds of the type $(\mu-H)Os_3(\mu_3,\eta^1-CR)(CO)_{10}$ (Formula III) and $[Os_3(\mu_3,\eta^1-COC(O)CH_3)(CO)_{10}]M$

*43 $(\mu-H)Os_3(\mu_3,\eta^1-CH)(CO)_{10}$	from $[(\mu-H)Os_3(\mu_2,\eta^1-CHOCH_3)(CO)_{10}]^-$ (Section 3.1.6.2.1; in situ generated from $(\mu-H)Os_3(\mu_2,\eta^1-COCH_3)(CO)_{10}$ and $LiBH(C_2H_5)_3$) by protonation with CF_3CO_2H in CH_2Cl_2 (CD_2Cl_2) at -60 °C (70%), or by protonation of $(\mu-H)Os_3(\mu_2,\eta^1-CHNC_5H_4CH_3-4)$-$(CO)_{10}$ (Section 3.1.6.2.1) [24]
	bright yellow, slightly air-sensitive crystals from pentane; stable at room temperature [24]

References on pp. 169/71

Table 5 (continued)

No. compound	method of preparation (yield in %) properties and remarks
	^{1}H NMR (CD_2Cl_2, -60 °C): -18.13 (d, μ-H; J(H,H) = 1.6), 14.16 (d, CH; J(H,^{13}C) = 166) [24] ^{13}C NMR (CD_2Cl_2/C_7D_8, 3:1, -100 °C): 168.0 (s, CO-f), 169.0 (s, CO-c), 171.1 (d, CO-b; ^{2}J(C,μ-H) = 10), 171.8 (s, CO-d), 173.4 (s, CO-a), 176.7 (s, CO-e), 219 (μ_3,η^1-CH); for assignment, see Formula III; optimum resolution revealed that the resonance of CO-a is a singlet overlapping a shorter doublet due to coupling (50%) with the methylidyne carbon; the unusual position of the resonance of axial CO-f at -100 °C was attributed to the distortion caused by steric interaction between CO-f and μ_3,η^1-CH; at ca. -60 °C the resonances of axial CO-e and CO-f broaden and coalesce with one of the remaining signals due to localized three-fold exchanges at the $Os(CO)_4$ center; the low activation energy G^+ of ca. 9 kcal/mol (at -60 °C?) was also attributed to the distortion, particularly to the pseudo seven-coordinate geometry around Os(1), see Fig. 54, p. 163 [24] IR (cyclohexane): 2002, 2012, 2023, 2030, 2066, 2068, 2110 (all CO) [24] mass spectrum: $[M]^+$ [24]
44 $[Os_3(\mu_3,\eta^1\text{-}COCOCH_3)(CO)_{10}]M$ $M = N(P(C_6H_5)_3)_2$, $P(C_6H_5)_4$	as an intermediate in the preparation of No. 39 from $[Os_3(\mu\text{-}CO)(CO)_{10}]M_2$ and CH_3COCl in CH_2Cl_2 (not isolated) [36, 51] yellow [51]; no further data given
45 $(\mu\text{-}H)Os_3(\mu_3,\eta^1\text{-}CCH_3)(CO)_{10}$	by protonation of No. 50 with $HBF_4\cdot O(C_2H_5)_2$ in ether, followed by addition of pentane to remove $[N(P(C_6H_5)_3)_2]BF_4$; purification by column chromatography on alumina with CH_2Cl_2/hexane (1:2) as eluant (17%) [51] yellow powder [51] ^{1}H NMR (CD_2Cl_2): -16.91 (s, μ-H), 4.54 (s, CH_3) [51] ^{13}C NMR (CD_2Cl_2/CH_2Cl_2, 1 °C): 168.9 (s, CO-a), 173.1 (d, CO-b; J(C, μ-H) = 8.5), 174.1 (s, CO-d,e,f), 174.3 (s, CO-c), 297.7 (s, μ_3,η^1-C); for assignment, see Formula III; the μ_3,η^1-C resonance ranged from $\delta =$ 293.3 ppm at -90 °C to 299.3 ppm at $+30$ °C, this temperature dependence is characteristic for capping carbons [51]

References on pp. 169/71

Table 5 (continued)

No. compound	method of preparation (yield in %) properties and remarks

45 (continued)

IR (pentane): 2000, 2012, 2023, 2032, 2068, 2085, 2112 (all CO) [51]

mass spectrum: $[M]^+$ [51]

*46 $(\mu-H)Os_3(\mu_3,\eta^1-CCH_2CH(CH_3)_2)(CO)_{10}$

from $(\mu-H)_2Os_3(CO)_{10}$ and 3,3–dimethylcyclo-propene in hexane at 25 °C for 16 h, followed by extraction of the residue with hexane (80 to 86%, along with ca. 10% of $(\mu-H)Os_3-(\mu_2,\eta^1-CCHC(CH_3)_2)(CO)_{10}$, Section 3.1.6.2.2) [28, 63]

yellow to orange crystals from hexane/ether at − 20 °C [28, 63]

^{1}H NMR (CDCl$_3$): − 16.73 (s, μ–H), 1.10 or 1.15 (d, CH$_3$; ^{3}J(H,H) = 6.6), 1.79 (t of septets, CH; ^{3}J(H,H) = 5.61, 6.6), 4.51 (d, CH$_2$; ^{3}J(H,H) = 5.61) [28, 63]

^{13}C NMR (CD$_2$Cl$_2$): 23.4 (q, CH$_3$; ^{1}J(C,H) = 128), 36.8 (d, CH; ^{1}J(C,H) = 128), 73.1 (t, CH$_2$; ^{1}J(C,H) = 128), 169.7 (d, CO–a(c); ^{2}J(C,μ–H) = 2.9), 174.0 (d, CO–b; ^{2}J(C,μ–H) = 11.2), 174.3 (d, CO–c(a); ^{2}J(C,μ–H) = 3.4), 174.6 (s, CO–d,e,f), 319.5 (dt, μ_3,η^1–C; ^{2}JC,H) = 5, 5.4) [28, 63]; for assignment, see Formula III; selective μ–^{1}H decoupling resulted in a collapse of the pseudo-quintet of the μ-carbyne resonance to a triplet at δ = 319.5 ppm [63]

variable-temperature ^{13}C NMR spectroscopy between − 80 and + 40 °C showed a shift of the carbyne carbon resonances from δ = 312.0 to δ = 320.54 ppm, respectively, implying a possible equilibrium in solution between the μ_3,η^1–C geometry of the ^{1}L ligand at low temperature and the μ_2,η^1–C as the thermodynamically favored geometry at higher temperatures; thus, the donor interaction between the carbyne carbon and the Os(CO)$_4$ group is weaker in comparison to that of No. 43 [63]

the existence of only four CO resonances indicates that the molecule possesses a time-averaged plane symmetry [28, 63]

IR (hexane): 1989, 2001, 2013, 2023, 2059, 2105 (all CO) [28, 63]

mass spectrum: $[M]^+$ [63]

References on pp. 169/71

Table 5 (continued)

No. compound	method of preparation (yield in %) properties and remarks
*47 $(\mu\text{-H})Os_3(\mu_3,\eta^1\text{-}CC_6H_5)(CO)_{10}$	from $(\mu\text{-H})Os_3(\mu_2,\eta^1\text{-}COCH_3)(CO)_{10}$ (Section 3.1.6.2.2) and C_6H_5Li (ca. 1:2.5) in ether at 0 °C for 15 to 30 min, followed by treatment with a large excess of $CF_3SO_3CH_3$ at 25 °C for 60 h [38] or at 35 °C for 72 h [45]; work-up by addition of n-pentane/H_2O and subjecting the evaporated organic layer to TLC with n-pentane as eluant (50 to 40%, along with 30 to 41% of No. 53, depending on reaction conditions) [38, 45] by methylation of $(\mu\text{-H})Os_3(=C(OCH_3)_2)(\mu_3,\eta^1\text{-}CC_6H_5)(CO)_9$ ("Organoosmium Compounds" B 5, Section 3.1.5.2, in preparation; intermediate in the above reaction by methylation at 0 °C for 30 min) with $CF_3SO_3CH_3$ as before (along with $Os_3(\mu_3,\eta^1\text{-}CC_6H_5)(\mu_3,\eta^1\text{-}COCH_3)(CO)_9$, No. 53) [37] red, air-stable crystals from CH_2Cl_2/acetone, m. p. 144 to 146 °C; stable up to 120 °C in the solid state and in solution [38] ^1H NMR (CCl_4): -16.6 (s, μ-H), 7.2 to 7.4 (m, H-β,γ of C_6H_5), 7.6 to 7.8 (m, H-α of C_6H_5) [38] ^{13}C NMR (CD_2Cl_2, 17 °C): 169.8 (s, CO-a), 174.0 (d, CO-b; J(H,C) = 11), 175.0 (s, CO-d,e,f), 175.9 (s, CO-c), 314.2 (s, μ_3,η^1-C); (CD_2Cl_2, < -100 °C): 170 (CO-a), 172.0 (CO-f), 173.2 (CO-b,d), 176 (CO-c), 179.7 (CO-e), 304.2 (μ_3,η^1-C); for assignment, see Formula III, spectra depicted [38] variable-temperature ^{13}C NMR spectroscopy revealed a shift of the μ_3,η^1-C resonance from $\delta = 304.2$ ppm at -115 °C to $\delta = 314.2$ ppm at 17 °C; similarly below -100 °C the $Os(CO)_4$ carbonyl resonance at $\delta = 175.0$ ppm splits to give a 1:2:1 pattern with $\delta = 172.0$ (CO-f), 173.2 (s, CO-d), and 179.7 ppm (CO-e); from the coalescence at ca. -70 °C, an approximate value of $\Delta G^{\neq}$ ca. 8.8 kcal/mol for the axial-equatorial exchange was estimated [38] IR (cyclohexane): 1995, 2000, 2023, 2063, 2107 (all CO) [38] mass spectrum: $[M]^+$ [38]

References on pp. 169/71

Table 5 (continued)

No. compound	method of preparation (yield in %) properties and remarks

48 $(\mu\text{-H})Os_3(\mu_3,\eta^1\text{-}CC_6H_4CH_3\text{-}4)(CO)_{10}$ — from $(\mu\text{-H})_2Os_3(\mu_3,\eta^1\text{-}CC_6H_4CH_3\text{-}4)(CO)_9W(\eta^5\text{-}C_5H_5)(CO)_3$ (Section 3.1.6.4.1) in refluxing toluene under 1 atm of CO for 15 min; isolated by TLC with pentane/CH_2Cl_2 (8:1) as eluant (87%, along with $(\eta^5\text{-}C_5H_5)WH(CO)_3$) [42]

for formation from $(\mu\text{-H})_2Os_3(\mu_3,\eta^1\text{-}CC_6H_4CH_3\text{-}4)(CO)_9W(\eta^5\text{-}C_5H_5)(CO)_3$ (Section 3.1.6.4.1) by reaction with $(\eta^5\text{-}C_5H_5)W(\equiv CC_6H_4CH_3\text{-}4)(CO)_2$, see No. 36 [42]

orange-red crystalline solid from acetone [42]

^{1}H NMR (C_6D_6): -16.27 (s, μ-H), 1.85 (s, CH_3), 6.70 to 7.77 (m, C_6H_4) [42]

IR (cyclohexane): 1987, 1995, 2020, 2057, 2123 (all CO) [42]

mass spectrum: $[M]^+$ [42]

other compounds with an μ_3,η^1-bridging ^{1}L ligand

*49

by UV irradiation of $(\mu\text{-H})Os_3(\mu_2,\eta^1\text{-}CHCH=N(C_2H_5)_2)(CO)_{10}$ (anti isomer; see Section 3.1.6.2.1) with a high-pressure Hg lamp in hexane for 2.5 h; separated with TLC using hexane/CH_2Cl_2 (4:1) as eluant (58%, along with $(\mu\text{-H})_2Os_3(\mu_3,\eta^2\text{-}CHCN(C_2H_5)_2)(CO)_9$ in 21% yield) [53, 79]

yellow crystals from CH_2Cl_2/hexane [53]

^{1}H NMR $(CDCl_3)$: -19.31 (s, μ-H), 1.28 (t, CH_3; $J(H,H)=7.2$), 3.53, 3.61 (q's, each CH_2; $J(H,H)=7.2$), 9.10 (s, CH) [53, 79]

IR (hexane): 1957, 1971, 1982, 2009, 2018, 2037, 2064, 2096 (all CO) [53, 79]

isomerization in hexane at 68 °C for 3 h gave also $(\mu\text{-H})_2Os_3(\mu_3,\eta^2\text{-}CHCN(C_2H_5)_2)(CO)_9$ (see above) in ca. 70% yield [53, 79]

50

VIII [51]

orange oil from ether/pentane [51]

^{1}H NMR (CD_2Cl_2): 4.12 (s, CH_3) [51]

^{13}C NMR $(CD_2Cl_2/CH_2Cl_2$, 32 °C): 167.7 (s, $\mu_3,\eta^1\text{-}C$), 184.7 (s, 10 CO); $(CD_2Cl_2/CH_2Cl_2$, -90 °C): 164.8 (s, $\mu_3,\eta^1\text{-}C$), 177.1 (s, 9 CO), 249.3 (s, $\mu\text{-}CO$); at 32 °C the terminal CO groups coalesce with μ-CO to give one resonance at $\delta=184.7$ ppm [51]

IR (CH_2Cl_2): 1971, 2024 (all CO) [51]

References on pp. 169/71

Table 5 (continued)

No. compound	method of preparation (yield in %) properties and remarks
	protonation with $HBF_4 \cdot O(C_2H_5)_2$ in ether yielded No. 45; methylation with $CF_3SO_3CH_3$ in CH_2Cl_2 gave No. 52 [51]

compounds of the type $Os_3(\mu_3,\eta^1-CR)(\mu_3,\eta^1-CR')(CO)_{9-n}D_n$ (Formula IV) and $[(\mu-H)Os_3(\mu_3,\eta^1-CC_6H_5)(\mu_3,\eta^1-COCH_3)(CO)_9]^+$

51 $Os_3(\mu_3,\eta^1-CC_6H_5)_2(CO)_9$	from $(\mu-H)Os_3(\mu_3,\eta^1-CC_6H_5)(CO)_{10}$ (No. 47) and C_6H_5Li (ca. 1:2) in ether at 0 °C for 15 to 30 min, followed by treatment with a large excess of $CF_3SO_3CH_3$ and stirring at 25 °C for 46 h; purified by TLC with n-pentane as eluant (29%, along with 32% of unreacted No. 47 and five other unidentified by-products) [52]
	dark red, air-stable crystals from CH_2Cl_2/CH_3OH, m. p. 139 to 140 °C [52]
	^1H NMR (CD_3CN, 17 °C): 7.32 (t, H-γ of C_6H_5), 7.40 (t, H-β of C_6H_5), 8.00 (d, H-α of C_6H_5); J(H,H) = 7 [52]
	^{13}C NMR ($CDCl_3$, 17 °C): 174.9 (s, 9 CO), 275.7 (s, μ_3,η^1-C), variable-temperature ^{13}C NMR spectra revealed no temperature dependence for the μ_3,η^1-C carbon resonances; similarly, the carbonyl singlet remained sharp down to low temperatures due to the threefold rotation on each $Os(CO)_3$ group [52]
	IR (cyclohexane): 1992, 2002, 2022, 2057, 2064 (all CO) [52]
	mass spectrum: $[M]^+$ [52]
52 $Os_3(\mu_3,\eta^1-CCH_3)(\mu_3,\eta^1-COCH_3)(CO)_9$	
	by methylation of $[Os_3(\mu_3,\eta^1-CCH_3)(\mu-CO)-(CO)_9][N(P(C_6H_5)_3)_2]$ (No. 50) with $CF_3SO_3CH_3$ in CH_2Cl_2 for 75 min, followed by extraction of the crude product with hexane and purification by column chromatography on alumina with hexane as eluant (27%, along with $Os_3(\mu_3,\eta^2-C(CH_3)=COCH_3)(\mu-CO)-(CO)_9$ in small amounts) [51]
	yellow powder [51]
	^1H NMR (CD_2Cl_2): 4.36 (s, CCH$_3$; $^2J(H,^{13}C) = 6$), 4.67 (s, OCH$_3$) [51]
	^{13}C NMR (CD_2Cl_2/CH_2Cl_2, 30 °C): 174.3 (s, 9 CO), 252.5 (s, μ_3,η^1-CCH_3), 312.5 (s, μ_3,η^1-COCH_3); the carbonyl singlet remains unchanged down to −90 °C [51]

References on pp. 169/71

Table 5 (continued)

No. compound	method of preparation (yield in %) properties and remarks

52 (continued)

IR (pentane): 1994, 2008, 2022, 2062 (all CO) [51]

mass spectrum: $[M]^+$ [51]

*53　$Os_3(\mu_3,\eta^1\text{-}CC_6H_5)(\mu_3,\eta^1\text{-}COCH_3)(CO)_9$

for formation, see No. 47 [37, 38, 45]

obtained by deprotonation of No. 54 with 10% $NaHCO_3$ solution in $CDCl_3$ (quantitative) [45]

orange-red air-stable crystals from ether/hexane, m. p. 137 to 139 °C [45]

^1H NMR (acetone-d_6): 4.62 (s, CH_3), 7.21 (t, H-γ of C_6H_5), 7.31 (t, H-β of C_6H_5), 7.81 (d, H-α of C_6H_5); J(H,H) = 7 [37, 45]; deshielding of the methyl protons in comparison to No. 6 (p. 128) may result from the resonance form $Os_3^- = C = {}^+OCH_3$; spectrum depicted [45]

^{13}C NMR (CD_2Cl_2): 173.9 (s, 9 CO), 234.6 (s, $\mu_3,\eta^1\text{-}CC_6H_5$), 319.4 (s, $\mu_3,\eta^1\text{-}COCH_3$) [37, 45]; the carbonyl resonance remains unchanged down to -110 °C indicating a low energy barrier for the threefold exchange of the CO's at the $Os(CO)_3$ centers; the $\mu_3,\eta^1\text{-}CCH_3$ resonance is temperature independent whereas the $\mu_3,\eta^1\text{-}CC_6H_5$ resonance is shifted from 234.6 ppm at 20 °C to 228.9 ppm at -80 °C; the upfield shift of the $\mu_3,\eta^1\text{-}CC_6H_5$ resonance in comparison to that of No. 51 is attributed to the presence of resonance interaction $Os_3^- = C = {}^+OCH_3$; spectra at -80 and $+20$ °C depicted [45]

IR (cyclohexane): 1989, 2003, 2024, 2060, 2065 (all CO); spectrum depicted [37, 45]

mass spectrum: $[M]^+$ [37, 45]

54

by protonation of No. 53 with CF_3SO_3H in $CDCl_3$ (not isolated, but based on ^1H and ^{13}C NMR spectroscopy) [45]

deep red solution, stable at 25 °C for ca. 24 h [45]

^1H NMR ($CDCl_3$, 17 °C): -15.56 (s, μ-H), 4.80 (s, CH_3), 7.42 to 7.53 (m, H-β,γ of C_6H_5), 7.69 (d, H-α of C_6H_5; J(H,H) = 7.9) [45]

^{13}C NMR (CD_2Cl_2, -75 °C): 162.2 (d, CO-a; ^{2}J(H,C) = 8), 165.2 (s, CO-b), 166.1 (s, CO-c), 167.1 (s, CO-d), 266.1 (s, $\mu_3,\eta^1\text{-}CC_6H_5$), 305.0 (s, $\mu_3,\eta^1\text{-}COCH_3$); the $\mu_3,\eta^1\text{-}CC_6H_5$ resonance is significantly shifted downfield com-

Table 5 (continued)

No. compound	method of preparation (yield in %) properties and remarks

pared to that of No. 53 being effected by the incorporated hydride ligand; at low temperature, the hindered threefold exchange at the two Os(CO)$_3$ centers being bridged by the hydride ligand resulted in a splitting of the CO resonance; the average resonance of 165.4 ppm for the nine CO groups is ca. 8.5 ppm upfield from that for No. 53, apparently due to the positive charge on the Os$_3$-moiety; spectrum depicted [45]

treatment with a 10% NaHCO$_3$ solution re-formed No. 53 [45]

55 Os$_3(\mu_3,\eta^1$-CC$_6$H$_5)(\mu_3,\eta^1$-COCH$_3$)(CO)$_8$P(C$_6$H$_5)_3$

for formation, see No. 56 (20%) [45]

orange crystals from CH$_2$Cl$_2$/CH$_3$OH [45]

^{1}H NMR (acetone-d$_6$, -20 °C): 3.88 (s, CH$_3$), 7.14 (t, H-γ of CC$_6$H$_5$), 7.29 (t, H-β of CC$_6$H$_5$), 7.40 to 7.53 (m, P(C$_6$H$_5)_3$), 7.94 (d, H-α of CC$_6$H$_5$); J(H,H) = 7; the high-field shift of the methyl protons in contrast to No. 53 results from the shielding effect of the phosphane ligand [45]

^{13}C NMR (CDCl$_3$, -60 °C): 174.8 (s, CO of Os(CO)$_3$), 183.5 (d, CO of Os(CO)$_2$; ^{2}J(^{13}C,^{31}P) = 3), 231.5 (d, μ_3,η^1-**C**C$_6$H$_5$; ^{2}J(^{13}C,^{31}P) = 21), 318.9 (s, μ_3,η^1-**C**OCH$_3$); the strong μ_3,η^1-^{13}C-^{31}P coupling indicates for the phosphane ligand a position trans to the Os-CC$_6$H$_5$ vector; spectrum depicted [45]

IR (cyclohexane): 1944, 1974, 1988, 2003, 2016, 2052, 2077 (all CO); spectrum depicted [45]

mass spectrum: [M]$^+$ [45]

56 Os$_3(\mu_3,\eta^1$-CC$_6$H$_5)(\mu_3,\eta^1$-COCH$_3$)(CO)$_7$(P(C$_6$H$_5)_3)_2$

from Os$_3(\mu_3,\eta^1$-CC$_6$H$_5)(\mu_3,\eta^1$-COCH$_3$)(CO)$_9$ (No. 53) and P(C$_6$H$_5)_3$, ca. 1:5, in n-decane at 120 °C for 7 h, followed by removal of the solvent and isolation by TLC with n-pentane as eluant (54%, along with 20% of No. 55) [45]

orange air-stable crystals from CH$_2$Cl$_2$/CH$_3$OH [45]

^{1}H NMR (acetone-d$_6$, 17 °C): 3.64 (s, CH$_3$), 7.01 to 7.10 (m, H-β,γ of CC$_6$H$_5$), 7.32 to 7.51 (m, P(C$_6$H$_5)_3$), 7.72 (d, H-α of CC$_6$H$_5$; J(H,H) = 7); the high-field shift of the methyl protons in

References on pp. 169/71

Table 5 (continued)

No. compound	method of preparation (yield in %) properties and remarks
56 (continued)	contrast to No. 53 results from the shielding effect of the phosphane ligands [45]

^{13}C NMR (CDCl$_3$, 17 °C): 176.9 (s, CO of Os(CO)$_3$), 183.6 (s, CO of Os(CO)$_2$), 184.1 (d, CO of Os(CO)$_2$; ^{2}J(C,P) = 8), 237.2 (t, μ_3,η^1-**CC$_6$H$_5$**; ^{2}J(^{13}C,^{31}P) = 10), 317.7 (s, μ_3,η^1-**COCH$_3$**); a decrease in μ_3,η^1-^{13}C-^{31}P coupling compared to that of No. 55 indicates static or dynamic structural distortion; spectrum depicted [45]

IR (cyclohexane): 1934, 1944, 1970, 1980, 2000, 2011, 2051 (all CO); spectrum depicted [45]

mass spectrum: [M]$^+$ [45]

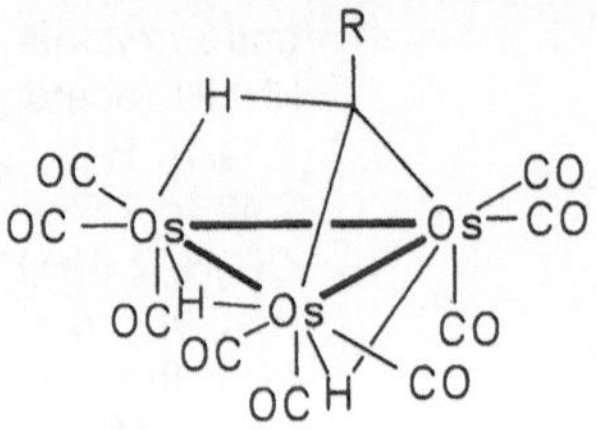

V

*Further information:

(μ-H)$_3$Os$_3$(μ$_3$,η^1-CH)(CO)$_9$ (Table **5**, No. **1**) crystallizes in the triclinic space group P$\bar{1}$ − C$_i^1$ (No. 2) with a = 9.399(3), b = 11.665(4), c = 15.651(7) Å, α = 112.32(3)°, β = 90.11(3)°, γ = 97.80(3)°; Z = 4, D$_c$ = 3.55 g/cm^3, R = 0.044 at − 73 °C. These data are consistent with the results of neutron diffraction at − 262 °C. The asymmetric unit contains two crystallographically independent molecules **A** and **B**, each having C$_{3v}$ molecular symmetry within the experimental error; the structure of Molecule **A** is shown in **Fig. 46**. The nearly equilateral Os$_3$ triangle of each molecule is symmetrically capped by the methylidyne ligand and each of the edges is nearly symmetrically bridged by a hydride ligand (the positions of the hydride ligands were calculated, only the methylidyne hydrogen could be located by neutron diffraction); the C–H vector is essentially perpendicular to the Os$_3$ plane. The Os–CO distances fall into two distinct sets, one being in trans position to the hydride ligands, and the other being in trans position to the methylidyne group; the average bond distances amounted to 1.911 and 1.964 Å, respectively. The differences in the bond lengths are due to the π–acid character of the CH ligand, which competes with the CO ligands for metal electron density. The methylidyne carbon is best described as sp hybridized with substantial metal–carbon overlap involving the 2pπ orbitals of the carbon [30].

The mean-square displacement amplitudes (thermal parameters) obtained from neutron diffraction at − 262 °C were determined for the terminally bound CO groups. The relationships between the mean-square displacement amplitudes of Os, C, and O atoms are investi-

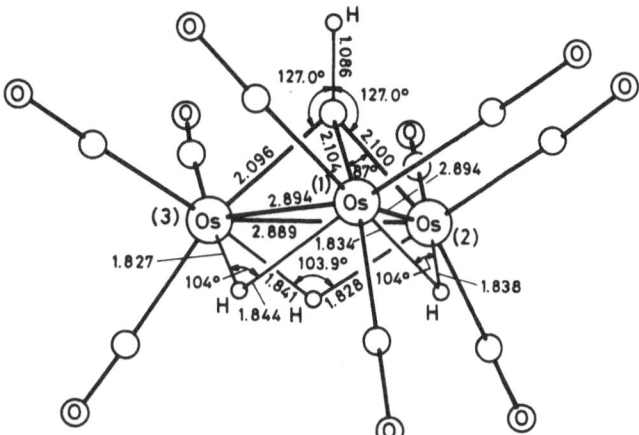

Fig. 46. Molecular structure of Molecule **A** of $(\mu\text{-H})_3Os_3(\mu_3,\eta^1\text{-CH})(CO)_9$ (No. 1) with selected bond distances (in Å) and bond angles [30].

gated by applying the rigid–body test along the Os–C, C≡O and Os···O vectors on the basis of simple vibrational motion models. The differences in the mean–square displacement amplitudes were calculated for each CO group; the averaged values are $\Delta_{Os\text{-}C} = 16(3) \times 10^4$ Å^2, $\Delta_{C≡O} = 6(4) \times 10^4$ Å^2, and $\Delta_{Os\cdots O} = 15(3) \times 10^4$ Å^2; the low $\Delta_{C≡O}$ value near zero indicates the rigid–bond character of the C≡O bond. The results revealed that the C atoms of terminally bound CO groups slide along the Os···O vectors towards the O atoms on simply passing from isotropic to anisotropic refinement, resulting in a lengthening of the Os–C bonds and a shortening of the C≡O bonds, while the Os···O distances remain unchanged. Both, the sliding effect and the mean–square displacement amplitude values probably reflect the deviation of the electron density distribution around the C atoms nuclei from the ideal spherical atom model. The bonding electron density is best described as an anisotropic thermal ellipsoid [54].

The formation of No. 1 from No. 5 and $Re_2(CO)_{10}$ under UV irradiation (see Preparation Method Ib, p. 122) proceeded exclusively in toluene, whereas in toluene-d_8 ring addition became competitive with D abstraction, as expected for a significant primary kinetic isotope effect. ^{1}H NMR determinations of $(\mu\text{-H})_3Os_3(\mu_3,\eta^1\text{-CD})(CO)_9$ in C_6D_6 showed a methylidyne signal after a few days, indicating only a slow hydrogen scrambling between the metal and carbon sites at room temperature [58]. Above 80 °C the site exchange can be observed by spin saturation transfer. Selective inversion–recovery measurements of the relaxation rates T_1 for the CH and μ-H sites at ambient temperature are 13.0 and 2.08 s; the respective nuclear Overhauser enhancements (NOE) were 32 and 2%. Variable-temperature measurements of the T_1 and NOE values below the onset of site exchange at ca. 80 °C are also consistent with dipole–dipole relaxation; the NOE values are temperature independent. Around ca. 80 °C the effects of exchange became apparent for the CH site, but complete spin saturation of the CH nuclei occurred at ca. 130 °C; for the faster relaxing μ-H site, measurements on the μ-H nuclei were possible between 115 and 152 °C. The rate constant k for the CH site exchange was 3.6×10^{-3} s^{-1} at 80 °C; the activation parameters were evaluated as $\Delta H^{\neq} = 24.0 \pm 1.7$ kcal/mol and $\Delta S^{\neq} = 2.0 \pm 1.8$ cal·mol^{-1}·K^{-1}. The CH/μ-H site exchange was proposed to proceed via the intermediate $(\mu\text{-H})_2Os_3(\mu_3,\eta^2\text{-HCH})(CO)_9$ having an agostic bonded hydrogen (see Formula V, p. 150; R = H). The same intermediate is probably formed by thermal conversion of $(\mu\text{-H})_2Os_3(\mu_2,\eta^1\text{-CH}_2)(CO)_{10}$ (Section 3.1.6.2.1) into No. 1 [50]; compare also Preparation Method VII, p. 122.

The CH/μ-H site exchange via the Os–H–C intermediate is also characterized by a change in the reaction enthalpy which has been calculated from Mullay's electronegativities ($\Delta H_R = 18.8$) and from Allred/Rochow's electronegativities ($\Delta H_R = 11.9$) for the C and Os atoms [69].

Reaction with $t\text{-C}_4H_9OCl$ in CO-saturated CCl_4 or $CBrCl_3$ at 45 °C yielded Nos. 4 and 5, respectively [58].

Treatment of the solid title compound with conc. H_2SO_4, followed by dilution with H_2O resulted in $(\mu\text{-H})_2Os_3(CO)_9(\mu_3,\eta^3\text{-O}_3SO)$ [43]; reaction with CF_3SO_3H at 25 °C presumably yielded $(\mu\text{-H})_2Os_3(CO)_9(O_3SCF_3)_2$; the structure has not yet been characterized [43, 75]. Dissolving in CF_3SO_3H, followed by subsequent reaction with phosphonic or arsonic acids REO_3H_2 (E = P, As; R = CH_3, C_6H_5) in CH_3OH resulted in $(\mu\text{-H})_2Os_3(CO)_9(\mu_3,\eta^3\text{-O}_3ER)$ [80].

Similarly, treatment with CF_3CO_2H at 75 °C gave $(\mu\text{-H})_2Os_3(CO)_9(\mu_2,\eta^2\text{-O}_2CCF_3)$-$(\eta^1\text{-O}_2CCF_3)$; the oxo ligands probably adopt a trans configuration [75].

$(\mu\text{-H})_3Os_3(\mu_3,\eta^1\text{-CCl})(CO)_9$ (Table 5, No. 4) crystallizes in the orthorhombic space group $Pnma-C_{2h}^{16}$ (No. 62) with a = 17.628(3), b = 14.593(3), c = 6.682(1) Å; Z = 4, D_c = 3.371 g/cm³, R = 0.059. The nearly equilateral Os_3 triangle is symmetrically capped by the μ_3,η^1-CCl unit; see **Fig. 47**. The bridging hydrogens were not observed crystallographically [81].

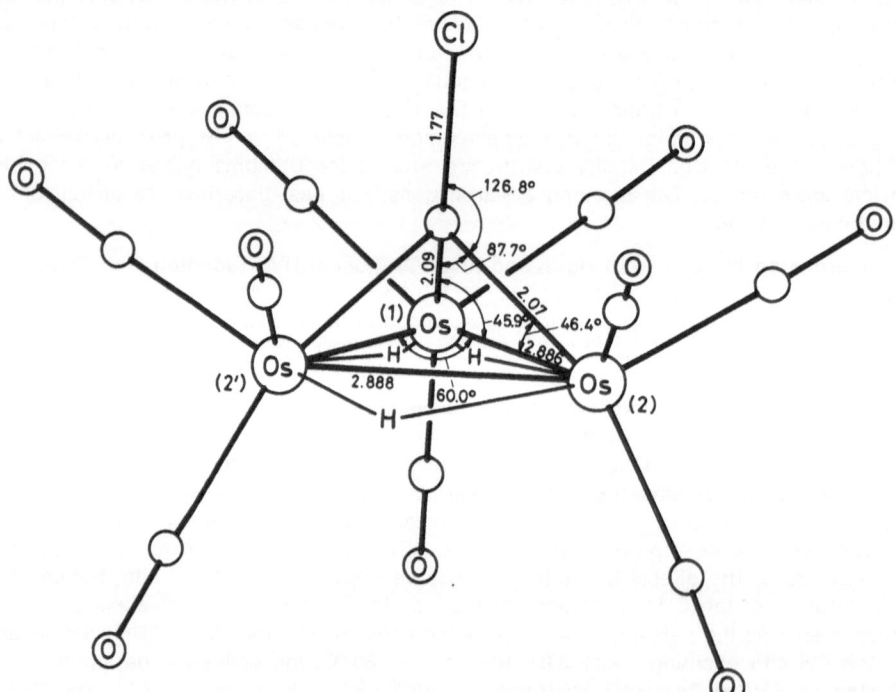

Fig. 47. Molecular structure of $(\mu\text{-H})_3Os_3(\mu_3,\eta^1\text{-CCl})(CO)_9$ (No. 4) with selected bond distances (in Å) and bond angles [81].

References on pp. 169/71

$(\mu\text{-H})_3Os_3(\mu_3,\eta^1\text{-CBr})(CO)_9$ (Table 5, No. 5) crystallizes in the orthorhombic space group $Pnma-C_{2h}^{16}$ (No. 62) with $a = 17.605(4)$, $b = 14.527(2)$, $c = 6.906(1)$ Å; $Z = 4$, $D_c = 3.451$ g/cm^3, $R = 0.053$. The structure is similar to that of No. 4; compare also Fig. 47. The bridging hydrogens were not observed crystallographically. Selected bond distances and angles [81]:

atoms	distance (Å)	atoms	angle (°)
Os(1)-Os(2)	2.885(1)	Os(1)-Os(2)-Os(1')	60.03(2)
Os(1)-Os(1')	2.886(1)	Os(1')-Os(1)-Os(2)	59.99(1)
Os(1)-μ_3,η^1-C	2.09(1)	Os(1)-Os(2)-μ_3,η^1-C	46.3(3)
Os(2)-μ_3,η^1-C	2.09(2)	Os(1')-Os(1)-μ_3,η^1-C	46.3(3)
μ_3,η^1-C-Br	1.91(1)	Os(1)-μ_3,η^1-C-Os(2)	87.4(5)
average Os-C	1.94	Os(1)-μ_3,η^1-C-Os(1')	87.5(6)
average C-O	1.12	Os(2)-μ_3,η^1-C-Br	127.1(8)

The title compound is a suitable starting material for the preparation of other functionalized μ_3,η^1-CR complexes.

Alkylidyne complexes (R = H, alkyl, allyl, aryl) were obtained by treatment of No. 5 with an excess of AgSbF$_6$ and nucleophile precursors of the type RSi(CH$_3$)$_3$, HSi(C$_2$H$_5$)$_3$, or SnR$_4$ generally in CH$_2$Cl$_2$ at room or reflux temperature, or by treatment with AgSbF$_6$/C$_6$H$_5$R' (R' = H, D, CH$_3$) at reflux temperature [65], or with AlCl$_3$/C$_6$D$_6$ [60]; see also Preparation Method Ia, p. 122.

Acyl groups were introduced by reaction with AgSbF$_6$/CO [65] or AlCl$_3$/CO [20, 78] leading to the intermediate $[(\mu\text{-H})_3Os_3(\mu_3,\eta^1\text{-CCO})(CO)_9]^+$ (No. 42) before treatment with a nucleophilic reagent [20, 65, 78]. Compound No. 3 bearing the μ_3,η^1-CF moiety was obtained by reaction with AgBF$_4$/[N(C$_2$H$_5$)$_4$]BF$_4$ in CH$_2$Cl$_2$ at 0 °C [57]; see also Preparation Method Ia, p. 122.

The reactions of No. 5, as described above, probably proceeded via the intermediate $[(\mu\text{-H})_3Os_3(\mu_3,\eta^1\text{-C})(CO)_9]^+$ formed by Br$^-$ abstraction by the Lewis acid (AgSbF$_6$, AgBF$_4$, and AlCl$_3$ under CO) and which then added either a nucleophile, or initially CO followed by nucleophile addition; the latter reactions with CO proceeded also via the intermediate, $[(\mu\text{-H})_3Os_3(\mu_3,\eta^1\text{-CCO})(CO)_9]^+$ (No. 42) [57, 65, 78]. Br$^-$ abstraction with AlCl$_3$ or TlPF$_6$ in the presence of allylsilane resulted in a mixture of products or in the recovery of No. 5 [65].

A radical intermediate, $(\mu\text{-H})_3Os_3(\mu_3,\eta^1\text{-C}^\cdot)(CO)_9$, produced by in situ-generated $^\cdot$Re(CO)$_5$ was proposed for the reactions upon UV irradiation under a CO atmosphere in the presence of Re$_2$(CO)$_{10}$. The radical Os$_3$ species on one hand abstracts a hydrogen from cyclohexane, toluene or C$_6$H$_5$F to give $(\mu\text{-H})_3Os_3(\mu_3,\eta^1\text{-CH})(CO)_9$ (No. 1), or a fluorine from C$_6$H$_5$F or C$_6$F$_6$ leading to $(\mu\text{-H})_3Os_3(\mu_3,\eta^1\text{-CF})(CO)_9$ (No. 3) [58, 76]. On the other hand, it adds to C$_6$F$_6$, C$_6$H$_5$F, and C$_6$H$_6$ to result in the corresponding μ_3,η^1-aryl clusters [76]. Therefore, photolysis in C$_6$F$_6$ resulted in a mixture of No. 3, $(\mu\text{-H})_3Os_3(\mu_3,\eta^1\text{-CC}_6F_5)$-(CO)$_9$ (No. 38), $\{(\mu\text{-H})_3Os_3(\mu_3,\eta^1\text{-CC(O)-})(CO)_9\}_2$, and $(\mu\text{-H})_2Os_3(\mu_2,\eta^1\text{-C}_6F_5)(CO)_9Re(CO)_5$ (Section 3.1.6.4.3), and photolysis in C$_6$H$_5$F gave a mixture of Nos. 1, 3, 34, and 37; a reaction mechanism, concerning the distribution of the reaction products was discussed [76]. The radical intermediate is also presumed in the conversion of No. 5 to No. 1 under the action of an excess of HSn(C$_4$H$_9$-n)$_3$/CO in heptane at 60 °C [58].

(μ-H)$_3$Os$_3$(μ$_3$,η^1-COCH$_3$)(CO)$_9$ (Table **5**, No. **6**) was formed as an equilibrium mixture with (μ-H)Os$_3$(μ$_2$,η^1-COCH$_3$)(CO)$_{10}$ (Section 3.1.6.2.2) upon treatment of the latter compound with a CO/hydrogen mixture (ratio 25:75) in decane at 110 °C for 5 d in an autoclave under 35 atm; the equilibrium constant [No. 6] [CO]/[(μ-H)Os$_3$(μ$_2$,η^1-COCH$_3$)(CO)$_{10}$][H$_2$] was calculated as 0.08±0.02 at 110 °C. Carbonylation of the equilibrium mixture in decane under 14 to 70 atm of CO regenerated (μ-H)Os$_3$(μ$_2$,η^1-COCH$_3$)(CO)$_{10}$ (Section 3.1.6.2.2), competing with the formation of Os(CO)$_5$ (or Os$_3$(CO)$_{12}$ at 100 °C), dimethyl ether, and H$_2$. The latter conversion is expected to be of zero order in CO pressure and of first order in cluster concentration, the first-order rate constant under CO atmosphere in decane represents the sum of the first-order rate constants for the H$_2$ and for the dimethyl ether elimination and ranges between ca. 2.1×10^{-6} s^{-1} at 100 °C (14 or 70 atm CO) and ca. 26×10^{-6} s^{-1} at 119 °C (68 atm CO). The relative amounts of the products indicate that the rate constant for H$_2$ elimination must be greater than that for dimethyl ether elimination by a factor of 2.5; the activation parameters were estimated as $\Delta H^{\pm} = 38$ kcal/mol and $\Delta S^{\pm} = 16$ cal·mol^{-1}·K^{-1} [33].

Reaction with BBr$_3$ in CH$_2$Cl$_2$ at 25 °C gave (μ-H)$_3$Os$_3$(μ$_3$,η^1-CBr)(CO)$_9$ (No. 5) [12]. Treatment with AlCl$_3$ (ca. 1:2) in CH$_2$Cl$_2$ under CO at 25 °C for 45 min, followed by addition of CH$_3$OH yielded (μ-H)$_3$Os$_3$(μ$_3$,η^1-CCO$_2$CH$_3$)(CO)$_9$ (No. 28), along with (μ-H)$_3$Os$_3$-(μ$_3$,η^1-CH)(CO)$_9$ (No. 1) and (μ-H)$_3$Os$_3$(μ$_3$,η^1-CCl)(CO)$_9$ (No. 4) [78].

Vacuum pyrolysis at 240 °C for 12 h resulted in a mixture of polynuclear clusters including 12% of H$_4$Os$_4$(CO)$_{12}$, 10% of Os$_6$(CO)$_{18}$, 50% of H$_2$Os$_7$C(CO)$_{10}$, 15% of Os$_8$C(CO)$_{22}$, and 3% of H$_2$Os$_{10}$C(CO)$_{24}$ [83].

(μ-H)$_3$Os$_3$(μ$_3$,η^1-CBCl$_2$)(CO)$_9$ (Table **5**, No. **9**) crystallizes in the monoclinic space group P2$_1$/n − C$_{2h}^5$ (No. 14) with a = 9.636(5), b = 13.810(6), c = 13.526(6) Å, β = 93.52(4)°; Z = 4, D$_c$ = 3.402 g/cm^3, R = 0.045 at 0 °C. The molecular structure, shown in **Fig. 48**, has approximate C$_s$ symmetry with a pseudo mirror plane passing through B, μ$_3$,η^1-C, Os(2), and the midpoint of the Os(1)-Os(3) edge. The short μ$_3$,η^1-C-B bond distance indicates some double bond character and is consistent with the boron being sp^2-hybridized. The μ$_3$,η^1-CBCl$_2$ unit is tilted only by 15° from perpendicularity with respect to the Os$_3$ plane. The hydrides were not observed crystallographically but were believed to bridge the three Os-Os bond edges. Average bond distances of Os-C and C≡O are 1.904 and 1.14 Å [48].

Hydrolysis or acid-catalyzed hydrolysis with glacial acetic acid in CH$_2$Cl$_2$ at −78 °C yielded No. 1 [48, 74]; see Preparation Method IIa, p. 122. Reaction with N(CH$_3$)$_3$, P(CH$_3$)$_3$, or P(C$_6$H$_5$)$_3$ in CD$_2$Cl$_2$ at −196 °C yielded Nos. 13 to 15 [74]; see Preparation Method IIb, p. 122. Treatment with KH/B(CH$_3$)$_3$ led to the K salt of No. 10 [74].

(μ-H)$_3$Os$_3$(μ$_3$,η^1-CCH$_3$)(CO)$_9$ (Table **5**, No. **18**). A mixture of partially deuterated [1] and perdeuterated [68] No. 18 was obtained from (μ-H)$_2$Os$_3$(μ$_3$,η^2-CCH$_2$)(CO)$_9$ with D$_2$ in refluxing n-octane [1, 68]; see also Preparation Method III, p. 122. Regioselective preparation of μ$_3$,η^1-CCD$_3$ or (μ-D)$_3$ labeled No. 18 could be achieved under the conditions of Preparation Method Ia, p. 122, by reaction of (μ-H)$_3$Os$_3$(μ$_3$,η^1-CBr)(CO)$_9$ (No. 5) with AgSbF$_6$/CO and DSi(C$_2$H$_5$)$_3$, or by treatment of deuterated No. 5 with AgSbF$_6$/Sn(CH$_3$)$_4$ [65]. Similarly, a μ$_3$,η^1-C^{13}CH$_3$ enriched complex could be obtained from No. 5 by reaction with AgSbF$_6$/^{13}CO and HSi(C$_2$H$_5$)$_3$ [65].

The title compound crystallizes in the orthorhombic space group Pnma − D$_{2h}^{16}$ (No. 62) with a = 17.55(3), b = 14.57(2), c = 6.76(1) Å. The structure with selected bond distances (in Å), shown in Formula VI (bond lengths in Å), was evaluated by X-ray powder photography in combination with nematic phase ^{1}H NMR spectroscopy. The Os-H bond distances and

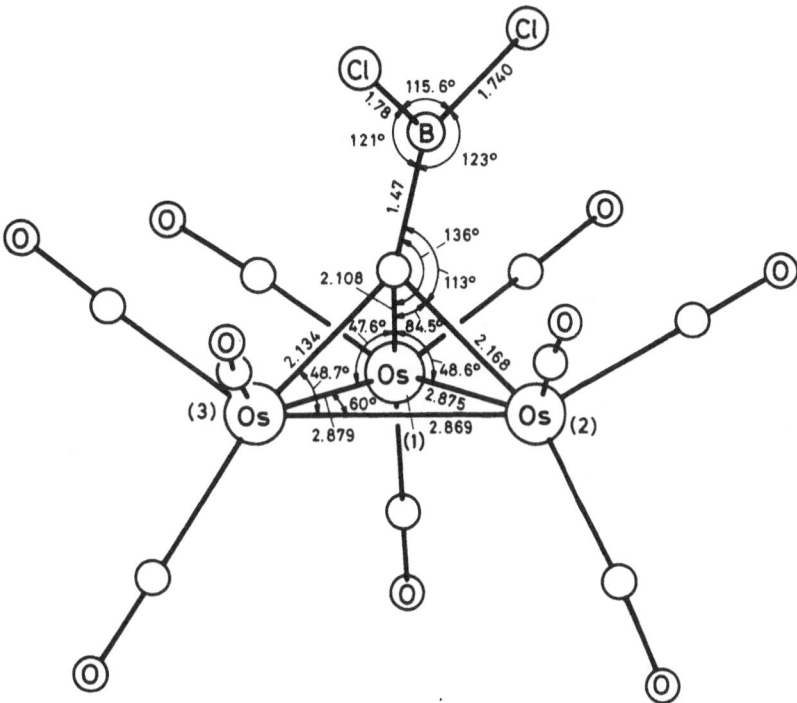

Fig. 18. Molecular structure of $(\mu\text{-H})_3Os_3(\mu_3,\eta^1\text{-CBCl}_2)(CO)_9$ (No. 9) with selected bond distances (in Å) and bond angles [48].

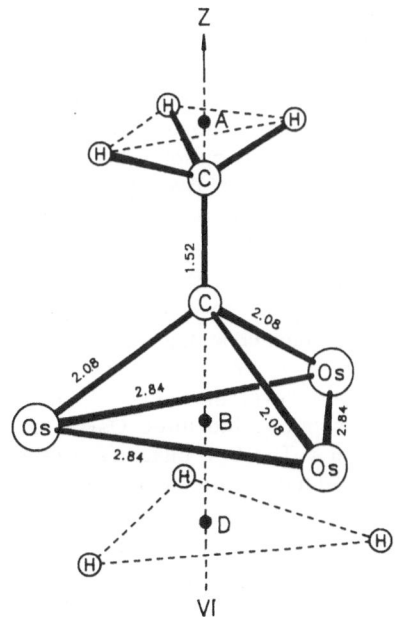

References on pp. 169/71

the Os–H–Os bond angle were calculated as 1.82 Å and 103°, respectively. The complex is isostructural to $(\mu-H)_3Ru_3(\mu_3,\eta^1-CCH_3)(CO)_9$ [2].

In addition to the IR investigations of the $(\mu-H)_3Os_3(CO)_9$ core vibrations (see Table 5, p. 133) [27], the vibrations of the μ_3,η^1-CCH_3 moiety in the C–H stretching region were analyzed by determination of the effect of pressure (0 to 96 kbar) on the vibrational frequencies; the observed pressure shifts were used to assign previously uncharacterized μ_3,η^1-CCH_3 ligand vibrations and to confirm other existing assignments. Additionally, to distinguish between C–H and C–C vibrations, the effect of deuteration on the hydrocarbon vibrations is determined. The experimental results for both, No. 18 and the deuterated analogue $(\mu-D)_3Os_3(\mu_3,\eta^1-CCD_3)(CO)_9$, are listed in the following table; ν_0 corresponds to the bands recorded in KBr at atmospheric pressure, the pressure shifts $d\nu/dp$ were observed in Nujol, the relationship $\Delta\nu/\nu_0$ reveals the percent change in ca. 60 kbar, and the designations (sym) and (asym) refer to the nature of $\nu(C–H)$ or $\delta(C–H)$ relative to the C_3 axis [68]; assignment of the overtones from [27]:

vibrations	$(\mu-H)_3Os_3(\mu_3,\eta^1-CCH_3)(CO)_9$			$(\mu-D)_3Os_3(\mu_3,\eta^1-CCD_3)(CO)_9$	
	ν_0 [cm^{-1}]	$d\nu/dp$ [cm^{-1}/kbar]	$\Delta\nu/\nu_0$	ν_0 [cm^{-1}]	$\nu(H)/\nu(D)$
$\nu_{as}(C–H)$	2918 m	+0.73	+2.05	2182 m	1.34
$\nu_s(C–H)$	2894 m	+0.97	+2.80		
$\delta_{as}(CH_3)$	1438 w			1028 w	1.40
$\delta_s(CH_3)$	1361 m			1002 w	1.36
$\nu(C–C)$	1147 m	+0.43	+3.30	1169 m	0.98
$\rho(CH_3)$	1037 w	+0.30	+2.02	848 s	1.22
overtones of $\delta(CH_3)$	2701 w	+0.46	+1.07		
	2833 w	+0.72	+1.48		
a_1 and two e modes	2863 w	+0.71	+1.53		

The vibration at 1037 cm^{-1} is a methyl rocking motion parallel to the Os_3 plane; for the perdeuterated complex the band is shifted to 848 cm^{-1}, whereas the $\nu(C–C)$ band at 1147 cm^{-1} is only slightly perturbed and shifted to 1169 cm^{-1} for μ_3,η^1-CCD_3. Generally, with increasing pressure all bands are shifted to higher frequencies; the pressure dependencies of the vibrational bands in terms of $d\nu/dp$ followed the order $\nu(C–H) > \nu(C–C) > \delta(C–H)$ or $\gamma(C–H)$; formulation of the shifts in terms of $\Delta\nu/\nu_0$ gave the order $\nu(C–C) > \nu(C–H)$ $\delta(C–H)$. For $\nu(C–C)$, the magnitude of its shift to higher energy with increasing pressure is proportional to the initial value at atmospheric pressure [27, 68].

$(\mu-H)_3Os_3(\mu_3,\eta^1-CCO_2H)(CO)_9$ (Table 5, No. 27) crystallizes in the monoclinic space group $P2_1/c-C_{2h}^5$ (No. 14) with a = 8.440(2), b = 12.869(3), c = 16.8011(4) Å, β = 100.47(2)°; Z = 4, D_c = 3.268 g/cm^3, R = 0.031. The complex crystallizes as the hydrogen-bonded dimer having C_i point symmetry; the structure of No. 27 and the packing unit of the dimer are shown in **Fig. 49** and **Fig. 50**, respectively. However, the hydride ligands were not observed crystallographically; they are believed to bridge the three Os–Os bond edges according to the typical distances around ca. 2.876 Å. The C–C distance is relatively short indicating some double bond character. Similar relations were found for the crystal structure of benzoic acid. Average bond distances of Os–CO and C≡O are 1.91 and 1.13 Å [49].

$(\mu-H)_3Os_3(\mu_3,\eta^1-CC_6H_5)(CO)_9$ (Table 5, No. 34). $(\mu-H)_3Os_3(\mu_3,\eta^1-CC_6D_5)(CO)_9$ was obtained in 90% yield by treatment of No. 5 with $AlCl_3/C_6D_6$ under the conditions similar to those of Preparation Method Ia, p. 122 [60]. No deuteration occurred upon heating of

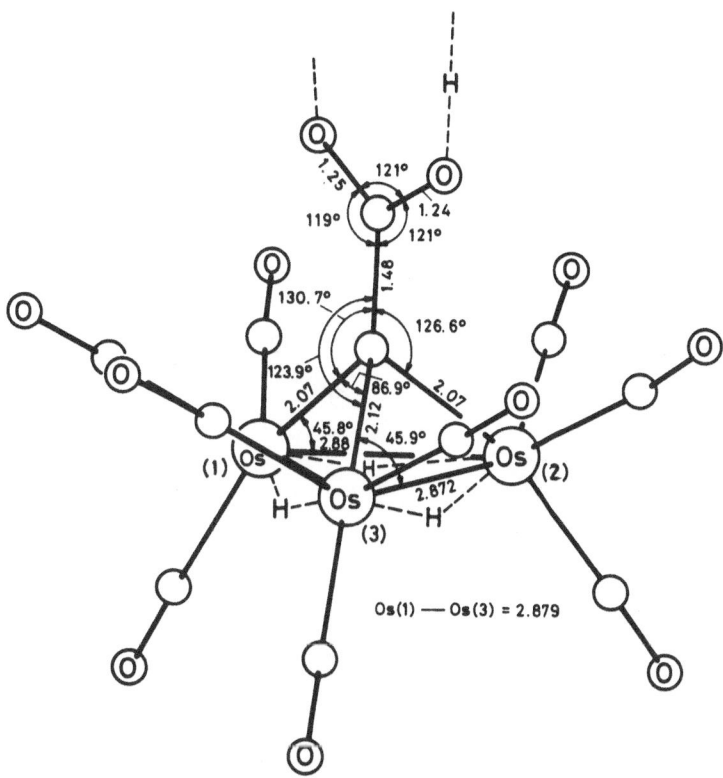

Fig. 49. Molecular structure of $(\mu\text{-H})_3\text{Os}_3(\mu_3,\eta^1\text{-CCO}_2\text{H})(\text{CO})_9$ (No. 27) with selected bond distances (in Å) and bond angles [49].

No. 34 in toluene-d_8 in a sealed tube to 140 °C for 40 min in the presence or absence of CO [56].

The title compound crystallizes in the monoclinic space group $\text{P2}_1/n - \text{C}_{2h}^5$ (No. 14) with a = 16.658(4), b = 15.636(4), c = 16.677(6) Å, β = 113.95°; Z = 8, D_c = 3.062 g/cm^3, R = 0.040. The structure is similar to that of Nos. 4 and 5; compare also Fig. 47, p. 152. The bridging hydrogens were not located crystallographically. The asymmetric unit contains two independent molecules. Selected bond distances and bond angles [81]:

atoms	distances (Å)	atoms	angles (°)
Os(1)-Os(2)	2.882(1)	Os(1)-Os(3)-Os(2)	60.18(2)
Os(1)-Os(3)	2.877(1)	Os(2)-Os(1)-μ_3,η^1-C	48.8(3)
Os(2)-Os(3)	2.870(1)	Os(1)-Os(3)-μ_3,η^1-C	47.5(4)
Os(1)-μ_3,η^1-C	2.125(10)	Os(3)-Os(2)-μ_3,η^1-C	47.0(4)
Os(2)-μ_3,η^1-C	2.145(10)	Os(1)-μ_3,η^1-C-Os(2)	84.9(5)
Os(3)-μ_3,η^1-C	2.110(10)	Os(1)-μ_3,η^1-C-Os(3)	85.7(5)
μ_3,η^1-C-C	1.49(2)	Os(2)-μ_3,η^1-C-C	127.6(9)
average Os-CO	1.92		
average C-O	1.13		

References on pp. 169/71

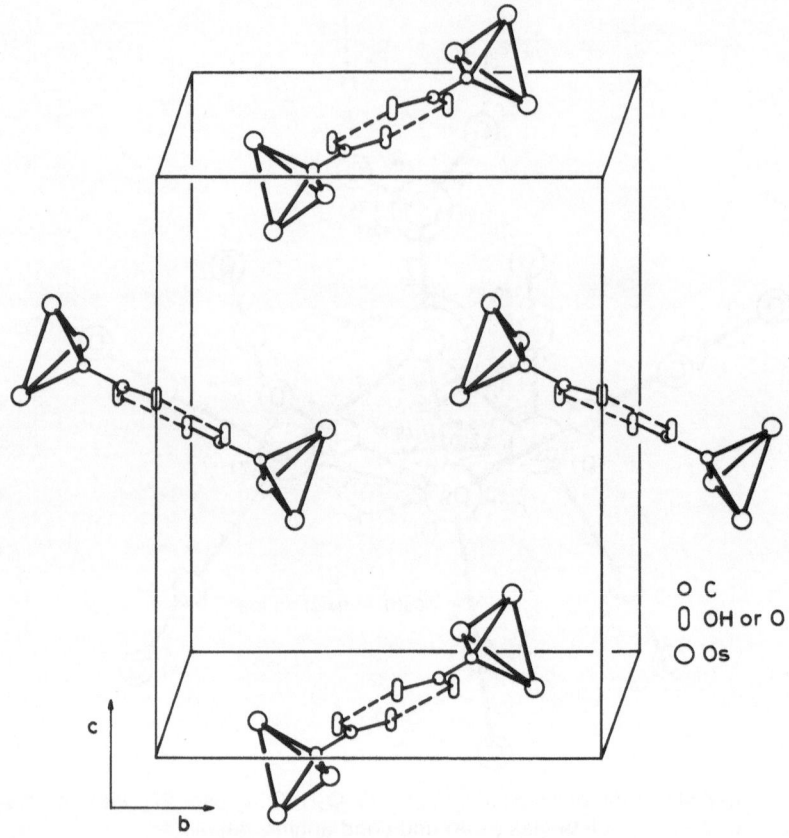

○ C
◗ OH or O
◯ Os

Fig. 50. Packing of the hydrogen-bonded dimers $\{(\mu-H)_3Os_3(\mu_3,\eta^1-CCO_2H)(CO)_9\}_2$ of No. 27 [49].

H/D exchange between the metal-hydride sites and the ortho C–D site of the phenyl ring in $(\mu-H)_3Os_3(\mu_3,\eta^1-CC_6D_5)(CO)_9$ was observed by ^1H NMR studies in tetrachloroethane-d_2 at 125 °C, and was indicated by a slow intensity decrease of the hydride signal and a growing of the peak for the ortho-C–H site. The reduced hydride signal shows a negative isotope effect on the chemical shift indicating the presence of $\mu-H_3$ ($\delta = -18.687$ ppm), $\mu-H_2D$ ($\delta = -18.678$ ppm), and $\mu-HD_2$ ($\delta = -18.671$ ppm) isotopomers; ($\mu-H_2D$) is the favored one. The isotope equilibrium constants were calculated from the relative ^1H NMR intensities of the two sites after H/D exchange is complete; $K[(\mu-H)_3Os_3(\mu_3,\eta^1-CC_6D_5)(CO)_9/(\mu-D)_3Os_3(\mu_3,\eta^1-CC_6H_3D_2)(CO)_9] = 1.28$, but $K[(\mu-D)_3Os_3(\mu_3,\eta^1-CC_6H_5)(CO)_9/(\mu-H)_3Os_3-(\mu_3,\eta^1-CC_6H_2D_3)(CO)_9] = 1.43$; the difference is assumed to be due to experimental uncertainty. For the H/D exchange a mechanism is discussed with intermediates bearing agostic Os–H–C and/or Os–D–C bonds (see Formula V, p. 150; $R = C_6H_5$). Kinetic studies conducted by following the appearance of the new C–H signal in the NMR spectrum revealed a first order rate constante of $k = 0.0285(10)$ min^{-1} at 125 °C; $\Delta G^{\ddagger}_{398\,K} = 29.6$ kcal/mol. The same reaction but at 110 °C in a sealed tube under CO (1 atm) led to a rate constant of $k = 0.0066(4)$ min^{-1} which gives $\Delta G^{\ddagger}_{383\,K} = 29.5(2)$ kcal/mol, indicating that the reaction is not affected by CO and therefore does not occur via reversible loss of CO to form coordinatively unsaturated species [56].

References on pp. 169/71

Pyrolysis at 220 °C in an evacuated sealed glass tube gave an isomeric mixture of $(\mu-H)_2Os_3(\mu_3,\eta^2-C_6H_3CH_3-4)(CO)_9$ and $(\mu-H)_2Os_3(\mu_3,\eta^2-C_6H_3CH_3-3)(CO)_9$ in a ratio of 4:1. ^1H NMR determinations of the deuterated products obtained from deuterated No. 34 revealed a deuterium scrambling in all sites of both isomers [60].

$(\mu-H)_3Os_3(\mu_3,\eta^1-CC_6F_5)(CO)_9$ (Table 5, No. 38) crystallizes in the monoclinic space group $P2_1/c-C_{2h}^5$ (No. 14) with a = 12.127(10), b = 9.386(5), c = 18.298(15) Å, β = 98.92(6)°; Z = 4, D_c = 3.243 g/cm³, R = 0.058. The compound consists of an equilateral triangle of three Os atoms with the averaged bond distances of 2.87 Å; see **Fig. 51**. The Os_3 plane is symmetri-

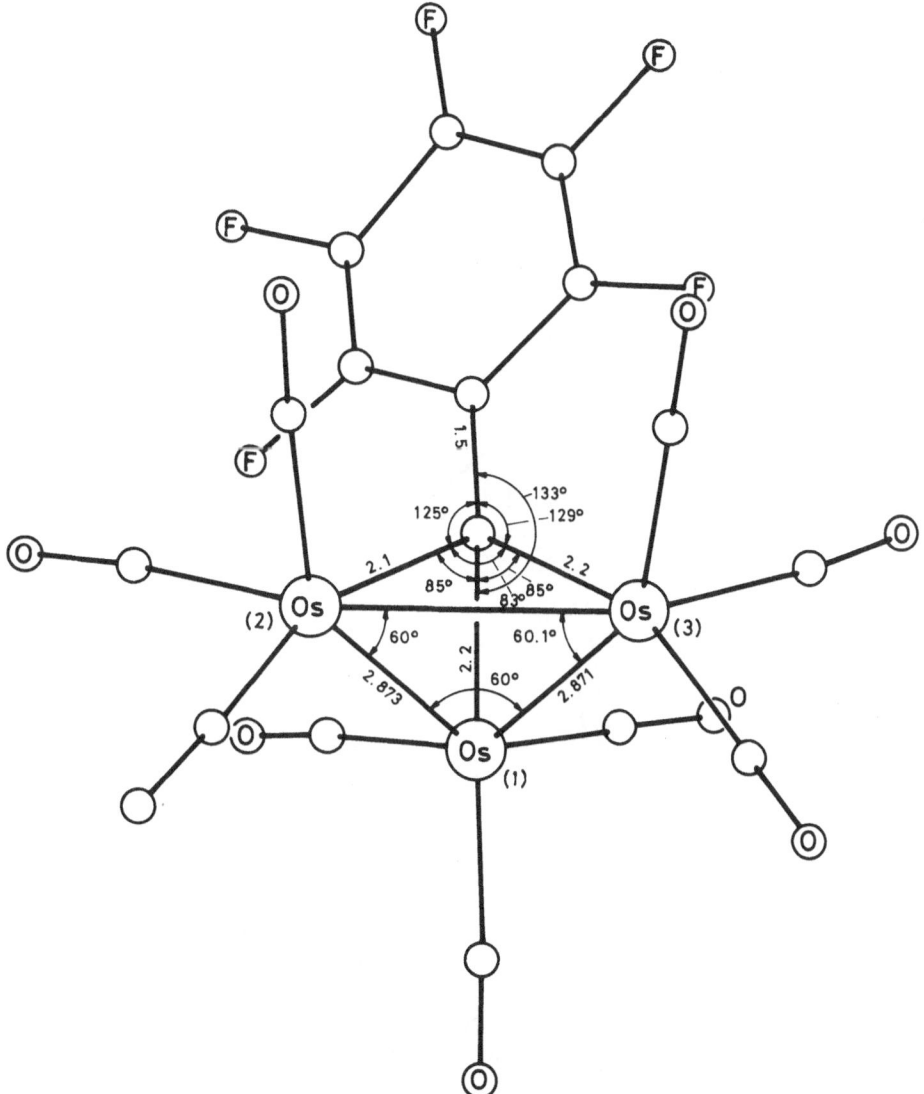

Fig. 51. Molecular structure of $(\mu-H)_3Os_3(\mu_3,\eta^1-CC_6F_5)(CO)_9$ (No. 38) with selected bond distances (in Å) and bond angles [76].

References on pp. 169/71

cally bridged by the μ_3-CC_6F_5 unit with the C_6F_5 ring being essentially perpendicular to the Os_3 plane. The hydride ligands were not observed crystallographically, but they are assumed to bridge the Os–Os bond edges [76].

[$Os_3(\mu_3,\eta^1$-CCO)(CO)$_9$]M$_2$ (M = N(P(C$_6$H$_5$)$_3$)$_2$, P(C$_6$H$_5$)$_4$; Table **5**, No. **39**). The P(C$_6$H$_5$)$_4$ salt crystallizes in the triclinic space group P$\bar{1}$ – C$_i^1$ (No. 2) with a = 12.762(2), b = 20.447(5), c = 10.779(2) Å, α = 97.68(2)°, β = 110.33(2)°, γ = 89.17(2)°; Z = 2, D$_c$ = 1.963 g/cm^3, R = 0.027 at –120 °C. Two cation-dianion combinations are found in the unit cell; the cation-anion contacts are small. The structure, shown in **Fig. 52**, revealed an approximate equilateral Os_3 core with relatively short Os–Os bond distances and an asymmetrically capping μ_3,η^1-CCO unit. The μ_3,η^1-CCO ligand is approximately linear and tilted over the Os(3) center by an angle of 26° between the least-squares line through C–C–O and the perpendicular to the plane of the Os_3 triangle. The μ_3,η^1-CCO moiety is formulated as a μ_3,η^1-C=C=O ketenylidene. Average bond distances of Os–CO and C≡O are 1.895 and 1.152 Å. Further considerations concerning the bonding mode of the μ_3,η^1-CCO ligand were performed by extended Hückel calculations with respect to the analogous Fe and Ru compounds [51].

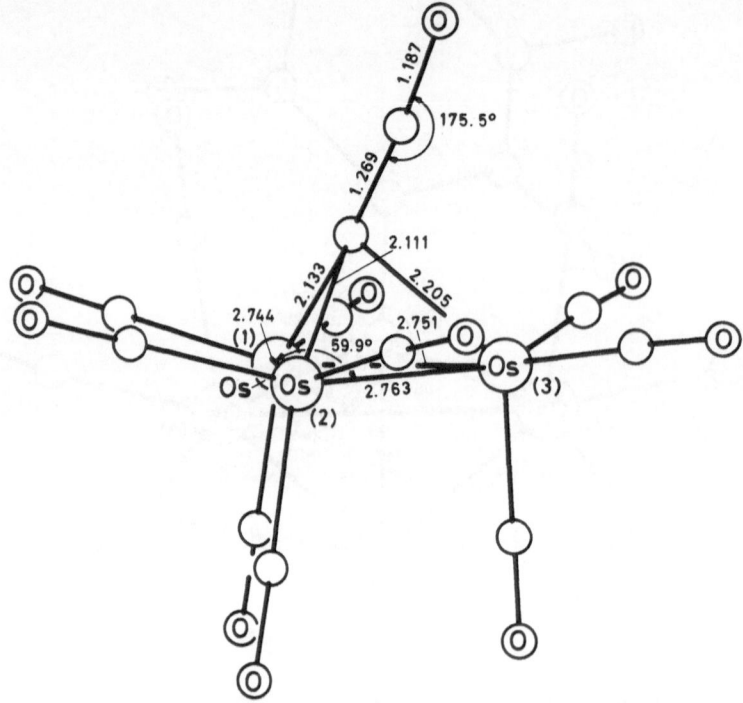

Fig. 52. Molecular structure of the dianion of [$Os_3(\mu_3,\eta^1$-CCO)(CO)$_9$][P(C$_6$H$_5$)$_4$]$_2$ (No. 39) with selected bond distances (in Å) and bond angles [51].

Successive protonation with HCl or CF$_3$SO$_3$H in CH$_2$Cl$_2$ yielded Nos. 40, 41 and 42; methylation with CF$_3$SO$_3$CH$_3$ resulted in No. 50, whereas similar methylation, but under a CO atmosphere, led to $Os_3(\mu_3,\eta^2$-C(CH$_3$)=COCH$_3$)(μ-CO)(CO)$_9$ [51].

The reaction with [Mn(CO)$_3$(NCCH$_3$)$_3$][PF$_6$] in acetone at 35 °C yielded [$Os_3(\mu_4$-C)(μ-CO)(CO)$_9$Mn(CO)$_3$][N(P(C$_6$H$_5$)$_3$)$_2$] (Section 3.1.6.4.3) [71]. Reaction with Ni(1,5-C$_8$H$_{12}$-c)$_2$ in refluxing THF under ca. 0.55 atm of CO gave $[Os_3(\mu_6-C)(CO)_9Ni_4(\mu-CO)_2(CO)_4][N(P(C_6H_5)_3)_2]_2$ (Section 3.1.6.4.3) [82].

$(\mu-H)_2Os_3(\mu_3,\eta^1-CCO)(CO)_9$ (Table 5, No. 41) crystallizes in the monoclinic space group $P2_1/m-C_{2h}^2$ (No. 11) with a = 9.2721(15), b = 14.2272(21), c = 12.6005(19) Å, β = 92.423(13)°; Z = 4, D_c = 3.46 g/cm³, R = 0.065. The crystal contains two sites for molecules, each lying about a crystallographic mirror plane. The crystallographic asymmetric unit thus consists of two independent half-molecules, Molecule **A** and **B**. Molecule **A** is ordered having what appears to be approximate C_{3v} symmetry at the $Os(CO)_3$ group centered on Os(1) while that on Os(2) has only approximate C_s symmetry. The symmetrically capping nearly linear μ_3,η^1-CCO ligand is in an upright position and is best described as a $\mu_3,\eta^1-C^--C\equiv O^+$ moiety. The bridging hydrides were located crystallographically and span the equivalent Os(1)-Os(2) and Os(1)-Os(2') bonds; the structure of an ordered molecule, Molecule **A**, is shown in **Fig. 53**. Average bond distances of Os-CO and C≡O are 1.909 and 1.152 Å, respectively [25, 67].

For the disordered Molecule **B** the hydrides could not be found crystallographically due to the anomalous short Os(1)-Os(2) and Os(1)-Os(2') bond distances (see the following table). These anomalous Os-Os bonds are explained as a result of the disorder in which

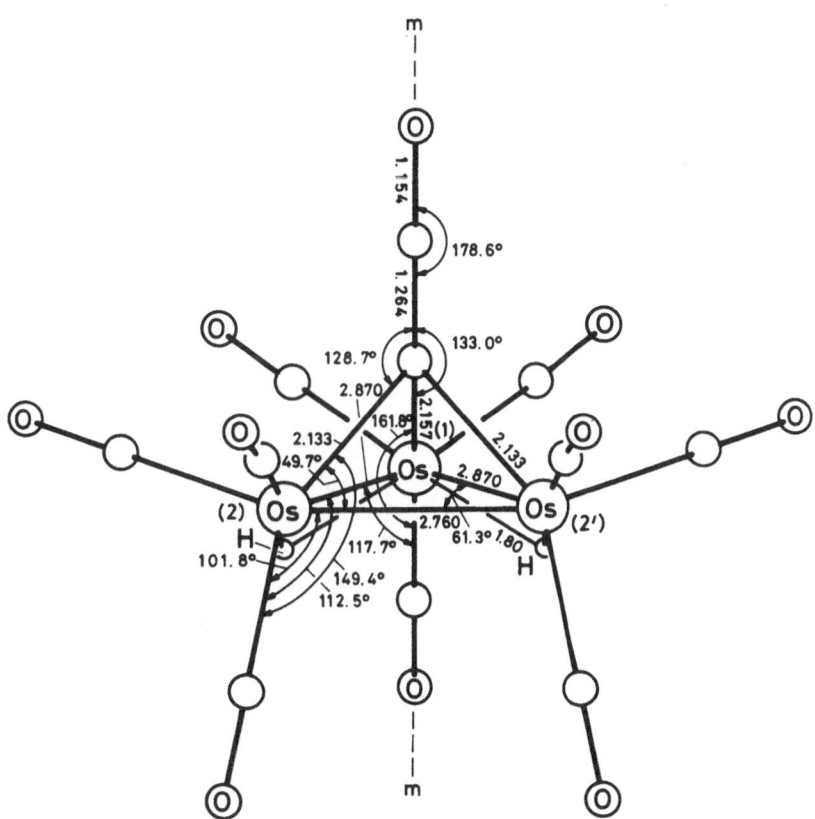

Fig. 53. Molecular structure of Molecule **A** of $(\mu-H)_2Os_3(\mu_3,\eta^1-CCO)(CO)_9$ (No. 41) with selected bond distances (in Å) and bond angles [67].

the individual molecules are aligned with their microscopically precise molecular mirror planes deviating from the macroscopically precise crystallographic mirror plane; this is probably affected by rotations about the Os_3 centroid$-$CCO axis by a pseudo-C_3^1 or pseudo-C_3^2 operation; see Scheme 1 (μ_3,η^1-CCO and CO omitted). Further structural differences for Molecule **B**, and the consistence with the model are provided by additional observations which are discussed in detail [67].

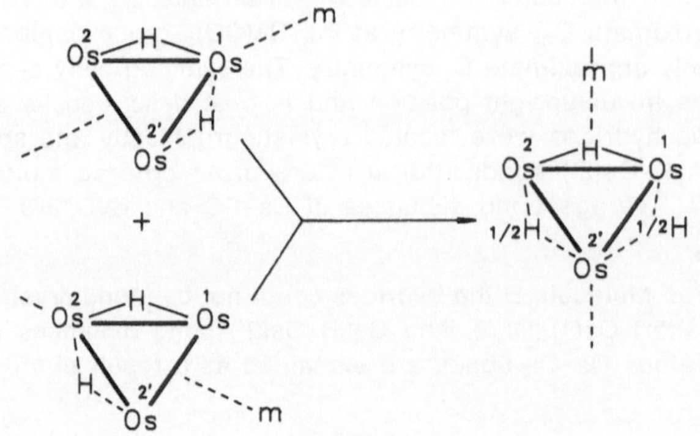

Scheme 1

For comparison with the ordered molecules, some selected bond distances and bond angles of the disordered Molecule **B** are given in the following table [67]:

atoms	distance (Å)	atoms	angle (°)
Os(1)-Os(2)	2.813(1)	Os(1)-Os(2)-Os(2')	59.26(3)
Os(1)-Os(2')	2.813(1)	Os(2)-Os(1)-Os(2')	61.48(3)
Os(2)-Os(2')	2.875(1)	Os(2)-Os(1)-C(1)	48.5(5)
Os(1)-C(1)	2.170(25)	Os(1)-Os(2)-C(1)	49.8(5)
Os(2)-C(1)	2.130(18)	Os(2')-Os(2)-C(1)	47.5(4)
Os(2')-C(1)	2.130(18)	C(11)-Os(1)-C(1)	150.1(13)
C(1)-C(2)	1.285(38)	C(22)-Os(2)-C(1)	160.3(10)
C(2)-O(1)	1.194(35)	Os(1)-C(1)-C(2)	130.4(17)
		Os(2)-C(1)-C(2)	130.1(17)
		Os(2)-Os(1)-C(11)	107.5(12)
		Os(1)-Os(2)-C(22)	113.5(8)
		Os(2')-Os(2)-C(22)	117.6(8)
		C(1)-C(2)-O(1)	176.6(25)

Slow conversion to $(\mu\text{-H})_3Os_3(\mu_3,\eta^1\text{-CCO}_2H)(CO)_9$ (No. 27) occurred in moisture [25]; a similar reaction is observed by treatment with gaseous HCl in undried CH_2Cl_2 (CD_2Cl_2) at room temperature; under these conditions small amounts of $(\mu\text{-H})_3Os_3(\mu_3,\eta^1\text{-CCl})(CO)_9$ (No. 4) and $(\mu\text{-H})_3Os_3(\mu_3,\eta^1\text{-CCH}_2Cl)(CO)_9$ (No. 22) were obtained as the by-products [49]. Treatment with methanol at a higher temperature gave $(\mu\text{-H})_3Os_3(\mu_3,\eta^1\text{-CCO}_2CH_3)(CO)_9$ (No. 28) [20].

Protonation with HCl or with $HBF_4 \cdot O(C_2H_5)_2$ in CH_2Cl_2 between -80 and $+25\,°C$ gave cationic No. 42; see also Preparation Method VIII, p. 124 [20, 25, 51, 78].

Reaction with H_2 (1 atm) in refluxing heptane or toluene yielded $(\mu\text{-H})_3Os_3(\mu_3,\eta^1\text{-CH})$-$(CO)_9$ (No. 1) via $(\mu\text{-H})Os_3(CH)(CO)_9$ as a proposed intermediate [25].

Treatment with BCl_3 in CH_2Cl_2 led to $(\mu\text{-H})Os_3(\mu_2,\eta^1\text{-COCH}_2Cl)(CO)_{10}$ (Section 3.1.6.2.2) [72].

Reaction with $c\text{-}C_2H_4S$ in refluxing n-octane gave $(\mu\text{-H})Os_3(\mu_3,\eta^2\text{-CH=S})(CO)_9$ ("Organo-osmium Compounds" B 5, Section 3.1.6.1.5, in preparation) [46].

Heterometallic clusters of the types $(\mu\text{-H})_2Os_3(\mu_4\text{-C})(CO)_9Pt(CO)P(C_6H_{11}\text{-c})_3$ and $(\mu\text{-H})_2Os_3(\mu_5\text{-C})(CO)_9Pt_2(\mu\text{-CO})(P(C_6H_{11}\text{-c})_3)_2$ (Section 3.1.6.4.2) were obtained from the reaction with $(C_2H_4)_2PtP(C_6H_{11}\text{-c})_3$ in toluene at 25 °C [34].

$(\mu\text{-H})Os_3(\mu_3,\eta^1\text{-CH})(CO)_{10}$ (Table 5, No. **43**) crystallizes in the triclinic space group $P\bar{1} - C_i^1$ (No. 2) with $a = 9.404(2)$, $b = 9.300(2)$, $c = 11.002(3)$ Å, $\alpha = 94.76(2)°$, $\beta = 94.44(2)°$, $\gamma = 118.49(2)°$; $Z = 2$, $D_c = 3.44$ g/cm³, $R = 0.035$. The methylidyne ligand symmetrically bridges the Os(2)-Os(3) edge and shows strong bonding interactions with the Os(1) center. The C-H bond is bent 24° out of the Os(2)-$(\mu_3,\eta^1\text{-C})$-Os(3) plane on the same side as Os(1); however, the Os(1)···H' distance is 2.36(14) Å. The semi-triply bridging coordination of the CH ligand together with a close H'···C' contact causes significant distortions in the intraligand angles of the Os(CO)$_4$ group; see **Fig. 54**. The interactions between Os(1) and the CH moiety can be regarded as a two-electron donation from the Os(CO)$_4$ center to the sp²-hybridized CH carbon. The angle between the Os$_3$ and the Os(1)-$(\mu_3,\eta^1\text{-C})$-Os(2) plane is 69.7°. Both the bridging hydride as well as the ligand hydrogen H' were located crystallographically. Average bond distances of Os-CO and C≡O are 1.94 and 1.13 Å [24].

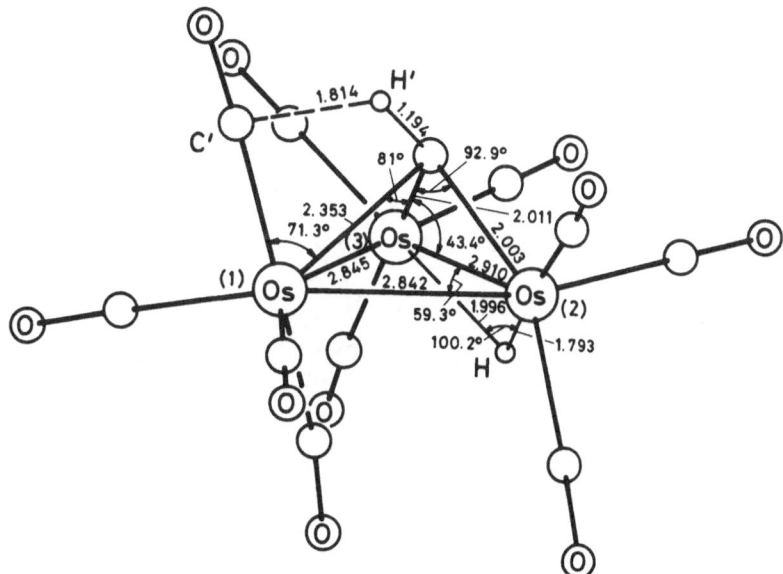

Fig. 54. Molecular structure of $(\mu\text{-H})Os_3(\mu_3,\eta^1\text{-CH})(CO)_{10}$ (No. 43) with selected interatomic distances (in Å) and bond angles [24].

Extended Hückel calculations suggested that the bridging methylidyne moiety should be treated only as a μ_2 bridge with weak Os(1)···H' bonding interactions, even though the Os···μ-C distance is in the bonding range [77].

References on pp. 169/71

Pyrolysis at 85 °C/0.1 Torr or heating in refluxing hexane yielded No. 41 [25]; $\Delta G^+ = 26$ kcal/mol at 60 °C [29].

The reaction with 4-methylpyridine at −60 °C gave $(\mu-H)Os_3(\mu_2,\eta^1-CHNC_5H_4CH_3-4)$-$(CO)_{10}$ (Section 3.1.6.2.1) [24].

Hydrogenation with $LiBH(C_2H_5)_3$ at −40 °C resulted in $[(\mu-H)Os_3(\mu_2,\eta^1-CH_2)(CO)_{10}]^-$ (Section 3.1.6.2.1), which upon protonation, led to $(\mu-H)Os_3(\eta^1-HCH_2)(CO)_{10}$ ("Organo-osmium Compounds" B 5, Section 3.1.2.2, in preparation) [24].

Treatment with $RHCN_2$ (R = H, $Si(CH_3)_3$) gave compounds of the type $(\mu-H)Os_3(\mu_2,\eta^2-CH=CHR)(CO)_{10}$ [24].

Reaction with $(C_2H_4)_2PtP(C_6H_{11}-c)_3$ in toluene at 80 °C yielded a mixture of $(\mu-H)_2Os_3(\mu_4-C)(CO)_9Pt(CO)P(C_6H_{11}-c)_3$ and $(\mu-H)_2Os_3(\mu_5-C)(CO)_9Pt_2(\mu-CO)(P(C_6H_{11}-c)_3)_2$ (both Section 3.1.6.4.2); No. 41 is believed to be an intermediate; see also p. 163 [34].

$(\mu-H)Os_3(\mu_3,\eta^1-CCH_2CH(CH_3)_2)(CO)_{10}$ (Table **5**, No. **46**). Partially deuterated $(\mu-D)Os_3$-$(\mu_3,\eta^1-CCHDCH(CH_3)_2)(CO)_{10}$ was obtained from $(\mu-D)_2Os_3(CO)_{10}$ and 3,3-dimethyl cyclopropene [28, 63].

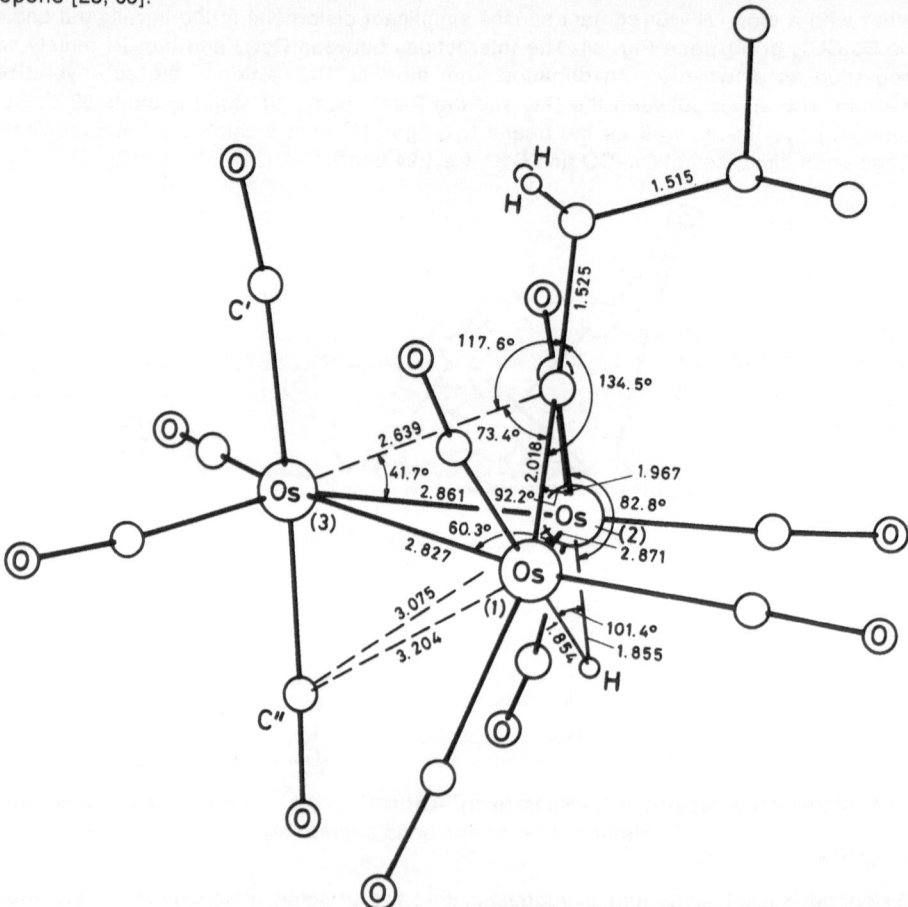

Fig. 55. Molecular structure of $(\mu-H)Os_3(\mu_3,\eta^1-CCH_2CH(CH_3)_2)(CO)_{10}$ (No. 46) with selected interatomic distances (in Å) and bond angles [28, 63].

References on pp. 169/71

The title compound crystallizes in the triclinic space group $P\bar{1} - C_i^1$ (No. 2) with a = 9.226(3), b = 9.333(5), c = 13.136(7) Å, α = 89.70(4)°, β = 74.18(4)°, γ = 78.00(3)°; Z = 2, D_c = 2.48 g/cm³, R = 0.071. The structure is shown in **Fig. 55**. The μ-alkylidyne ligand bridges the Os(1)-Os(2) edge almost symmetrically; the carbyne carbon interacts with the Os(3) center, although to a lesser extent than observed for No. 43. These interactions cause a distorted C_{2v} symmetry at Os(CO)$_4$ with an angle of 82.1° between the Os$_3$ and the Os(1)-(μ_3,η^1-C)-Os(2) planes. Thus, the distortion reduces the steric interactions between the μ_3,η^1-CCH$_2$CH(CH$_3$)$_2$ ligand and the axial carbonyl CO'. Therefore, symmetry in bonding is achieved by a shift of the axial CO'' towards the opposite face of the Os$_3$ core. The μ-hydride ligand is not observed crystallographically; its position is calculated and it is assigned to bridge the Os(1)-Os(2) edge as the μ-alkylidyne group due to the counterbalancing effects of metal-metal bond lengthening by μ-hydrides and shortening by μ-C moieties, hence the Os atoms describe an approximately equilateral triangle. Average bond distances of Os-CO and C≡O are 1.925 and 1.135 Å [28, 63].

Thermolysis in octane/hexane (1:1) at 100 °C gave a mixture of (μ-H)Os$_3$(μ_2,η^2-CH=CHCH(CH$_3$)$_2$)(CO)$_{10}$ and (μ-H)$_2$Os$_3$(μ_3,η^3-C=CHCH(CH$_3$)$_2$)(CO)$_9$. Monitoring of the reaction by ^{1}H NMR revealed the formation of the latter complex from the former one. In order to examine the nature of the hydrogen shift in this carbyne to μ-vinylidene rearrangement, the thermolysis of the deuterium labeled complex (μ-D)Os$_3$(μ_3,η^1-CCHDCH(CH$_3$)$_2$)(CO)$_{10}$ was determined. The mechanism, discussed in detail, indicates an initial shift of a proton from the methylene group to the electron-deficient μ-C carbon without direct assistance of the metal core [28, 63].

In contrast to the thermolysis, photolysis in hexane at 20 °C for 15 min under UV light yielded (μ-H)$_2$Os$_3$(μ_3,η^3-C=CHCH(CH$_3$)$_2$)(CO)$_9$ directly [63].

(μ-H)Os$_3$(μ_3,η^1-CC$_6$H$_5$)(CO)$_{10}$ (Table 5, No. 47). (μ-D)Os$_3$(μ_3,η^1-CC$_6$H$_5$)(CO)$_{10}$ was prepared from (μ-D)Os$_3$(μ_2,η^1-COCH$_3$)(CO)$_{10}$ as described before; see p. 145 [38].

The title compound crystallizes in the monoclinic space group $P2_1/m - C_{2h}^2$ (No. 11) with a = 8.1205(11), b = 14.2658(26), c = 8.9165(18) Å, β = 98.701(14)°; Z = 2, D_c = 3.06 g/cm³, R = 0.022. The dibridged Os(2)-Os(2') bond is slightly longer than the other two metal-metal bonds; the structure is shown in **Fig. 56**. The capping benzylidyne ligand lies over the Os$_3$ plane rather than external to it. The interatomic Os(1)···(μ_3,η^1-C) distance of 2.586 Å is indicative of some direct interaction. Thus, this bonding mode can be characterized as "semi-triply bridging". The bridging Os(2)-(μ_3,η^1-C)-Os(2') and Os(2)-H-Os(2') systems are close to being coplanar, whereas the phenyl group is bent away from the Os$_3$ plane as proved by the dihedral angle of 163.25° between the Os(2)-(μ_3,η^1-C)-Os(2') plane and the phenyl ring. The molecular mirror plane through Os(1)···(μ_3,η^1-C)-C'/C'' prevents any rotation about the (μ_3,η^1-C)-C' axis. One of the axial carbonyl ligands at Os(1) is repulsed by the C$_6$H$_5$ group in such a way that the entire Os(CO)$_4$ group is rotated about its Os(1) center. The phenyl/Os(CO)$_4$ repulsion also causes the equatorial ligands at Os(1) to fall below the Os$_3$ plane. The angle between the Os$_3$ and the Os(1)···(μ_3,η^1-C)-Os(2) plane is 78.17°. Average bond distances of Os-CO and C≡O are 1.929 and 1.136 Å [38].

Pyrolysis under vacuum at 170 °C/0.1 Torr gave the ortho-metalated complex (μ-H)Os$_3$(μ_3,η^3-CHC$_6$H$_4$)(CO)$_9$ in 83% yield. Similar pyrolysis of (μ-D)Os$_3$(μ_3,η^1-CC$_6$H$_5$)-(CO)$_{10}$ led to a 1:1 mixture of (μ-D)Os$_3$(μ_3,η^3-CHC$_6$H$_4$)(CO)$_9$ and (μ-H)Os$_3$(μ_3,η^3-CDC$_6$H$_4$)-(CO)$_9$; a mechanism for the rearrangement is discussed. Thermolysis in xylene at 135 °C resulted in ca. 30% of (μ-H)Os$_3$(μ_3,η^3-CHC$_6$H$_4$)(CO)$_9$ in addition to various other products [38].

References on pp. 169/71

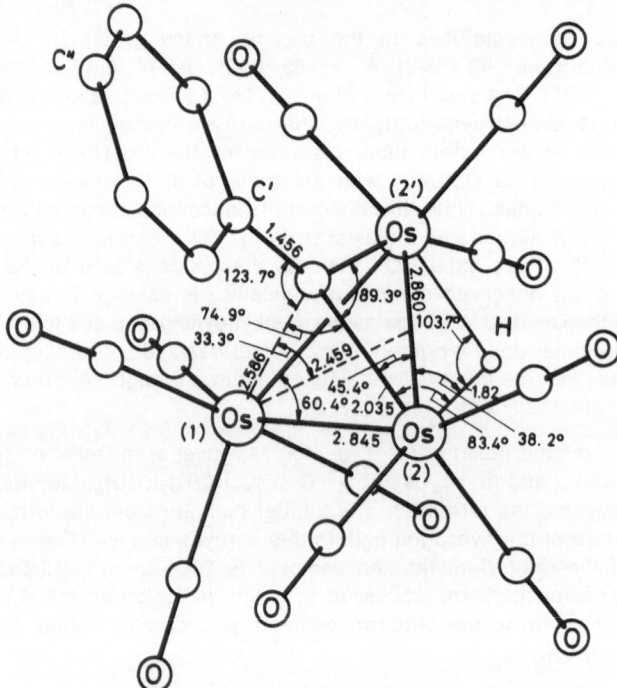

Fig. 56. Molecular structure of $(\mu\text{-H})Os_3(\mu_3,\eta^1\text{-}CC_6H_5)(CO)_{10}$ (No. 47) with selected interatomic distances (in Å) and bond angles [38].

Hydrogenation in decane at 150 °C for 20 min (or at 110 °C for 90 min) gave $(\mu\text{-H})_3Os_3(\mu_3,\eta^1\text{-}CC_6H_5)(CO)_9$ (No. 34) with the loss of one carbonyl group; see also Preparation Method III, p. 122 [38].

Reaction with C_6H_5Li in ether at 0 °C followed by treatment with $CF_3SO_3CH_3$ at 25 °C yielded $Os_3(\mu_3,\eta^1\text{-}CC_6H_5)_2(CO)_9$ (No. 51); the reaction probably proceeded via the intermediate $Os_3(\eta^1\text{=}C(OCH_3)C_6H_5)(\mu_3,\eta^1\text{-}CC_6H_5)(CO)_8$, which upon prolonged exposure to $CF_3SO_3CH_3$, resulted either in No. 51 or reformed No. 47 [52].

Sequential treatment with two equivalents of sodium benzophenone ketyl in THF at 25 °C followed by protonation with $HBF_4 \cdot O(C_2H_5)_2$ yielded $(\mu\text{-H})Os_3(\mu_3,\eta^2\text{-}C\equiv CC_6H_5)(CO)_9$; the reaction is believed to proceed via $[(\mu\text{-H})Os_3(C_2(O)C_6H_5)(CO)_9]^{2-}$ which is then doubly protonated at the oxygen followed by loss of H_2O [44].

Treatment with $C_6H_5C\equiv CC_6H_5$ (ca. 1:2) in refluxing n-heptane for 4.5 h yielded a separable mixture of $(\mu\text{-H})Os_3(\mu_3,\eta^3\text{-}(CC_6H_5)_3)(CO)_9$ (Formula VII), two isomers of

VII   VIII

Gmelin Handbook
Os-Org. Comp. B6

Os$_3$(μ_2,η^3-C$_3$(C$_6$H$_5$)$_3$)(μ_2,η^2-CC$_6$H$_5$CHC$_6$H$_5$)(CO)$_8$ (one of the isomers is shown in Formula VIII), and two isomers of Os$_3$(H(CC$_6$H$_5$)$_5$)(CO)$_9$ of unknown structure [55].

(μ-H)$_2$Os$_3$(μ_3,η^1-C-CH=N(C$_2$H$_5$)$_2$)(CO)$_9$ (Table 5, No. **49**) crystallizes in the monoclinic space group P2$_1$/n $-$ C$^5_{2h}$ (No. 14) with a $=$ 10.233(2), b $=$ 14.834(4), c $=$ 14.538(2) Å, β $=$ 99.88(2)°; Z $=$ 4, D$_c$ $=$ 2.82 g/cm^3, R $=$ 0.036. The short μ_3,η^1-C-CH and CH-N distances (see **Fig. 57**) as well as the non-equivalence of the C$_2$H$_5$ groups exhibiting hindered rotation about the CH-N bond (see ^{1}H NMR spectrum) indicate partial double-bond characters for both bonds; thus, the ligand can be formulated as a dimethyl amino-substituted vinylidene. The alkylidyne hydrogen was not located in the structural analysis but its location was indicated by the characteristic low-field resonance in the ^{1}H NMR spectrum; the position was calculated assuming idealized trigonal planar geometry. The geometry at the nitrogen is planar. Both hydride ligands were located crystallographically. Average bond distances of Os-CO and C≡O are 1.86 and 1.16 Å [53, 79].

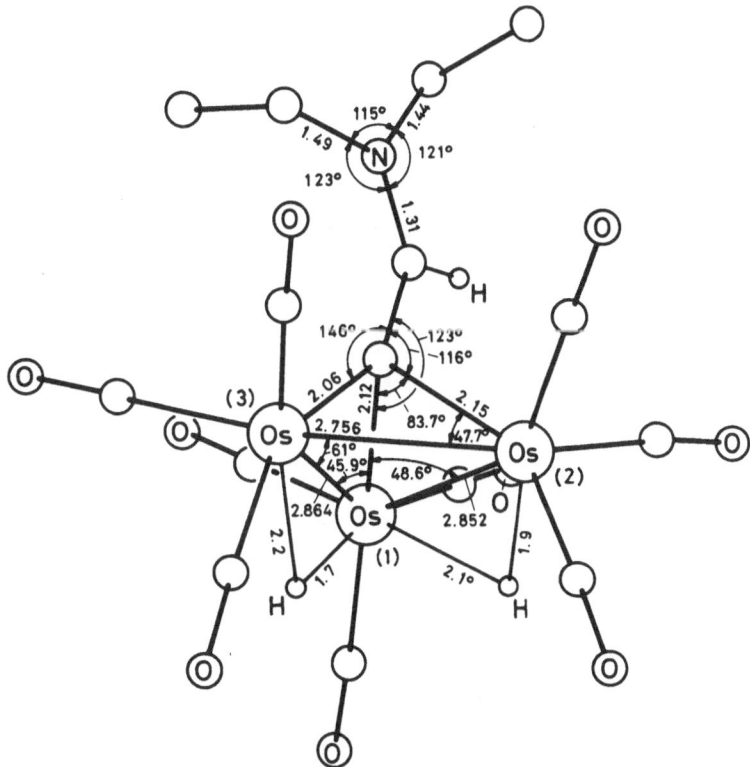

Fig. 57. Molecular structure of (μ-H)$_2$Os$_3$(μ_3,η^1-C-CH=N(C$_2$H$_5$)$_2$)(CO)$_9$ (No. 49) with selected bond distances (in Å) and bond angles [53, 79].

Os$_3$(μ_3,η^1-CC$_6$H$_5$)(μ_3,η^1-COCH$_3$)(CO)$_9$ (Table 5, No. **53**) crystallizes in the monoclinic space group P2$_1$/c $-$ C$^5_{2h}$ (No. 14) with a $=$ 10.6679(19), b $=$ 14.2218(12), c $=$ 14.2097 Å, β $=$ 99.713(13)°; Z $=$ 4, D$_c$ $=$ 2.98 g/cm^3, R $=$ 0.029. The equilateral Os$_3$ core is capped on one face by a μ_3-CC$_6$H$_5$ ligand and on the opposite face by the μ_3-COCH$_3$ ligand; see **Fig. 58**. The (μ_3-C)···(μ_3-C) distance of 2.688 Å indicates no direct interaction between the two alkylidyne groups. Apparently the μ_3,η^1-CC$_6$H$_5$ ligand is less strongly bounded to the Os$_3$ core

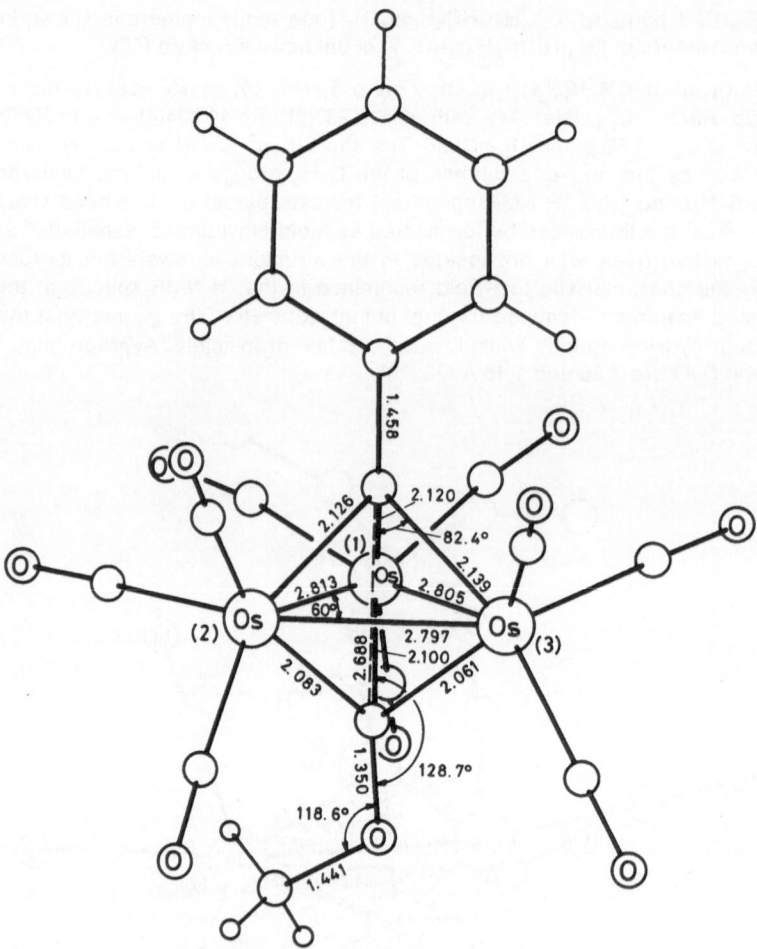

Fig. 58. Molecular structure of $Os_3(\mu_3,\eta^1\text{-}CC_6H_5)(\mu_3,\eta^1\text{-}COCH_3)(CO)_9$ (No. 53) with selected interatomic distances (in Å) and bond angles [45].

than the $\mu_3,\eta^1\text{-}COCH_3$ unit, indicated by the longer Os–μ_3–C bond distances. Average bond distances of Os–CO and C≡O are 1.922 and 1.132 Å [45].

Protonation with CF_3SO_3H in $CDCl_3$ gave cationic No. 54, based on NMR spectra [45].

Hydrogenation in toluene at 110 °C led to a mixture of $(\mu\text{-}H)_3Os_3(\mu_3,\eta^1\text{-}COCH_3)(CO)_9$ (No. 6; see also Preparation Method III, p. 122) and $(\mu\text{-}H)_2Os_3(\mu_3,\eta^2\text{-}C(C_6H_5)=COCH_3)(CO)_9$ in 12 and 55%, respectively, under 1 atm of H_2, or in 54 and 23% yield, respectively, under 4 atm of H_2 [44].

Reduction of the title compound with 2 equivalents of sodium benzophenone ketyl in THF at ca. 25 °C, followed by protonation with 5 equivalents of $HBF_4 \cdot O(C_2H_5)_2$ gave $(\mu\text{-}H)Os_3(\mu_3,\eta^2\text{-}C\equiv CC_6H_5)(CO)_9$ in 95% yield; ^{13}C labeled experiments indicate a possible intermediate alkylidyne-carbonyl coupling [44].

References on pp. 169/71

Reaction with C_6H_5Li in ether at 0 °C followed by treatment of $CF_3SO_3CH_3$ led to $Os_3(\eta^1\text{=}C(OCH_3)C_6H_5)(\mu_3,\eta^1\text{-}CC_6H_5)(\mu_3,\eta^1\text{-}COCH_3)(CO)_8$ ("Organoosmium Compounds" B 5, Section 3.1.5.2, in preparation) [64].

Substitution of carbonyl ligands occurred upon treatment with an excess of $P(C_6H_5)_3$ in refluxing toluene [44] or in n-decane at 120 °C to give a mixture of Nos. 55 and 56 [45].

References:

[1] Deeming, A. J.; Underhill, M. (J. Chem. Soc. Chem. Commun. **1973** 277/8).

[2] Yesinowski, J. P.; Bailey, D. (J. Organomet. Chem. **65** [1974] C 27/C 29).

[3] Bryan, E. G.; Jackson, W. G.; Johnson, B. F. G.; Kelland, J. L.; Lewis, J.; Schorpp, K. T. (J. Organomet. Chem. **108** [1976] 385/91).

[4] Forster, A.; Johnson, B. F. G.; Lewis, J.; Matheson, T. W. (J. Organomet. Chem. **104** [1976] 225/9).

[5] Azam, A. K.; Deeming, A. J. (J. Chem. Soc. Chem. Commun. **1977** 472/3).

[6] Calvert, R. B.; Shapley, J. R. (J. Am. Chem. Soc. **99** [1977] 5225/6).

[7] Azam, K. A.; Deeming, A. J. (J. Mol. Catal. **3** [1977/78] 207/12).

[8] Calvert, R. B.; Shapley, J. R. (J. Am. Chem. Soc. **100** [1978] 6544).

[9] Yin, C. C.; Deeming, A. J. (J. Organomet. Chem. **144** [1978] 351/5).

[10] Keister, J. B. (J. Chem. Soc. Chem. Commun. **1979** 214/5).

[11] Brown, S. C.; Evans, J. (J. Organomet. Chem. **194** [1980] C 53/C 56).

[12] Keister, J. B.; Horling, T. L. (Inorg. Chem. **19** [1980] 2304/7).

[13] Azam, K. A.; Deeming, A. J.; Rothwell, I. P. (J. Chem. Soc. Dalton Trans. **1981** 91/8).

[14] Brown, S. C.; Evans, J. (J. Mol. Catal. **11** [1981] 143/9).

[15] Johnson, B. F. G.; Lewis, J.; Raithby, P. R.; Sankey, S. W. (J. Organomet. Chem. **231** [1982] C 65/C 67).

[16] Shapley, J. R.; Sievert, A. C.; Churchill, M. R.; Wasserman, H. J. (J. Am. Chem. Soc. **103** [1981] 6975/7).

[17] Steinmetz, G. R.; Geoffroy, G. L. (J. Am. Chem. Soc. **103** [1981] 1278/9).

[18] Arce, A. J.; Deeming, A. J. (J. Chem. Soc. Chem. Commun. **1982** 364/5).

[19] Brown, S. C.; Evans, J. (J. Chem. Soc. Dalton Trans. **1982** 1049/54).

[20] Sievert, A. C.; Strickland, D. S.; Shapley, J. R.; Steinmetz, G. R.; Geoffroy, G. L. (Organometallics **1** [1982] 214/5).

[21] Churchill, M. R.; Wasserman, H. J. (J. Organomet. Chem. **248** [1983] 365/73).

[22] Keister, J. B.; Payne, M. W.; Muscatella, M. J. (Organometallics **2** [1983] 219/25).

[23] Morrison, E. D.; Steinmetz, G. R.; Geoffroy, G. L.; Fultz, W. C.; Rheingold, A. L. (J. Am. Chem. Soc. **105** [1983] 4104/5).

[24] Shapley, J. R.; Cree-Uchiyama, M. E.; St. George, G. M.; Churchill, M. R.; Bueno, C. (J. Am. Chem. Soc. **105** [1983] 140/2).

[25] Shapley, J. R.; Strickland, D. S.; St. George, G. M.; Churchill, M. R.; Bueno, C. (Organometallics **2** [1983] 185/7).

[26] Boyar, E.; Deeming, A. J. (J. Organomet. Chem. **276** [1984] C 45/C 48).

[27] Evans, J.; McNulty, G. S. (J. Chem. Soc. Dalton Trans. **1984** 79/85).

[28] Green, M.; Orpen, A. G.; Schaverien, C. J. (J. Chem. Soc. Chem. Commun. **1984** 37/9).

[29] Holmgren, J. S.; Shapley, J. R. (Organometallics **3** [1984] 1322/3).

[30] Orpen, A. G.; Koetzle, T. F. (Acta Crystallogr. B **40** [1984] 606/12).

[31] Shore, S. G.; Jan, D.-Y.; Hsu, W.-L.; Hsu, L.-Y.; Kennedy, S.; Huffman, J. C.; Wang, T.-C. L.; Marshall, A. G. (J. Chem. Soc. Chem. Commun. **1984** 392/4).

[32] Steinmetz, G. R.; Morrison, E. D.; Geoffroy, G. L. (J. Am. Chem. Soc. **106** [1984] 2559/64).

[33] Bavaro, L. M.; Keister, J. B. (J. Organomet. Chem. **287** [1985] 357/67).

[34] Farrugia, L. J.; Miles, A. D.; Stone, F. G. A. (J. Chem. Soc. Dalton Trans. **1985** 2437/47).

[35] Holmgren, J. S.; Shapley, J. R.; Belmonte, P. A. (J. Organomet. Chem. **284** [1985] C 5/C 8).

[36] Sailor, M. J.; Shriver, D. F. (Organometallics **4** [1985] 1476/8).

[37] Shapley, J. R.; Yeh, W.-Y.; Churchill, M. R.; Li, Y.-J. (Organometallics **4** [1985] 1898/1900).

[38] Yeh, W.-Y.; Shapley, J. R.; Li, Y.-J.; Churchill, M. R. (Organometallics **4** [1985] 767/72).

[39] Barreto, R. D.; Fehlner, T. P.; Hsu, L.-Y.; Jan, D.-Y.; Shore, S. G. (Inorg. Chem. **25** [1986] 3572/81).

[40] Bower, D. K.; Keister, J. B. (J. Organomet. Chem. **312** [1986] C 33/C 36).

[41] Boyar, E.; Deeming, A. J.; Kabir, S. E. (J. Chem. Soc. Chem. Commun. **1986** 577/9).

[42] Chi, Y.; Shapley, J. R.; Churchill, M. R.; Li, Y.-J. (Inorg. Chem. **25** [1986] 4165/70).

[43] Keiter, R. L.; Strickland, D. S.; Wilson, S. R.; Shapley, J. R. (J. Am. Chem. Soc. **108** [1986] 3846/7).

[44] Yeh, W.-Y.; Shapley, J. R. (J. Organomet. Chem. **315** [1986] C 29/C 31).

[45] Yeh, W.-Y.; Shapley, J. R.; Ziller, J. W.; Churchill, M. R. (Organometallics **5** [1986] 1757/63).

[46] Adams, R. D.; Babin, J. E.; Tasi, M. (Organometallics **6** [1987] 1717/27).

[47] Deeming, A. J.; Felix, M. S. B.; Bates, P. A.; Hursthouse, M. B. (J. Chem. Soc. Chem. Commun. **1987** 461/3).

[48] Jan, D.-Y.; Hsu, L.-Y.; Workman, D. P.; Shore, S. G. (Organometallics **6** [1987] 1984/5).

[49] Krause, J.; Jan, D.-Y.; Shore, S. G. (J. Am. Chem. Soc. **109** [1987] 4416/8).

[50] Vander Velde, D. G.; Holmgren, J. S.; Shapley, J. R. (Inorg. Chem. **26** [1987] 3077/8).

[51] Went, M. J.; Sailor, M. J.; Bogdan, P. L.; Brock, C. P.; Shriver, D. F. (J. Am. Chem. Soc. **109** [1987] 6023/9).

[52] Yeh, W.-Y.; Shapley, J. R.; Ziller, J. W.; Churchill, M. R. (Organometallics **6** [1987] 1/7).

[53] Adams, R. D.; Tanner, J. T. (Organometallics **7** [1988] 2241/3).

[54] Braga, D.; Koetzle, T. F. (Acta Crystallogr. B **44** [1988] 151/5).

[55] Churchill, M. R.; Ziller, J. W.; Shapley, J. R.; Yeh, W.-Y. (J. Organomet. Chem. **353** [1988] 103/17).

[56] Kneuper, H.-J.; Shapley, J. R. (New J. Chem. **12** [1988] 479/ 80).

[57] Kneuper, H.-J.; Strickland, D. S.; Shapley, J. R. (Inorg. Chem. **27** [1988] 1110/1).

[58] Strickland, D. S.; Wilson, S. R.; Shapley, J. R. (Organometallics **7** [1988] 1674/6).

[59] Wermer, J. R.; Jan, D.-Y.; Getman, T. D.; Moher, E.; Shore, S. G. (Inorg. Chem. **27** [1988] 4274/6).

[60] Yeh, W.-Y.; Kneuper, H.-J.; Shapley, J. R. (Polyhedron **7** [1988] 961/5).

[61] Boyar, E.; Deeming, A. J.; Felix, M. S. B.; Kabir, S. E.; Adatia, T.; Bhusate, R.; McPartlin, M.; Powell, H. R. (J. Chem. Soc. Dalton Trans. **1989** 5/12).

[62] Chi, Y.; Chen, B.-F.; Wang, S.-L.; Chiang, R.-K.; Hwang, L.-S. (J. Organomet. Chem. **377** [1989] C 59/C 64).

[63] Green, M.; Orpen, A. G.; Schaverien, C. J. (J. Chem. Soc. Dalton Trans. **1989** 1333/40).

[64] Yeh, W.-Y.; Wilson, S. R.; Shapley, J. R. (J. Organomet. Chem. **371** [1989] 257/65).

[65] Akita, M.; Shapley, J. R. (Organometallics **9** [1990] 2209/11).

[66] Bower, D. K.; Keister, J. B. (Organometallics **9** [1990] 2321/7).

[67] Churchill, M. R.; Bueno, C. (J. Organomet. Chem. **396** [1990] 327/38).

[68] Coffer, J. L.; Drickamer, H. G.; Park, J. T.; Roginski, R. T.; Shapley, J. R. (J. Phys. Chem. **94** [1990] 1981/5).

[69] Fehlner, T. P. (Polyhedron **9** [1990] 1955/63).

[70] Feighery, W. G.; Allendoerfer, R. D.; Keister, J. B. (Organometallics **9** [1990] 2424/6).

[71] Jensen, M. P.; Henderson, W.; Johnston, D. H.; Sabat, M.; Shriver, D. F. (J. Organomet. Chem. **394** [1990] 121/43).

[72] Krause, J. A.; Workman, D. P.; Shore, S. G. (Acta Crystallogr. C **46** [1990] 2086/8).

[73] Workman, D. P.; Deng, H.-B.; Shore, S. G. (Angew. Chem. **102** [1990] 328/30; Angew. Chem. Int. Ed. Engl. **29** [1990] 309).

[74] Workman, D. P.; Jan, D.-Y.; Shore, S. G. (Inorg. Chem. **29** [1990] 3518/25).

[75] Frauenhoff, G. R.; Wilson, S. R.; Shapley, J. R. (Inorg. Chem. **30** [1991] 78/85).

[76] Hadj–Bagheri, N.; Strickland, D. S.; Wilson, S. R.; Shapley, J. R. (J. Organomet. Chem. **410** [1991] 231/9).

[77] Jemmis, E. D.; Prasad, B. V. (Organometallics **10** [1991] 3613/20).

[78] Strickland, D. A.; Shapley, J. R. (J. Organomet. Chem. **401** [1991] 187/97).

[79] Adams, R. D.; Tanner, J. T. (Appl. Organomet. Chem. **6** [1992] 449/62).

[80] Frauenhoff, G. R.; Liu, J.-C.; Wilson, S. R.; Shapley, J. R. (J. Organomet. Chem. **437** [1992] 347/61).

[81] Jan, D.-Y.; Workman, D. P.; Hsu, L.-Y.; Krause, J. A.; Shore, S. G. (Inorg. Chem. **31** [1992] 5123/31).

[82] Karet, G. B.; Espe, R. L.; Stern, C. L.; Shriver, D. F. (Inorg. Chem. **31** [1992] 2658/60).

[83] Amoroso, A. J.; Johnson, B. F. G.; Lewis, J.; Li, C.-K.; Raithby, P. R.; Wong, W.-T. (J. Organomet. Chem. **444** [1993] C 55/C 56).

3.1.6.4 Heterometallic Compounds

3.1.6.4.1 Os₃Mo and Os₃W Compounds

The tetranuclear Os₃M compounds (M = Mo, W) dealt with in this section are represented by the Formulas I to X:

M = Mo, W

anti — Isomer 1

syn — Isomer 2 n = 0 or 5

anti — Isomer 3

I

II

III n = 0 or 5

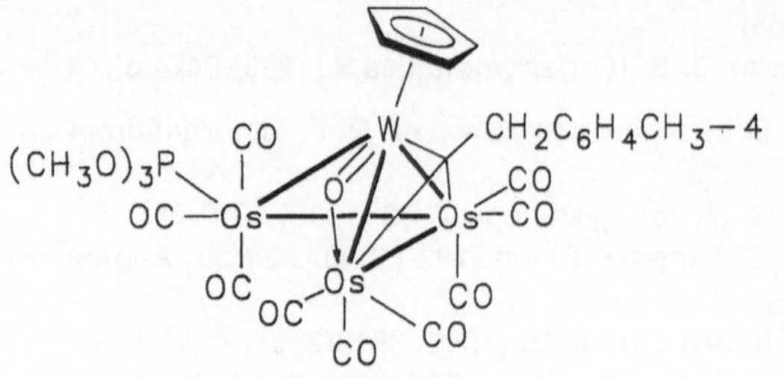

$(CH_3O)_3P$ $CH_2C_6H_4CH_3-4$

IV

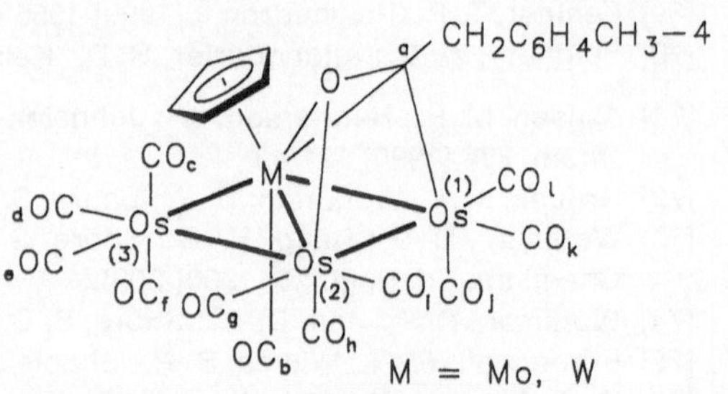

$CH_2C_6H_4CH_3-4$

M = Mo, W

V

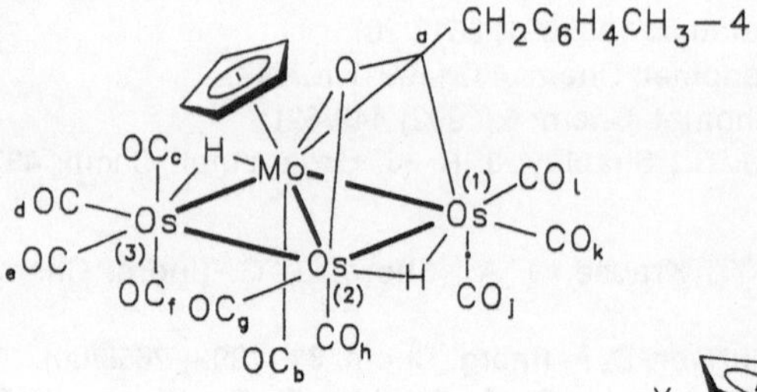

$CH_2C_6H_4CH_3-4$

VI

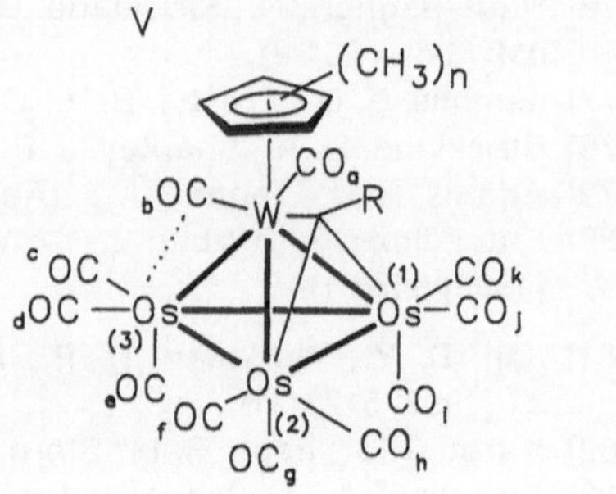

$(CH_3)_n$

n = 0 or 5

VII

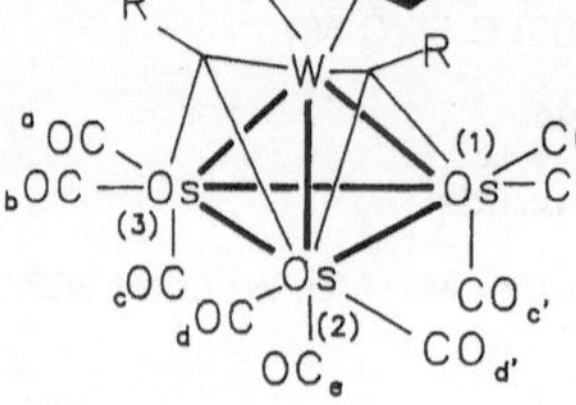

VIII

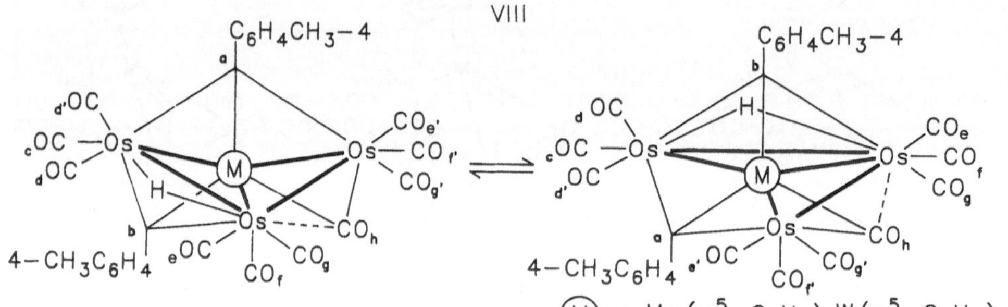

$C_6H_4CH_3-4$

$4-CH_3C_6H_4$

$C_6H_4CH_3-4$

$4-CH_3C_6H_4$

$M = Mo(\eta^5-C_5H_5), W(\eta^5-C_5H_5)$

IX

$C_6H_4CH_3-4$

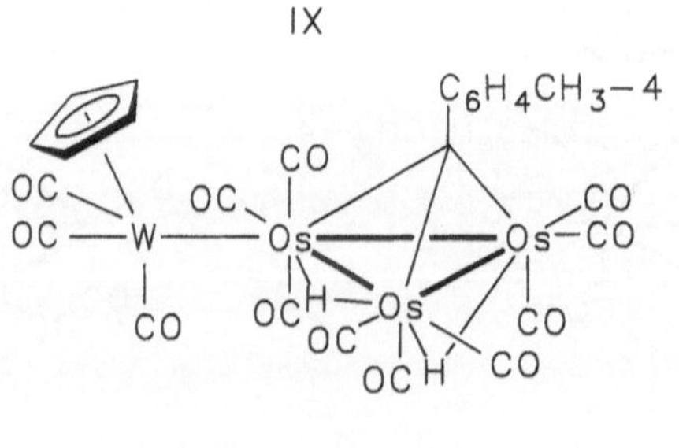

X

References on p. 208

In compounds of type I (X=H) to III, VII, and VIII the metal frameworks adopt a pseudo-tetrahedral arrangement associated with 60 outer-valence electrons (as expected for a tetrahedral arrangement). In most of the molecules each metal atom is in a different chemical environment resulting in chiral Os_3M compounds [7 to 10, 18, 19, 21] with the exception of those of type VIII [6]. The μ-oxo ligand in compounds of Formula I to III acts as a four-electron donor; the metal-oxygen distances are consistent with W=O double and Os-O single bonds [8, 9].

In compounds of type I (X=Cl, Br, SC_6H_5), IV and IX, the metal frameworks are arranged in a "butterfly" configuration [6, 8, 12, 16]; in those of type V and VI an essentially planar triangulated rhomboidal Os_3M skeleton is found [2, 5, 7, 19, 21]. For both metal arrangements typically 62 outer-valence electrons are expected. Most of these molecules are chiral result-ing from the different coordination environments of each metal atom in a molecule [5, 12].

The acyl compounds of type V having an activated CO bond, are discussed as possible intermediates in chain growth reactions on metal surfaces in the Fischer-Tropsch synthesis [5, 7, 8, 19, 21].

$(μ-H)_2Os_3(μ_3,η^1-CC_6H_4CH_3-4)(CO)_9W(η^5-C_5H_5)(CO)_3$ (No. 36; see Formula X) is the only complex where the metal-atom core takes up an equatorial spiked triangular geometry which is associated with 64 outer-valence electrons for the compound [11].

The tetrametallic core for compounds of type I either adopts a tetrahedral, or a butterfly configuration, depending on the presence of bridging μ-H or μ-X ligands (X=Cl, Br, SC_6H_5). An Os(1)-Os(2) bond is observed for the compounds bearing a μ-H ligand [19, 21], whereas in the case of a μ-Cl ligand the interatomic distance of 3.747 Å is out of the bonding range [12].

Compounds of type I having a μ-H ligand, adopt the configuration of Isomers 1, 2, and/or 3, whereas only Isomers 1 or 2 were observed for compounds with μ-X ligands. The syn/anti denotations reflect the orientation of R in the μ-alkylidene ligand with respect to the triply edge-bridged Os(1)-M-Os(2) face [12, 19, 21]. In [18] the syn/anti denotation is related to the $R/η^5-C_5H_5$ locations and thus is contrary to that of [12, 19, 21].

In the ^1H NMR spectra of Nos. 1 [21] and 6 to 12 [12], all of structure I, the μ-$CHCH_2$ hydrogens exhibit 1:1:1 AMX patterns with three separate resonances [12].

Structures not confirmed by X-ray analysis can generally only be considered as propos-als, based on spectroscopic and analytical data.

Some compounds of Table 6 were prepared by the following methods:

Method I: Starting from $(η^5-C_5H_5)M(≡CC_6H_4CH_3-4)(CO)_2$, M=Mo, W.

   a. The reaction of $(μ-H)_2Os_3(CO)_{10}$ with $(η^5-C_5H_5)M(≡CC_6H_4CH_3-4)(CO)_2$ (ca. 1:1 [5, 11] or 1:6 [20]) led to mixtures of $Os_3(μ_3,η^2-OCCH_2C_6H_4CH_3-4)-(CO)_{10}M(η^5-C_5H_5)(CO)$ (Formula V) and $(μ-H)Os_3(μ_3,η^1-CC_6H_4CH_3-4)_2-(μ_3-CO)(CO)_9M(η^5-C_5H_5)$ (Formula IX) [2, 5, 11, 19, 20].

   The preparations were performed by treatment of the Os_3 compound with a cold (−30 °C) CH_2Cl_2 solution of the metal-alkylidyne complex in CH_2Cl_2 at ca. 22 °C [19], or at 5 °C (M=Mo [20]) or 0 °C (M=W [2, 5, 11]) for 2 to 5 h, followed by warming-up and stirring for additional 10 to 20 h at room temperature. The reaction mixtures were separated by TLC with CH_2Cl_2/petroleum ether (1:4 [20] or 3:2 [2, 5, 11]) as eluant.

References on p. 208

$(\mu-H)_2Os_3(\mu_3,\eta^1-CC_6H_4CH_3-4)(CO)_9W(\eta^5-C_5H_5)(CO)_3$ (No. 36, Formula X) was only identified as a by-product in the preparation of Nos. 24 and 35 [11].

b. The reaction of $(\mu-H)_2Os_3(\mu_3,\eta^1-CC_6H_4CH_3-4)(CO)_9W(\eta^5-C_5H_5)(CO)_3$ (No. 36; Formula X) with $(\eta^5-C_5H_5)W(\equiv CC_6H_4CH_3-4)(CO)_2$ (ca. 1:1.5) led either to $Os_3(\mu_3,\eta^1-CC_6H_4CH_3-4)_2(CO)_9W(\eta^5-C_5H_5)H$ (No. 32, see Formula VIII), or to $(\mu-H)Os_3(\mu_3,\eta^1-CC_6H_4CH_3-4)_2(\mu_3-CO)(CO)_9W(\eta^5-C_5H_5)$ (No. 35, see also Formula IX), depending on the reaction conditions: Upon refluxing in benzene for 25 min No. 32 was formed in more than 50% yield in addition to a 1:1 mixture of $(\mu-H)_3Os_3(\mu_3,\eta^1-CC_6H_4CH_3-4)(CO)_9$ and $(\mu-H)Os_3-(\mu_3,\eta^1-CC_6H_4CH_3-4)(CO)_{10}$ (both Section 3.1.6.3), but in CH_2Cl_2 at room temperature for ca. 40 h No. 35 (ca. 26% yield), traces of No. 32, and the two above mentioned Os_3 carbyne complexes of Section 3.1.6.3 were obtained. The reaction mixtures were separated by TLC with CH_2Cl_2/pentane as eluant [11].

Method II: Compounds of type I, II, III, and VIII can be prepared by thermal rearrangement of a suitable precursor complex upon thermolysis in toluene at reflux temperature between 15 min and 19 h, followed by work-up by TLC with CH_2Cl_2/pentane (ca. 1:1, ca. 2:3, or 3:2) as eluant; the precursors are given in the table. The preparations were performed

a. by direct thermolysis of the precursor without initial decarbonylation [4, 6, 7, 12, 14, 16, 19, 21], or

b. by initial decarbonylation with one equivalent of $(CH_3)_3NO$ in CH_2Cl_2/CH_3CN at room temperature for ca. 20 min [4, 7, 14, 19, 21].

Method III: Starting from $Os_3(\mu_3,\eta^1-CCH_2C_6H_4CH_3-4)(CO)_9(\mu-O)W(\eta^5-C_5H_5)$ (No. 17, Formula III).

a. Compounds of type I can be prepared by reaction with H_2 [7], gaseous HX (X = Cl, Br), or an excess of HSC_6H_5 [12] in toluene at reflux temperature [7, 12] or at room temperature [12] for ca. 1 h, followed by purification by TLC with CH_2Cl_2/pentane (2:3) [7] or (1:1 or 3:4) [12] as eluant.
Treatment of No. 17 with BCl_3 (1 atm) in CH_2Cl_2 at room temperature for 5 min, followed by washing with H_2O before TLC, led to No. 7 in 89% yield [12].

b. Compounds of type III and IV bearing a terminal PRR'_2 ligand, were prepared by treatment with PRR'_2 (R = CH_3, R' = C_6H_5; R, R' = OCH_3) in refluxing toluene for ca. 20 min. Work-up by TLC with CH_2Cl_2/pentane (2:1) as eluant [16].

Table 6

Os$_3$Mo and Os$_3$W Compounds.

An asterisk preceding the compound number indicates further information at the end of the table, pp. 193/208.

Explanations, abbreviations, and units on p. X.

No. compound	method of preparation (yield) properties and remarks

compounds of the type Os$_3$(μ_2,η^1-CHR)(CO)$_8$D(μ-X)(μ-O)M(η^5-C$_5$H$_{5-n}$(CH$_3$)$_n$) (Formula I)

*1 (μ-H)Os$_3$(μ_2,η^1-CHCH$_2$C$_6$H$_4$CH$_3$-4)(CO)$_9$(μ-O)Mo(η^5-C$_5$H$_5$)

Isomer 1

IIb (from No. 25 in 56 to 59% yield as a non-chromatographically separable mixture with No. 3, along with 18% of No. 2) [19, 21]

the equilibrium mixture of No. 1 and No. 3 crystallizes as pure No. 3 from CH$_2$Cl$_2$/petroleum ether [21]

^{1}H NMR (acetone-d$_6$): − 19.3 (d, μ-H; ^{3}J(μ-H,CH) = 2.0), 2.28 (s, CH$_3$), 3.33 (dd, 1 H, CH$_2$; J(H,H) = 9.6, 14.4), 4.02 (dd, 1 H, CH$_2$; J(H,H) = 5.8, 14.4), 5.93 (ddd, CH; J(H,H) = 2.0, 5.8, 9.6), 6.15 (s, C$_5$H$_5$), 7.17 to 7.33 (m, C$_6$H$_4$) [21]; (CDCl$_3$): − 19.2 (s, μ-H), 2.38 (s, CH$_3$), 3.37 (dd, 1 H, CH$_2$; J(H,H) = 8.7, 14.5), 3.71 (dd, 1 H, CH$_2$; J(H,H) = 6.4, 14.5), 5.70 (t, CH; J(H,H) = 8.7), 5.89 (s, C$_5$H$_5$), 7.14 to 7.27 (m, C$_6$H$_4$) [19, 21]; spectra of a mixture of Nos. 1 and 3 depicted [21]

the μ-H/CH coupling observed in acetone-d$_6$ revealed that the hydride bridges the Os(1)-Os(2) edge adjacent to the ^{1}L ligand; interestingly, no coupling was observed in CDCl$_3$; compare also No. 3 [21]

^{13}C NMR (CD$_2$Cl$_2$, − 20 °C): 135.8 (d, μ_2,η^1-C; ^{1}J(C,H) = 134.9), 168.9 (d, 1 CO trans to μ-H; ^{2}J(C,μ-H) = 12.9), 169.5 (d, 1 CO trans to μ-H; ^{2}J(C,μ-H) = 9.6), 170.6, 173.7, 177.3, 181.9, 183.5, 184.3, 189.7 (each 1 CO); the spectrum of a mixture of Nos. 1 and 3 is depicted [21]

IR (cyclohexane; mixture with No. 3): 1930, 1955, 1963, 1987, 2003, 2012, 2028, 2033, 2053, 2061, 2087 [21]

*2 (μ-H)Os$_3$(μ_2,η^1-CHCH$_2$C$_6$H$_4$CH$_3$-4)(CO)$_9$(μ-O)Mo(η^5-C$_5$H$_5$)

Isomer 2

IIa (61 to 70% from a mixture of Nos. 1 and 3) [19, 21]; see also No. 1 [19, 21]

by treatment of a mixture of Nos. 1 and 3 with aqueous HCl in acetone between 45 and 56 °C for 4 to 14 h [19, 21], followed by neutralization with K$_2$CO$_3$ and filtration; purification by TLC with petroleum ether/CH$_2$Cl$_2$ (3:2) as eluant (72%) [21]

References on p. 208

Table 6 (continued)

No. compound	method of preparation (yield) properties and remarks
*2 (continued)	dark brown crystals from $CHCl_3/CH_3OH$ at $-15\,°C$ [21] ^1H NMR (acetone-d_6): -18.4 (d, μ-H; $J(\mu$-H,CH$) = 1.5$), 2.37 (s, CH_3), 2.79 (dd, 1 H, CH_2; $J(H,H) = 10.8, 12.9$), 4.62 (dd, 1 H, CH_2; $J(H,H) = 6.2, 12.9$), 5.59 (s, C_5H_5), 7.04 (ddd, CH; $J(H,H) = 1.5, 6.2, 10.8$), 7.21 to 7.27 (m, C_6H_4) [21]; (CDCl$_3$): -18.6 (s, μ-H), 2.40 (s, CH_3), 2.52 (t, 1 H, CH_2; $J(H,H) = 11.7$), 4.74 (dd, 1 H, CH_2; $J(H,H) = 5.9, 12.2$), 5.38 (s, C_5H_5), 6.59 (dd, CH; $J(H,H) = 5.9, 11.4$), 7.14 to 7.23 (m, C_6H_4) [19, 21]; see also No. 1 ^{13}C NMR (CDCl$_3$, $-10\,°C$): 140.5 (d, μ_2,η^1-C; $^1J(C,H) = 140.6$), 168.3 (d, 1 CO trans to μ-H; $^2J(C,\mu$-H$) = 9.8$), 169.7 (d, 1 CO trans to μ-H; $^2J(C,\mu$-H$) = 13.1$), 170.2, 175.2, 176.1, 182.4, 185.1, 185.2, 189.0 (each 1 CO) [19, 21]; in toluene-d_8 the CO resonances at 182.4, 185.1, and 189.0 ppm broaden at higher temperatures and coalesce at 100 °C to a broad single peak due to localized three-fold CO exchange processes on the unique Os(3) atom; see Formula I [21] IR (cyclohexane): 1941, 1955, 1990, 2000, 2012, 2028, 2060, 2086 (all CO) [19, 21] mass spectrum: [M]$^+$ [19, 21]
*3 $(\mu$-H)Os$_3(\mu_2,\eta^1$-CHCH$_2$C$_6$H$_4$CH$_3$-4)(CO)$_9(\mu$-O)Mo(η^5-C$_5$H$_5$) Isomer 3	for formation, see No. 1 [19, 21] dark brown crystals from CH_2Cl_2/petroleum ether; see also No. 1 [21] ^1H NMR (acetone-d_6): -16.6 (s, μ-H), 2.32 (s, CH_3), 3.39 (dd, 1 H, CH_2; $J(H,H) = 9.4, 14.1$), 4.44 (dd, 1 H, CH_2; $J(H,H) = 5.4, 14.1$), 6.16 (s, C_5H_5), 6.83 (dd, CH; $J(H,H) = 5.4, 9.4$), 7.17 to 7.33 (m, C_6H_4) [21]; (CDCl$_3$): -16.6 (s, μ-H), 2.39 (s, CH_3), 3.52 (dd, 1 H, CH_2; $J(H,H) = 8.6, 14.4$), 4.01 (dd, 1 H, CH_2; $J(H,H) = 6.4, 14.4$), 5.90 (s, C_5H_5), 6.67 (t, CH; $J(H,H) = 7.5$), 7.17 to 7.27 (m, C_6H_4) [19, 21]; the absence of the μ-H/CH coupling in CDCl$_3$ as well as in acetone-d_6 revealed that the hydride bridges the Os(2)-Os(3) edge; compare also No. 1 [21] ^{13}C NMR (CD$_2$Cl$_2$, $-20\,°C$): 146.3 (d, μ_2,η^1-C; $^1J(C,\mu$-H$) = 132.5$), 170.2 (d, 1 CO trans to

References on p. 208

Table 6 (continued)

No. compound	method of preparation (yield) properties and remarks

μ–H; $^2J(C,H) = 12.0$), 179.3 (d, 1 CO trans to
μ–H; $^2J(C,H) = 8.0$), 168.5, 174.7, 177.7, 182.2,
184.3, 184.9, 185.9 (each 1 CO) [21]
IR: see No. 1
mass spectrum: $[M]^+$ [19, 21]

4 (μ-H)Os$_3$(μ_2,η^1-CHCH$_3$)(CO)$_9$(μ-O)W(η^5-C$_5$(CH$_3$)$_5$)
Isomer 2

by solid state pyrolysis of (μ-H)Os$_3$(CO)$_{10}$-
{μ_2,η^2-C(O)CH$_2$W(η^5-C$_5$(CH$_3$)$_5$)(CO)$_3$}
("Organoosmium Compounds" B 5, Section
3.1.6.1.2.3, in preparation) at 195 °C for 10
min, followed by extraction with CH$_2$Cl$_2$ and
separation by TLC with CH$_2$Cl$_2$/hexane (10%,
along with 9% of No. 16, 3% of No. 26, and
3% of No. 27) [23]
by hydrogenation of No. 16 in refluxing toluene
for 2 h under H$_2$ atmosphere, followed by
TLC purification as before (45%) [23]
orange solid from CH$_2$Cl$_2$/heptane [23]
^{1}H NMR (CDCl$_3$): -18.11 (s, μ–H), 2.05 (s,
C$_5$(CH$_3$)$_5$), 2.77 (d, CCH$_3$), 4.78 (q, CH);
$^3J(H,H) = 7.6$; ^{1}H nuclear Overhauser effect
investigations revealed that the structure is
in accordance with Isomer 2 of Formula I
[23]
^{13}C NMR (CD$_2$Cl$_2$): 11.9 (C$_5$(**CH$_3$**)$_5$), 37.9
(CH**CH$_3$**), 112.9 (**C$_5$**(CH$_3$)$_5$), 128.9 (μ_2,η^1-C;
J(C,W) = 97), 170.5, 171.2, 173.1, 174.1, 175.9,
183.9, 185.4, 187.0, 192.8 (all CO) [23]
IR (cyclohexane): 1928, 1949, 1986, 1992, 2002,
2008, 2017, 2056, 2084 [23]
FAB mass spectrum: $[M]^+$ [23]

*5 (μ-H)Os$_3$(μ_2,η^1-CHC$_6$H$_4$CH$_3$-4)(CO)$_9$(μ-O)W(η^5-C$_5$H$_5$)
Isomer 2

from Os$_3$(μ_3,η^1-CC$_6$H$_4$CH$_3$-4)$_2$(CO)$_9$W(η^5-
C$_5$H$_5$)H (No. 32) and H$_2$ (1 atm) in refluxing
xylene for 6 h, followed by evaporation of
the solvent and TLC with hexane/CH$_2$Cl$_2$
(3:2) as eluant (12%); the source of the
μ-oxo ligand is unknown; the formation of
the syn isomer is believed to be due to its
greater thermodynamic stability with respect
to the unobserved anti-configurated complex
which would be destabilized because of the
unfavorable steric repulsion between the
tolyl substituent and the η^5-C$_5$H$_5$ ligand [18]
red-orange crystals from CH$_2$Cl$_2$/CH$_3$OH [18]

                                          References on p. 208

Table 6 (continued)

No. compound	method of preparation (yield) properties and remarks

*5 (continued)

^{1}H NMR (CDCl$_3$): -18.05 (s, μ-H), 2.39 (s, CH$_3$), 5.52 (s, CH), 5.94 (s, C$_5$H$_5$), 6.45 to 7.05 (m, C$_6$H$_4$) [18]
^{13}C NMR (CDCl$_3$, -60 °C): 169.1, 169.7, 169.9, 173.2, 176.4, 180.9, 181.4, 184.4, 194.5 (each 1 CO) [18]
IR (cyclohexane): 1940, 1958, 1991, 2007, 2014, 2020, 2028, 2065, 2090 (all CO) [18]
mass spectrum: [M]$^+$ [18]

*6 (μ-H)Os$_3$(μ_2,η^1-CHCH$_2$C$_6$H$_4$CH$_3$-4)(CO)$_9$(μ-O)W(η^5-C$_5$H$_5$)
Isomers 1+3

the solid state structure corresponds to Isomer 3 which equilibrates in solution with Isomer 1 [7, 9, 12]
IIIa (more than 90%) [7]
from No. 24 under the conditions of Preparation Method IIIa (high yields) [7]
dark red [9] or orange crystals from CH$_3$OH/CH$_2$Cl$_2$/toluene at $+10$ °C [7]
^{1}H NMR (CD$_2$Cl$_2$): Isomer 1: -18.87 (s, μ-H), 2.37 (s, CH$_3$), 3.44 (dd, 1 H, CH$_2$; J(H,H) = 8.7, 14.7), 3.71 (dd, 1 H, CH$_2$; J(H,H) = 6.4, 14.7), 4.95 (dd, CH; J(H,H) = 6.4, 8.7), 6.11 (s, C$_5$H$_5$), 7.08 (s, C$_6$H$_4$); Isomer 3: -16.93 (s, μ-H), 2.38 (s, CH$_3$), 3.60 (dd, 1 H, CH$_2$; J(H,H) = 8.5, 14.5), 4.06 (dd, 1 H, CH$_2$; J(H,H) = 6.4, 14.5), 5.80 (dd, CH; J(H,H) = 6.4, 8.5), 6.08 (s, C$_5$H$_5$), 7.1 to 7.3 (m, C$_6$H$_4$) [7]
IR (cyclohexane): 1938, 1958, 1987, 2004, 2009, 2025, 2031, 2050, 2059, 2086 (all CO) [7]
mass spectrum (Isomer 3): [M]$^+$ [7]
pyrolysis of a mixture of the Isomers 1 and 3 in refluxing xylene led to the syn isomer, Isomer 2 (not characterized) [12, 13]

*7 Os$_3$(μ_2,η^1-CHCH$_2$C$_6$H$_4$CH$_3$-4)(CO)$_9$(μ-Cl)(μ-O)W(η^5-C$_5$H$_5$)
Isomer 1

IIIa (67 or 89% from HCl or BCl$_3$/H$_2$O, respectively) [12]
orange–red crystals from CH$_2$Cl$_2$/hexane [12]
^{1}H NMR (CD$_2$Cl$_2$, 18 °C): 2.38 (s, CH$_3$), 4.01 (dd, 1 H, CH$_2$; ^{2}J(H,H) = 14.3, ^{3}J(H,H) = 9.0), 4.49 (dd, 1 H, CH$_2$; ^{2}J(H,H) = 14.3, ^{3}J(H,H) = 5.5), 6.13 (s, C$_5$H$_5$), 6.33 (dd, CH; ^{3}J(H,H) = 5.5, 9.0), 7.22, 7.30 (d's, both C$_6$H$_4$; ^{3}J(H,H) = 7) [12]
^{13}C NMR (CD$_2$Cl$_2$, 18 °C): 135.8 (CH; ^{1}J(C,H) = 130, ^{1}J(C,W) = 100), 173.1, 174.4,

Table 6 (continued)

No. compound	method of preparation (yield) properties and remarks

175.1, 175.7, 182.1, 183.3, 186.3, 186.7, 189.5
(each 1 CO); spectra of ^{13}C enriched samples depicted; the resonances at $\delta = 173.1$,
175.1, and 175.7 ppm are associated with the
CO's at Os(2), see Formula I [12]
IR (cyclohexane): 1942, 1956, 1970, 1987, 1997,
2008, 2030, 2064, 2090 (all CO) [12]
FD mass spectrum: $[M]^+$ [12]

8 $Os_3(\mu_2,\eta^1\text{-}CHCH_2C_6H_4CH_3\text{-}4)(CO)_9(\mu\text{-}Cl)(\mu\text{-}O)W(\eta^5\text{-}C_5H_5)$
   Isomer 2   IIa (71%, from No. 7) [12]
orange-red viscous oil, attempted crystallization failed [12]
^1H NMR (CDCl$_3$, 18 °C): 2.48 (s, CH$_3$), 3.87 (t, 1
H, CH$_2$; J(H,H) $\approx$ 12), 5.22 (dd, 1 H, CH$_2$;
$^2J(H,H) = 12.9$, $^3J(H,H) = 5.7$), 5.63 (s, C$_5$H$_5$),
6.81 (dd, CH; $^3J(H,H) = 5.7$, 11.3), 7.34, 7.49
(d's, both C$_6$H$_4$; $^3J(H,H) = 7$) [12]
^{13}C NMR (CDCl$_3$, 18 °C): 133.5 (CH;
$^1J(C,H) = 140$, $^1J(C,W) = 100$), 172.6, 173.4,
174.9, 175.0, 183.3, 183.6, 185.0, 187.1, 189.0
(each 1 CO) [12]
IR (cyclohexane): 1942, 1956, 1970, 1987, 1997,
2008, 2030, 2064, 2090 (all CO) [12]
FD mass spectrum: $[M]^+$, $[M-CO]^+$ [12]

9 $Os_3(\mu_2,\eta^1\text{-}CHCH_2C_6H_4CH_3\text{-}4)(CO)_8(PCH_3(C_6H_5)_2)(\mu\text{-}Cl)(\mu\text{-}O)W(\eta^5\text{-}C_5H_5)$
   Isomer 1   from No. 7 and P(C$_6$H$_5$)$_2$CH$_3$ as described for
Preparation Method IIIb; no isomerization
was observed [12]
orange-red crystalline solid from
CH$_2$Cl$_2$/pentane [12]
^1H NMR (CDCl$_3$, 18 °C): 2.09 (d, PCH$_3$;
$^2J(H,P) = 9.4$), 2.39 (s, CH$_3$), 4.19 (dd, 1 H,
CH$_2$; $^2J(H,H) = 15.0$, $^3J(H,H) = 8.1$), 4.38 (dd, 1
H, CH$_2$; $^2J(H,H) = 15.0$, $^3J(H,H) = 6.7$), 5.95 (t,
CH; J(H,H) $\approx$ 7), 6.04 (s, C$_5$H$_5$), 7.22, 7.30 (d's,
both C$_6$H$_4$; J(H,H) = 7), 7.54 (br s, C$_6$H$_5$) [12]
^{13}C NMR (CDCl$_3$, 18 °C): 131.8 (CH;
$^1J(C,H) = 128$, $^1J(C,W) = 94$), 174.9, 179.8,
182.9, 183.7, 184.1, 187.3, 189.5 (8 CO); spec-
trum depicted [12]
IR (cyclohexane): 1931, 1941, 1966, 1983, 2000,
2030, 2064 (all CO) [12]
FAB mass spectrum: $[M]^+$ [12]

                    References on p. 208

Table 6 (continued)

No. compound	method of preparation (yield) properties and remarks

10 $Os_3(\mu_2,\eta^1\text{-}CHCH_2C_6H_4CH_3\text{-}4)(CO)_9(\mu\text{-}Br)(\mu\text{-}O)W(\eta^5\text{-}C_5H_5)$
  Isomer 1

    IIIa (21%, along with 11% of No. 11) [12]
    orange crystalline solid from CH_2Cl_2/pentane [12]
    ^{1}H NMR (CD_2Cl_2, 18 °C): 2.37 (s, CH_3), 4.13 (dd, 1 H, CH_2; $^2J(H,H) = 14.8$, $^3J(H,H) = 8.4$), 4.47 (dd, 1 H, CH_2; $^2J(H,H) = 14.8$, $^3J(H,H) = 6.8$), 6.05 (s, C_5H_5), 6.53 (t, CH; $J(H,H) \approx 8$), 7.21, 7.30 (d's, both C_6H_4; $^3J(H,H) = 7$); spectrum depicted [12]
    IR (cyclohexane): 1942, 1956, 1970, 1987, 2005, 2030, 2064, 2090 (all CO) [12]
    FD mass spectrum: $[M]^+$ [12]

11 $Os_3(\mu_2,\eta^1\text{-}CHCH_2C_6H_4CH_3\text{-}4)(CO)_9(\mu\text{-}Br)(\mu\text{-}O)W(\eta^5\text{-}C_5H_5)$
  Isomer 2

    for formation, see No. 10 [12]
    orange oil, attempted crystallization failed [12]
    ^{1}H NMR ($CDCl_3$, 18 °C): 2.45 (s, CH_3), 3.96 (t, 1 H, CH_2; $J(H,H) \approx 12$), 5.13 (dd, 1 H, CH_2; $^2J(H,H) = 12.9$, $^3J(H,H) = 5.6$), 5.56 (s, C_5H_5), 6.77 (dd, CH; $^3J(H,H) = 5.6$, 11.3), 7.32, 7.47 (d's, both C_6H_4; $^3J(H,H) = 7$); spectrum depicted [12]
    IR (cyclohexane): 1942, 1956, 1970, 1987, 2005, 2030, 2064, 2090 (all CO) [12]
    FD mass spectrum: $[M]^+$, $[M-CO]^+$ [12]

12 $Os_3(\mu_2,\eta^1\text{-}CHCH_2C_6H_4CH_3\text{-}4)(CO)_9(\mu\text{-}SC_6H_5)(\mu\text{-}O)W(\eta^5\text{-}C_5H_5)$
  Isomer 1

    IIIa (59%, along with 22% of No. 13) [12]
    orange crystalline solid from CH_2Cl_2/pentane [12]
    ^{1}H NMR ($CDCl_3$, 18 °C): 2.34 (s, CH_3), 3.98 (dd, 1 H, CH_2; $^2J(H,H) = 14.8$, $^3J(H,H) = 7.9$), 4.24 (dd, 1 H, CH_2; $^2J(H,H) = 14.8$, $^3J(H,H) = 6.9$), 5.62 (t, CH; $J(H,H) \approx 7$), 6.04 (s, C_5H_5), 6.70 (d), 7.00 (m, both C_6H_5; $J(H,H) = 7$), 7.00, 7.31 (d's, both C_6H_4; $^3J(H,H) = 7$) [12]
    ^{13}C NMR ($CDCl_3$, 18 °C): 133.8 (CH; $^1J(C,H) = 127$, $^1J(C,W) = 96$), 172.5, 173.1, 176.3, 179.3, 179.6, 181.2, 186.0, 186.1, 186.6 (each 1 CO); spectrum depicted [12]
    IR (cyclohexane): 1948, 1955, 1963, 1985, 2000, 2006, 2026, 2056, 2081 (all CO) [12]
    FD mass spectrum: $[M]^+$ [12]

13 $Os_3(\mu_2,\eta^1\text{-}CHCH_2C_6H_4CH_3\text{-}4)(CO)_9(\mu\text{-}SC_6H_5)(\mu\text{-}O)W(\eta^5\text{-}C_5H_5)$
  Isomer 2

    for formation, see No. 12 [12]
    orange crystalline solid from CH_2Cl_2/CH_3OH [12]

References on p. 208

Table 6 (continued)

No. compound	method of preparation (yield) properties and remarks

^{1}H NMR (CDCl$_3$, 18 °C): 2.45 (s, CH$_3$), 3.65 (t, 1 H, CH$_2$; J(H,H) ≈ 12), 4.65 (dd, 1 H, CH$_2$; ^{2}J(H,H) = 12.8, ^{3}J(H,H) = 5.8), 5.57 (s, C$_5$H$_5$), 6.63 (dd, CH; ^{3}J(H,H) = 5.8, 11.6), 7.15 to 7.34 (m, C$_6$H$_5$), 7.15, 7.34 (d's, both C$_6$H$_4$; ^{3}J(H,H) = 7) [12]
^{13}C NMR (CDCl$_3$, 18 °C): 133.3 (CH; ^{1}J(C,H) = 138, ^{1}J(C,W) = 103), 173.2, 173.5, 175.7, 179.2, 180.3, 182.7 (each 1 CO), 185.6 (2 CO), 187.8 (1 CO); spectrum depicted [12]
IR (cyclohexane): 1940, 1957, 1968, 1990, 1999, 2008, 2026, 2057, 2081 (all CO) [12]
FD mass spectrum: [M]$^+$, [M − CO]$^+$ [12]

compounds of the type (μ-H)Os$_3$(μ$_2$,η^1-C=CHC$_6$H$_4$CH$_3$-4)(CO)$_8$D(μ-O)W(η^5-C$_5$H$_5$) (Formula II)

*14 (μ-H)Os$_3$(μ$_2$,η^1-C=CHC$_6$H$_4$CH$_3$-4)(CO)$_9$(μ-O)W(η^5-C$_5$H$_5$)

IIa (52% from No. 17) [7], IIa (36% from No. 24, along with No. 17) [7]
dark red crystals from CH$_2$Cl$_2$/hexane [10] or CH$_2$Cl$_2$/Nujol/hexane [7]
^{1}H NMR (CD$_2$Cl$_2$): − 17.18 (s, μ-H), 2.41 (s, CH$_3$), 5.82 (s, C$_5$H$_5$), 7.05 to 7.29 (m, C$_6$H$_4$), 8.20 (s, CH) [7]
IR (cyclohexane): 1964, 1988, 2005, 2032, 2054, 2089 (all CO) [7]
FD mass spectrum: [M]$^+$ [7]

15 (μ-H)Os$_3$(μ$_2$,η^1-C=CHC$_6$H$_4$CH$_3$-4)(CO)$_8$P(OCH$_3$)$_3$(μ-O)W(η^5-C$_5$H$_5$)

probably structure II, but neither evidence nor assignment given
as a by-product in the preparation of No. 19 from No. 22 under the conditions of Preparation Method IIa, or in the preparation of No. 22 by carbonylation of No. 19 in refluxing toluene under a CO atmosphere for 6 h (ca. 2.5%) [16]
IR: 1949, 1952, 1964, 1979, 1988, 2018, 2034, 2061 (all CO) [16]
FD mass spectrum: [M]$^+$ [16]

compounds of the type Os$_3$(μ$_3$,η^1-CR)(CO)$_7$DD'(μ-O)W(η^5-C$_5$H$_{5-n}$(CH$_3$)$_n$) (Formula III)

*16 Os$_3$(μ$_3$,η^1-CCH$_3$)(CO)$_9$(μ-O)W(η^5-C$_5$(CH$_3$)$_5$)

for formation, see No. 4 [23]
dark brown crystals from CH$_2$Cl$_2$/heptane [23]

References on p. 208

Table 6 (continued)

No. compound	method of preparation (yield) properties and remarks
*16 (continued)	^{1}H NMR (CD_2Cl_2): 2.18 (s, $C_5(CH_3)_5$), 3.60 (s, CCH_3) [23] ^{13}C NMR (CD_2Cl_2): 12.8 ($C_5(\textbf{CH}_3)_5$), 43.8 ($C\textbf{CH}_3$), 114.8 ($\textbf{C}_5(CH_3)_5$), 167.5, 179.4, 180.4 (each 1 CO), 183.0 (br, 3 CO), 183.1, 190.5, 196.7 (each 1 CO), 225.4 (μ_3,η^1-C; J(C,W) = 115) [23] IR (cyclohexane): 1943, 1951, 1962, 1983, 1989, 2005, 2022, 2038, 2073 [23] FAB mass spectrum: [M]$^+$ [23] hydrogenation in refluxing toluene gave No. 4 [23]
*17 $Os_3(\mu_3,\eta^1\text{-}CCH_2C_6H_4CH_3\text{-}4)(CO)_9(\mu\text{-}O)W(\eta^5\text{-}C_5H_5)$	IIa (from No. 24, along with No. 14) [7], IIb (80% from No. 24) [7] dark brown crystals from CH_2Cl_2/CH_3OH [7] ^{1}H NMR ($CDCl_3$): 2.36 (s, CH_3), 4.46 (d, 1 H, CH_2; J(H,H) = 15.2), 5.25 (d, 1H, CH_2; J(H,H) = 15.2), 5.75 (s, C_5H_5), 7.1 to 7.3 (m, C_6H_4) [7] ^{13}C NMR ($CDCl_3$?): 220.8 (μ_3-C) [7], 166.2 (CO-c), 175.0 (CO-c), 178.6 (CO-a), 179.6 (CO-c), 179.9 (CO-b), 182.5 (CO-a), 187.7 (CO-b), 188.7 (CO-a), 195.2 (CO-b); the designation CO-a, CO-b, and CO-c refers to the carbonyl ligands grouped in three sets on the basis of barriers to local exchange processes at each $Os(CO)_3$ center evaluated by variable ^{13}C NMR spectroscopy; the CO's of set c are assigned to the carbonyls attached to the Os-(μ-O) center; the local CO exchange barrier increases in the sequence CO-a < CO-b < CO-c [16] IR (cyclohexane): 1954, 1961, 1972, 1989, 1997, 2008, 2028, 2040, 2075 (all CO) [7] mass spectrum: [M]$^+$, [M − x CO]$^+$, x = 1 to 9 [7]
*18 $Os_3(\mu_3,\eta^1\text{-}CCH_2C_6H_4CH_3\text{-}4)(CO)_8P(C_6H_5)_2CH_3(\mu\text{-}O)W(\eta^5\text{-}C_5H_5)$	IIIb (50%); an unstable by-product, probably the addition product $\textbf{Os}_3(\mu_3,\eta^1\text{-}\textbf{CCH}_2\textbf{C}_6\textbf{H}_4\text{-}\textbf{CH}_3\text{-}4)(\textbf{CO})_9\textbf{P}(\textbf{C}_6\textbf{H}_5)_2\textbf{CH}_3(\mu\text{-}\textbf{O})\textbf{W}(\eta^5\text{-}\textbf{C}_5\textbf{H}_5)$, decomposed by TLC work-up upon contact with silica gel [16] red crystals from hot $CHCl_3$/heptane [16] ^{1}H NMR ($CDCl_3$, 18 °C): 2.24 (d, PCH_3; ^{2}J(H,P) = 9.2), 2.28 (s, CH_3), 3.35 (d, 1 H, CH_2;

Table 6 (continued)

No. compound	method of preparation (yield)
	properties and remarks

^{2}J(H,H) = 15.2), 4.71 (d, 1 H, CH$_2$;
^{2}J(H,H) = 15.2), 5.58 (s, C$_5$H$_5$), 6.86 to 7.06 (m,
C$_6$H$_4$), 7.42 to 7.50 (m, C$_6$H$_5$) [16]
^{13}C NMR (CDCl$_3$, −48 °C): 173.6 (CO-c), 176.2
(CO-a), 180.8 (CO-c), 181.7 (CO-b),
185.1 (CO-a), 189.3 (CO-b), 191.5 (CO-a),
196.2 (CO-b), 211.3 (μ$_3$-C; ^{1}J(C,W) = 110); the
designation CO-a, CO-b, and CO-c refers to
the carbonyl ligands grouped in three sets
on the basis of barriers to local exchange
processes at each Os(CO)$_3$ center evaluated
by variable ^{13}C NMR spectroscopy from −48
to +45 °C (spectra depicted); the CO's of set
c are assigned to the carbonyls attached to
the Os-(μ-O) center; set CO-a broadens at
−20 °C and coalesces at +18 °C, and set
CO-b loses intensity (broadens faster) relative to set CO-c; the local CO exchange barrier increases in the sequence CO-a<
CO-b<CO-c [16]
IR (cyclohexane): 1947, 1967, 1999, 2016, 2051
(all CO) [16]
FAB mass spectrum: [M]$^+$ [16]

19 Os$_3$(μ$_3$,η^1-CCH$_2$C$_6$H$_4$CH$_3$-4)(CO)$_8$P(OCH$_3$)$_3$(μ-O)W(η^5-C$_5$H$_5$)

IIa (64% from No. 22, along with moderate
yields of No. 15) [16], IIIb (21%, along with
32% of No. 22) [16]
red-brown solid CH$_2$Cl$_2$/pentane [16]
^{1}H NMR (CDCl$_3$, 18 °C): 2.33 (s, CH$_3$), 3.76 (d,
OCH$_3$; ^{2}J(H,P) = 11.7), 4.24, 5.15 (d's, each 1
H of CH$_2$; ^{2}J(H,H) = 15), 5.64 (s, C$_5$H$_5$), 7.13 to
7.25 (m, C$_6$H$_4$) [16]
^{13}C NMR (CDCl$_3$, −24 °C): 170.9 (d, CO-c;
^{2}J(C,P) = 10), 176.3 (CO-a), 181.4 (CO-c),
181.6 (CO-b), 184.7 (CO-a), 189.2 (CO-b),
190.8 (CO-a), 196.2 (CO-b), 214.1 (μ$_3$-C;
^{1}J(C,W) = 122); the designation CO-a, and
CO-b, and CO-c refers to the carbonyl li-
gands grouped in three sets on the basis of
barriers to local exchange processes at each
Os(CO)$_3$ center evaluated by variable ^{13}C
NMR spectroscopy; the CO's of set c are as-
signed to the carbonyls attached to the
Os-(μ-O) center; the local CO exchange bar-
riers increase in the sequence CO-a<
CO-b<CO-c [16]

References on p. 208

Table 6 (continued)

No. compound	method of preparation (yield) properties and remarks

19 (continued)

IR (cyclohexane): 1940, 1950, 1968, 1986, 2002, 2009, 2020, 2055 (all CO) [16]

FAB mass spectrum: $[M]^+$ [16]

carbonylation in refluxing toluene for 6 h under 1 atm of CO led to Nos. 15 and 22 in moderate yields [16]

thermolysis in the presence of $P(OC_2H_5)_3$ resulted in No. 20 [16]

20 $Os_3(\mu_3,\eta^1\text{-}CCH_2C_6H_4CH_3\text{-}4)(CO)_7P(OCH_3)_3P(OC_2H_5)_3(\mu\text{-}O)W(\eta^5\text{-}C_5H_5)$

probably structure III, but neither evidence nor assignment given

by thermolysis of No. 19 in the presence of $P(OC_2H_5)_3$ (only identified by mass spectrometry) [16]

compounds of the type $Os_3(\mu_3,\eta^1\text{-}CCH_2C_6H_4CH_3\text{-}4)(CO)_9D(\mu\text{-}O)W(\eta^5\text{-}C_5H_5)$ (Formula IV)

21 $Os_3(\mu_3,\eta^1\text{-}CCH_2C_6H_4CH_3\text{-}4)(CO)_{10}(\mu\text{-}O)W(\eta^5\text{-}C_5H_5)$

from No. 17 upon treatment with CO [22]

^{13}C NMR (no medium given): 220.7 (μ_3–C; J(C,W) = 105) [22]; no further data given

22 $Os_3(\mu_3,\eta^1\text{-}CCH_2C_6H_4CH_3\text{-}4)(CO)_9P(OCH_3)_3(\mu\text{-}O)W(\eta^5\text{-}C_5H_5)$

in solution a 3:1 mixture of two isomers was observed, probably corresponding to a different location of the phosphite ligand in the $Os(CO)_3D$ moiety

IIIb (32%, along with 21% of No. 19) [16]

by carbonylation of No. 19 in toluene for 6 h under 1 atm of CO (4%, along with small amounts of No. 15) [16]

^{1}H NMR (CDCl$_3$, 18 °C): major isomer: 2.37 (s, CH$_3$), 3.78 (d, OCH$_3$; ^{2}J(H,P) = 11), 5.30 (d, 1 H, CH$_2$; ^{2}J(H,H) = 15), 5.82 (s, C$_5$H$_5$), 6.27 (d, 1 H, CH$_2$; ^{2}J(H,H) = 15), 7.16 to 7.38 (m, C$_6$H$_4$); minor isomer: 3.81 (d, OCH$_3$; ^{2}J(H,P) = 11), 5.94 (s, C$_5$H$_5$), other signals not resolved [16]

^{13}C NMR (CD$_2$Cl$_2$, −45 °C): major isomer: 175.9, 179.5, 183.0, 183.7, 184.5 (trans, superimposed AB pattern; ^{2}J(C,C) ≈ 30), 185.0 (trans, superimposed AB pattern; ^{2}J(C,C) ≈ 30), 185.3 (^{2}J(C,P) = 11), 187.7, 188.0? (each 1 CO), 219.7 (μ_3–C; ^{1}J(C,W) = 109); signals at 179.5, 183.0 and 187.7 broaden above −45 °C and disappear at ca. 25 °C due to their fluxional behavior;

References on p. 208

Table 6 (continued)

No. compound	method of preparation (yield) properties and remarks

both the CO at 185.3 ppm and the phosphite
ligand are assigned positions mutually cis to
the carbonyls trans the $Os(CO)_3P(OCH_3)_3$
moiety; the spectrum closely resembles that
of No. 21 (?; no CO resonances given [15]);
minor Isomer: signals not resolved [16]
IR (cyclohexane): 1934, 1947, 1970, 1982, 1994,
2010, 2026, 2033, 2090 (all CO) [16]
FD mass spectrum: $[M]^+$ [16]
pyrolysis in toluene for 2 h yielded No. 19 and
No. 15 [16]

compounds of the types $Os_3(\mu_3,\eta^2\text{-}OCCH_2C_6H_4CH_3\text{-}4)(CO)_{10}M(\eta^5\text{-}C_5H_5)(CO)$ (Formula V)
and $(\mu\text{-}H)_2Os_3(\mu_3,\eta^2\text{-}OCCH_2C_6H_4CH_3\text{-}4)(CO)_9Mo(\eta^5\text{-}C_5H_5)(CO)$ (Formula VI, No. 25)

23 $Os_3(\mu_3,\eta^2\text{-}OCCH_2C_6H_4CH_3\text{-}4)(CO)_{10}Mo(\eta^5\text{-}C_5H_5)(CO)$

Ia (60%) [19], Ia (65%, along with 10% of No.
34 and 6% of $(\mu\text{-}H)Os_3(\mu_3,\eta^2\text{-}C_2(C_6H_4CH_3\text{-}$
$4)_2)(CO)_8Mo(\eta^5\text{-}C_5H_5)(CO)_2)$; a mechanism
for the formation was discussed [20]
dark red crystalline solid [20]
^1H NMR ($CDCl_3$): 2.36 (s, CH_3), 3.44 to 3.54 (AB
pattern, CH_2; J(H,H) = 13.3), 5.30 (s, C_5H_5),
7.14 to 7.21 (AB pattern, C_6H_4) [19, 20]
^{13}C NMR (CD_2Cl_2, −10 °C): 172.8 (CO-l), 175.2
(μ_3,η^2-CO-a; J(C,C) = 22.2), 177.2 (CO-e),
177.9 (CO-i), 180.1 (CO-d), 180.8 (CO-k),
184.2 (CO-h), 184.3 (CO-g), 186.1 (CO-j;
J(C,C) = 22.2), 189.0, 189.4 (both AB pattern of
^{13}C satellites, CO-f,c; J(C,C) = 35.0), 231.4
(CO-b); the CO resonances were assigned
on the basis of variable-temperature ^{13}C
NMR (spectra in CD_2Cl_2 from −10 to
+30 °C, and in $CDCl_3$ at +40 and +50 °C
depicted) and the C–C coupling constants; a
temperature increase from −10 to +10 °C
results in a broadening of the signals of
CO-j,k,l, while the signals of CO-d,e,f broad-
en at ca. 40 °C, due to threefold localized CO
exchanges at Os(1) and Os(3), respectively;
the CO signals which became broad with the
slowest rate are assigned to CO-g,h,i on the
Os(2) atom. Thus, the activation barrier for
CO exchange processes at the Os centers increases in the sequence Os(1) < Os(3) < Os(2)
[20]

References on p. 208

Table 6 (continued)

No. compound	method of preparation (yield) properties and remarks

23 (continued)

IR (CCl_4): 1978, 2011, 2022, 2032, 2064, 2095
(all CO) [19, 20]; spectrum depicted [20]
mass spectrum: $[M]^+$ [19, 20]
decarbonylation with $(CH_3)_3NO/CH_3CN$ followed
by reaction with H_2 (3 to 4 atm) in toluene at
80 °C yielded No. 25 [19, 21]

*24 $Os_3(\mu_3,\eta^2\text{-}OCCH_2C_6H_4CH_3\text{-}4)(CO)_{10}W(\eta^5\text{-}C_5H_5)(CO)$

Ia (21%, along with 22% of No. 35, and 33% of
isomeric $Os(\mu_3,\eta^2\text{-}C_2(C_6H_4CH_3\text{-}4)_2)(CO)_3$-
$W_2(\eta^5\text{-}C_5H_5)_2(CO)_4$ [2, 5], Ia (ca. 26%, along
with 9.3% of No. 35, 13.5% of the mono-
osmium complex, and 4.8% of No. 36) [11]
dark red crystals from pentane/CH_2Cl_2; from
pentane/CCl_4 at -10 °C crystallizing with 0.5
molecule CCl_4 [5]
^1H NMR (CD_2Cl_2): 2.35 (s, CH_3), 3.47 (s, CH_2),
5.43 (s, C_5H_5), 7.15 to 7.20 (m, C_6H_4) [2, 5]
^{13}C NMR (CD_2Cl_2, -30 °C): 61.2 (t, CH_2;
$^1J(C,H) = 127.2$), 163.3 ($\mu_3,\eta^2\text{-}CO\text{-}a$;
$^1J(C,C) = 58.2$), 173.4 (CO-l), 177.6 (CO-i ?),
177.7 (CO-e), 179.5 (CO-d), 181.7 (CO-k),
184.4 (CO-h?), 185.2 (CO-g?), 186.4 (CO-j;
$^2J(C,C) = 22.0$), 187.0, 187.2 (both superimposed AB pattern of ^{13}C satellites, CO-f,c;
$^2J(C,C) = 34.7$), 224.3 (CO-b; $^1J(C,W) = 155.4$);
a temperature increase from -30 to 0 °C (in
CD_2Cl_2) results in a broadening of the signals of CO-j,k,l, while the signals of CO-d,e,f
broaden at higher temperature, due to threefold localized CO exchanges at Os(1) and
Os(3), respectively; the CO signals which became broad with the slowest rate (at ca.
50 °C in $CDCl_3$) are assigned to CO-g,h,i on
the Os(2) atom. Thus, the activation barrier
for CO exchange processes at the Os
centers increases in the sequence
Os(1) < Os(3) < Os(2) [5]
IR (CCl_4): 1939, 1978, 2007, 2021, 2032, 2063,
2095 (all CO) [2, 5]
mass spectrum: $[M]^+$ [2, 5]

*25 $(\mu\text{-}H)_2Os_3(\mu_3,\eta^2\text{-}OCCH_2C_6H_4CH_3\text{-}4)(CO)_9Mo(\eta^5\text{-}C_5H_5)(CO)$

from No. 23 by decarbonylation with $(CH_3)_3NO$
(see Preparation Method IIb), followed by reaction with H_2 (3 to 4 atm) in toluene at
80 °C for 1 h; purified by TLC with petroleum

Table 6 (continued)

No. compound	method of preparation (yield) properties and remarks

ether/CH_2Cl_2 (3:1) as eluant (more than 86%) [19, 21]

orange crystalline solid from pentane/CH_2Cl_2 [21]

^{1}H NMR ($CDCl_3$): -17.6 (s, μ-H), -12.2 (s, μ-H), 2.35 (s, CH_3), 3.56 to 3.74 (AB pattern, CH_2; $J(H,H) = 13.1$), 5.33 (s, C_5H_5), 7.12 to 7.19 (m, C_6H_4) [19, 21]

^{13}C NMR (CD_2Cl_2, $-10\,^\circ C$): 168.0 (1 CO), 168.9 (d, 1 CO; $^2J(C,H) = 7.1$), 173.4 (μ_3,η^2-CO-a), 174.5 (1 CO), 175.9 (d, 1 CO; $^2J(C,H) = 11.8$), 176.1, 177.4, 180.2, 185.1, 187.2, 224.9 (each 1 CO) [19]; (CD_2Cl_2, $-10\,^\circ C$): 168.2 (CO-h, d, or g), 169.5 (d, CO-i; $^2J(C,H) = 7.1$), 173.6 (μ_3,η^2-CO-a), 175.3 (1 CO), 175.9 (d, CO-e; $^2J(C,H) = 11.8$), 176.5, 177.5 (both AB pattern of ^{13}C satellites, CO-c,f; $^2J(C,C) = 34.6$), 181.0 (CO-j; $^2J(C,C) = 23.4$), 185.9 (CO-g, d, or h), 187.8 (CO-d, g, or h), 224.9 (CO-b); spectra depicted [21]

IR (CCl_4): 1946, 1981, 2000, 2012, 2024, 2033, 2045, 2059, 2076, 2103, 2119 [19, 21]; the shift of the CO bands to higher frequency as compared to those of No. 23 reflects the electronic effect of the hydrides [19]

mass spectrum: [M]$^+$ [21]

treatment with $(CH_3)_3NO$ in CH_2Cl_2/CH_3CN and subsequent thermolysis in toluene at 100 °C yielded an isomeric mixture of Nos. 1, 2, and 3 [19, 21]; see also Preparation Method IIb

compounds of the type $Os_3(\mu_3,\eta^1\text{-CR})(CO)_9W(\eta^5\text{-}C_5H_{5-n}(CH_3)_n)(CO)_2$ (Formula VII)

*26 $Os_3(\mu_3,\eta^1\text{-CH})(CO)_9W(\eta^5\text{-}C_5(CH_3)_5)(CO)_2$

for formation, see No. 4 [23]

red crystals from CH_2Cl_2/heptane [23]

^{1}H NMR (CD_2Cl_2): 2.09 (s, $C_5(CH_3)_5$), 18.76 (s, CH) [23]

^{13}C NMR (CD_2Cl_2): 11.1 ($C_5(\mathbf{CH_3})_5$), 104.5 ($\mathbf{C_5}(CH_3)_5$), 180.7 (6 CO), 181.6 (3 CO), 219.2 (2 CO; $J(C,W) = 154$), 282.6 (μ_3,η^1-C; $J(C,W) = 76$); the CO ligands are believed to undergo highly fluxional exchange processes; the molecule probably possesses a time-average plane of symmetry [23]

IR (cyclohexane): 1922, 1968, 1977, 1983, 1993, 2003, 2033, 2038, 2078 [23]

FAB mass spectrum: [M]$^+$ [23]

References on p. 208

Table 6 (continued)

No. compound	method of preparation (yield) properties and remarks

27 Os$_3$(μ_3,η^1-CCH$_3$)(CO)$_9$W(η^5-C$_5$(CH$_3$)$_5$)(CO)$_2$

for formation, see No. 4 [23]
orange solid [23]
^{1}H NMR (CD$_2$Cl$_2$): 2.08 (s, C$_5$(CH$_3$)$_5$), 3.70 (s, CCH$_3$) [23]
IR (cyclohexane): 1824 (indicating the presence of bridging CO), 1950, 1965, 1974, 1980, 1986, 2003, 2027, 2034, 2075 [23]
FAB mass spectrum: [M]$^+$ [23]

28 Os$_3$(μ_3,η^1-CC$_5$H$_{11}$-n)(CO)$_9$W(η^5-C$_5$H$_5$)(CO)$_2$

by hydrogenation of Os$_3$(μ_4,η^2-CCC$_4$H$_9$-n)-(CO)$_9$W(η^5-C$_5$H$_5$)(CO)$_2$ with H$_2$ (1 atm) in refluxing toluene for 90 min (45%) [15]
solid [15]
^{1}H NMR (CDCl$_3$): 0.97 (t, CH$_3$), 1.46, 1.58, 2.06 (q's, each 1 CH$_2$), 3.26 (t, CH$_2$), 5.33 (s, C$_5$H$_5$) [15]
IR (cyclohexane): 1856, 1973, 1984, 1994, 2007, 2032, 2037, 2079 (all CO) [15]
FAB mass spectrum: [M]$^+$ [15]

*29 Os$_3$(μ_3,η^1-CC$_6$H$_4$CH$_3$-4)(CO)$_9$W(η^5-C$_5$H$_5$)(CO)$_2$

from Os$_3$(η^2-C$_8$H$_{14}$-c)$_2$(CO)$_{10}$ and (η^5-C$_5$H$_5$)-W($\equiv$CC$_6$H$_4$CH$_3$-4)(CO)$_2$ (ca. 1:1) in toluene at room temperature for 5 d [1, 3]; work-up by column chromatography on alumina with CH$_2$Cl$_2$/light petroleum (1:1) as eluant (16%) [3]
by treatment of Os$_3$(CO)$_{10}$(NCCH$_3$)$_2$ in toluene at 110 °C with a cold toluene solution (−20 °C) of (η^5-C$_5$H$_5$)W($\equiv$CC$_6$H$_4$CH$_3$-4)(CO)$_2$ followed by heating to reflux for 30 min, evaporation and extraction of the residue with CH$_2$Cl$_2$; purified by TLC with pentane/CH$_2$Cl$_2$, 3:2, as eluant; R$_f$=0.25 (42%, along with Os$_3$(CO)$_{12}$ in 36% yield) [16]
red-brown crystals from CH$_2$Cl$_2$/light petroleum [3, 16], m. p. 225 °C [3]
^{1}H NMR (CDCl$_3$): 2.23 (s, CH$_3$), 5.48 (s, C$_5$H$_5$), 7.08 to 7.40 (m, C$_6$H$_4$) [3]; (CD$_2$Cl$_2$): 2.27 (s, CH$_3$), 5.52 (s, C$_5$H$_5$), 7.12 to 7.46 (m, C$_6$H$_4$) [16]
^{13}C NMR (CD$_2$Cl$_2$/CH$_2$Cl$_2$): 21.7 (CH$_3$), 90.2 (C$_5$H$_5$), 129.7, 132.2 (C–both C-2,3 of C$_6$H$_4$), 138.3 (C-4 of C$_6$H$_4$), 167.5 (C-1 of C$_6$H$_4$),

References on p. 208

Table 6 (continued)

No. compound	method of preparation (yield) properties and remarks

180.6, 181.9 (11 CO), 264.7 (μ_3-C) [3]; (CD$_2$Cl$_2$, -80 °C): 175.8 (CO-e), 179.8 (br), 181.0 (each 3 CO, CO-f to k), 181.4 (CO-d), 187.9 (CO-c), 202.1 (CO-a, J(^{183}W,C) = 143.3), 218.4 (CO-b; J(^{183}W,C) = 131.9); at ca. -40 °C the two doubly bridged Os centers undergo fast threefold exchanges localized at each atom, and the activation barrier at one Os center is higher as indicated by the broad resonance at 179.8 ppm at -80 °C; with increasing temperature the resonances for CO-a,b broaden and those for CO-f to k and for CO-c to e broaden and coalesce at ca. 20 °C into single resonances, implying that between -40 and $+20$ °C the tungsten and Os(3) centers undergo slow and fast exchanges of the ligands, respectively; at 100 °C the spectrum exhibits only three resonances with 2:3:6 intensity ratios, revealing a time-averaged symmetrical configuration with a mirror plane through μ_3-C, W, and Os(3), which arises from localized ligand exchange on each metal center and equilibration of the six CO-f to k; the activation barriers for localized carbonyl exchange increase as Os(1), Os(2) < Os(3) < W; spectra at -80, -40, 0, $+20$, $+35$, and $+100$ °C depicted [16]

IR (methylcyclohexane): 1836, 1974, 1980, 1999, 2009, 2033, 2040, 2078 (all CO) [1, 3]; (CCl$_4$): 1975, 1982, 2000, 2007, 2033, 2039, 2078 [16]

FD mass spectrum: [M]$^+$ [3]

mass spectrum: [M]$^+$, [M $-$ x CO]$^+$, x = 1 to 11 [16]

compounds of the type Os$_3(\mu_3,\eta^1$-CR)(μ_3,η^1-CR$'$)(CO)$_9$W(η^5-C$_5$H$_5$)X (Formula VIII)

30 Os$_3(\mu_3,\eta^1$-CC$_6$H$_5)_2$(CO)$_9$W(η^5-C$_5$H$_5$)H

IIa (10 to 20% from (μ-H)Os$_3(\mu_2,\eta^2$-C$_6$H$_5$C=CC$_6$H$_5$)(CO)$_8$W(η^5-C$_5$H$_5$)(CO)$_2$) [4, 14]

IIb (60% from (μ-H)Os$_3(\mu_2,\eta^2$-C$_6$H$_5$C=CC$_6$H$_5$) (CO)$_8$W(η^5-C$_5$H$_5$)(CO)$_2$) [4, 14], along with 9% of the μ-oxo compound of type XI, p. 193 [14]

dark red solid [14]

^{1}H NMR (CD$_2$Cl$_2$): 4.00 (s, WH; ^{1}J(H,W) = 89), 5.42 (s, C$_5$H$_5$), 7.14 to 7.32 (m, C$_6$H$_5$) [14]

References on p. 208

Table 6 (continued)

No. compound	method of preparation (yield) properties and remarks

30 (continued)

^{13}C NMR (CD$_2$Cl$_2$): 268.1 (μ_3-C; ^{1}J(C,W) = 106) [4, 14]

IR (cyclohexane): 1981?, 1992, 2003, 2037, 2041, 2077 (all CO) [14]

FD mass spectrum: [M]$^+$ [14]

31 Os$_3$(μ_3,η^1-CC$_6$H$_5$)(μ_3,η^1-CC$_6$H$_4$CH$_3$-4)(CO)$_9$W(η^5-C$_5$H$_5$)H

IIa (10 to 20% from (μ-H)Os$_3$(μ_2,η^2-4-CH$_3$C$_6$H$_4$C=CC$_6$H$_5$)(CO)$_8$W(η^5-C$_5$H$_5$)(CO)$_2$) [4, 14]

IIb (61% from (μ-H)Os$_3$(μ_2,η^2-4-CH$_3$C$_6$H$_4$C=CC$_6$H$_5$)(CO)$_8$W(η^5-C$_5$H$_5$)(CO)$_2$) [4, 14], along with traces of the μ-oxo compound of type XI, p. 193 [14]

dark red crystalline solid [14]

^{1}H NMR (CD$_2$Cl$_2$): 2.36 (s, CH$_3$), 3.89 (s, WH; ^{1}J(H,W) = 89), 5.41 (s, C$_5$H$_5$), 7.04 to 7.31 (m, C$_6$H$_4$, C$_6$H$_5$) [14]

IR (cyclohexane): 1981, 1993, 2003, 2011, 2036, 2041, 2078 (all CO) [14]

FAB mass spectrum: [M]$^+$, [M − x CO]$^+$, x = 1 to 3 [14]

*32 Os$_3$(μ_3,η^1-CC$_6$H$_4$CH$_3$-4)$_2$(CO)$_9$W(η^5-C$_5$H$_5$)H

Ib (53%) [11]

IIa (10 to 20% from (μ-H)Os$_3$(μ_2,η^2-4-CH$_3$C$_6$H$_4$C=CC$_6$H$_4$CH$_3$-4)(CO)$_8$W(η^5-C$_5$H$_5$)(CO)$_2$) [4, 14], IIa (90% from No. 35) [6]

IIb (63% from (μ-H)Os$_3$(μ_2,η^2-4-CH$_3$C$_6$H$_4$C=CC$_6$H$_4$CH$_3$-4)(CO)$_8$W(η^5-C$_5$H$_5$)(CO)$_2$) [4, 14], along with 8% of the μ-oxo compound of type XI, p. 193 [14]

dark red crystals from hexane/CH$_2$Cl$_2$ [4, 6, 14]

^{1}H NMR (CD$_2$Cl$_2$): 2.36 (s, CH$_3$), 3.75 (s, WH; ^{1}J(H,W) = 89), 5.40 (s, C$_5$H$_5$), 7.03 to 7.11 (m, C$_6$H$_4$) [4, 6, 14]

^{13}C NMR (CD$_2$Cl$_2$, −95 °C): 172.3 (2 CO of CO-a,a',b,b',c,c'), 172.6 (CO-d,d'), 175.7 (CO-e), 179.6 185.5 (each 2 CO of CO-a,a',b,b',c,c'), 268.1 (μ_3-C; ^{1}J(C,W) = 106) [4, 6]; spectra at −90, −50, and +20 °C depicted; between −95 and −50 °C the three resonances assigned to CO-a,a',b,b',c,c', and between −50 to +20 °C the resonances assigned to CO-d,d' and CO-e broaden and coalesce being consistent with CO scrambling involving threefold localized CO ex-

Table 6 (continued)

No. compound	method of preparation (yield) properties and remarks

changes at each Os center; the activation barriers are in the order of $Os(1) = Os(1') < Os(2)$ [6]

IR (cyclohexane): 1980, 1991, 1999, 2031, 2036, 2072 (all CO) [4, 6, 14]

mass spectrum: [M]$^+$ [4, 6, 14], [M − x CO]$^+$, x = 1 to 9 [14]

33   $Os_3(\mu_3,\eta^1\text{-}CC_6H_4CH_3\text{-}4)_2(CO)_9W(\eta^5\text{-}C_5H_5)Cl$

from No. 32 and CCl_4 in the presence of ca. one equivalent of $t\text{-}C_4H_9OCl$ at 25 °C for 1 h, followed by TLC with pentane/CH_2Cl_2 (3: 2) as eluant [4, 6]

solid [6]

^{1}H NMR (CD_2Cl_2): 2.47 (s, CH_3), 5.56 (s, C_5H_5), 7.15 to 7.33 (m, C_6H_4) [4, 6]

IR (cyclohexane): 1980, 1999, 2041, 2046, 2080 (all CO) [4, 6]

FD mass spectrum: [M]$^+$ [4, 6]

compounds of the type $(\mu\text{-}H)Os_3(\mu_3,\eta^1\text{-}CC_6H_4CH_3\text{-}4)_2(\mu_3\text{-}CO)(CO)_9M(\eta^5\text{-}C_5H_5)$ (Formula IX)

34   $(\mu\text{-}H)Os_3(\mu_3,\eta^1\text{-}CC_6H_4CH_3\text{-}4)_2(\mu_3\text{-}CO)(CO)_9Mo(\eta^5\text{-}C_5H_5)$

Ia (10%, along with 65% of No. 23 and 6% of $(\mu\text{-}H)Os_3(\mu_3,\eta^2\text{-}C_2(C_6H_4CH_3\text{-}4)_2)(CO)_8$ $Mo(\eta^5\text{-}C_5H_5)(CO)_2$, isomeric with No. 34); a mechanism for the formation was discussed [20]

red crystalline solid [20]

^{1}H NMR ($CDCl_3$, −40 °C): −21.5 (s, μ-H), 2.33, 2.39 (s's, both CH_3), 5.35 (s, C_5H_5), 6.97 to 7.30 (m, C_6H_4); the two distinct CH_3 resonances broaden upon raising the temperature and coalesce at ca. 23 °C indicating a framework rearrangement; the free activation energy for the enantiomer–interchange process was evaluated as $\Delta G^{\neq} = 15.2 \pm 0.2$ kcal/mol at 23 °C [20]

IR (cyclohexane): 1718 (μ_3–CO), 1975, 1991, 2002, 2012, 2025, 2054, 2071, 2092 (all CO) [20]

mass spectrum: [M]$^+$ [20]

*35 $(\mu\text{-}H)Os_3(\mu_3,\eta^1\text{-}CC_6H_4CH_3\text{-}4)_2(\mu_3\text{-}CO)(CO)_9W(\eta^5\text{-}C_5H_5)$

Ia (22%, along with 21% of No. 24, and 33% of isomeric $Os(\mu_3,\eta^2\text{-}C_2(C_6H_4CH_3\text{-}4)_2)\text{-}(CO)_3W_2(\eta^5\text{-}C_5H_5)_2(CO)_4$ [5]; see also [2], Ia (9.3%, along with ca. 26% of No. 24, 13.5%

References on p. 208

Table 6 (continued)

No. compound	method of preparation (yield) properties and remarks

*35 (continued)

of the mono-osmium complex, and 4.8% of No. 36 [11], Ib (26%) [11]

dark red crystals from CH_2Cl_2/pentane [5, 6, 11]

^1H NMR (CD_2Cl_2): -21.57 (s, μ-H), 2.25, 2.34 (s's, both CH_3), 5.26 (s, C_5H_5), 6.97 to 7.11 (m, C_6H_4) [5]; in $CDCl_3$, the two distinct CH_3 resonances broaden above 25 °C and coalesce at ca. 50 °C indicating a framework rearrangement; the rate constant for the enantiomer-interchange process is 18 s^{-1} at 20 °C and 278 s^{-1} at 60 °C; the free activation energy for the enantiomer-interchange process was evaluated as $\Delta G^+ = 15.7 \pm 0.3$ kcal/mol at 50 °C [6]

^{13}C NMR ($CDCl_3$, -50 °C): 166.3 (d, CO-c; $^2J(C,H) = 12.3$), 170.5 (d, CO-f; $^2J(C,H) = 11.8$), 186.5 (s, μ_3-C-a; $^1J(C,W) = 106$), 209.0 (s, μ_3-C-b, $^1J(C,W) = 108$), 281.0 (μ_3-CO-h; $^1J(C,W) = 108$); other CO resonances not given; increasing the temperature to $+50$ °C caused broadening of all resonances with the exception of CO-c and CO-h; the results indicate a degenerate framework rearrangement that interchanges the two possible enantiomeric forms by reversible migration of the hydride and a CO exchange process in a pairwise manner: $a \rightarrow b$, $d \rightarrow d'$, $e \rightarrow e'$, $f \rightarrow f'$ and $g \rightarrow g'$; complete coalescence of μ_3-C-a,b could not be observed due to conversion into No. 32 above 70 °C [6]

IR (cyclohexane): 1703 (μ_3-CO), 1975, 1990, 1996, 2001, 2011, 2025, 2052, 2071, 2091 (all CO) [5]

FD mass spectrum: $[M]^+$ [5]

thermolysis in refluxing toluene for 1.5 h gave No. 32; see also Scheme 1, p. 207 [6]

a compound with a terminal bonded (η^5-C_5H_5)W(CO)$_3$ ligand (Formula X)

*36 $(\mu$-H$)_2$Os$_3(\mu_3,\eta^1$-CC$_6$H$_4$CH$_3$-4)(CO)$_9$W(η^5-C$_5$H$_5$)(CO)$_3$

Ia (4.8%; see also No. 35) [11]

yellow-orange crystals from CH_2Cl_2/CH_3OH [11]

^1H NMR ($CDCl_3$): -19.78, -17.50 (d's, both μ-H; $J(H,H) = 1.4$), 2.37 (s, CH_3), 5.45 (s, C_5H_5), 7.05 to 7.44 (m, C_6H_4) [11]

References on p. 208

Table 6 (continued)

No. compound	method of preparation (yield) properties and remarks

^{13}C NMR (CD$_2$Cl$_2$): 164.7 (1 CO; J(C,H) = 11.5), 167.1 (1 CO), 168.8 (1 CO; J(C,H) = 11.8), 169.0 (1 CO), 171.5 (1 CO; J(C,H) = 11.2), 174.0 (1 CO), 176.1 (1 CO; J(C,H) = 9.2), 183.7, 189.8 (each 1 CO; ^{2}J(C,C) = 30), 190.4 (μ$_3$–C), 215.9, 216.5, 222.3 (each 1 CO-W; ^{1}J(C,W) = 158, 159, and 146); the four CO resonances showing C,H couplings are located in trans positions to μ–hydrides; the resonances at 183.7 and 189.8 ppm showing large C,C couplings are assigned to the trans axial carbonyls on the Os center bearing the W(η^5-C$_5$H$_5$)(CO)$_3$ fragment [11]

IR (cyclohexane): 1909, 1921, 1986, 1944, 2010, 2024, 2052, 2067, 2075, 2098 (all CO) [11]

FAB mass spectrum: [M]$^+$ [11]

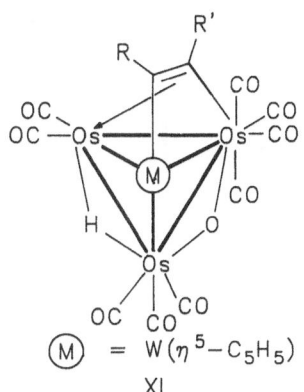

$\textcircled{M}$ = W(η^5–C$_5$H$_5$)

XI

* Further information:

(μ–H)Os$_3$(μ$_2$,η^1-CHCH$_2$C$_6$H$_4$CH$_3$-4)(CO)$_9$(μ-O)Mo(η^5-C$_5$H$_5$) (Table **6**, Nos. **1**, **2**, and **3**, corresponding to Isomers 1, 2, and 3 in Formula I, p. 171). Deuterated No. 2 (100% on μ–alkylidene carbon, 37% on μ–hydride) was obtained from a mixture of Nos. 1 and 3 upon treatment with DCl in acetone-d$_6$. The H/D exchange probably occurs independently on the two sites [19, 21].

No. 2 crystallizes in the orthorhombic space group Pmn2$_1$ – C$_{2v}^7$ (No. 31) with a = 12.438(7), b = 8.797(2), c = 11.846(4) Å; Z = 2, D$_c$ = 2.87 g/cm^3, R = 0.056. The solid state structure, depicted in **Fig. 59**, shows the CH$_2$C$_6$H$_4$CH$_3$-4 moiety on the alkylidene carbon to be oriented syn with respect to the Os(1)–Mo–Os(1') face and inclined towards the μ–oxo ligand. Both μ–oxo and μ$_2$,η^1-CHCH$_2$ are disordered out of the crystallographic mirror plane which passes through Mo and Os(2) and bisects the Os(1)–Os(1') bond, with each occupying the opposite side of the plane with a half occupancy. The C-1, C-4, and methyl carbons of

References on p. 208

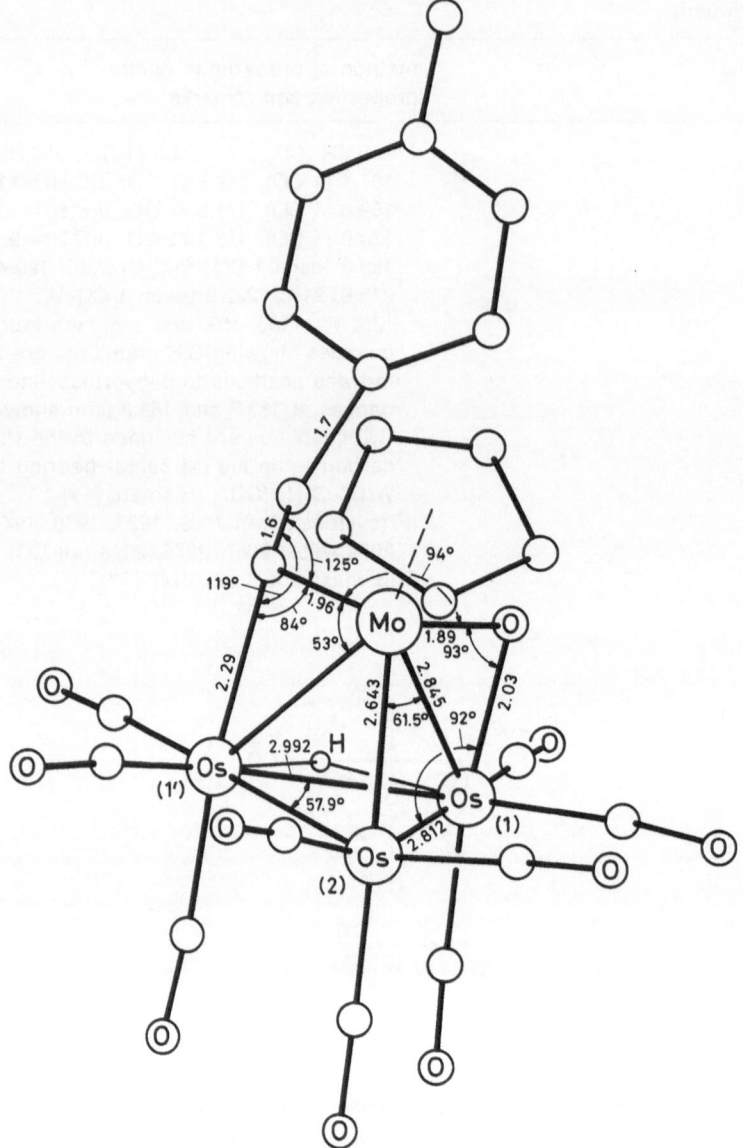

Fig. 59. Molecular structure of $(\mu\text{-H})Os_3(\mu_2,\eta^1\text{-CHCH}_2C_6H_4CH_3\text{-}4)(CO)_9(\mu\text{-O})Mo(\eta^5\text{-}C_5H_5)$ (No. 2; see also Formula I, Isomer 2) with selected bond distances (in Å) and bond angles [19, 21].

$C_6H_4CH_3$ are located nearly on this mirror plane and could be treated as ordered in the refinement. The μ-hydride was not observed crystallographically but is believed to span the Os(1)–Os(1′) bond edge. The compound as a whole is chiral; the crystal contains a disordered racemic mixture of two enantiomers. Average bond distances Mo–C (of η^5-C_5H_5), Os–CO and C≡O are 2.37, 1.88 and 1.18 Å, respectively [19, 21].

References on p. 208

Nos. 1 and 3 equilibrate in solution by rapid hydride migration between the edges of Os–Os bonds. The hydride migration rates were investigated using ^{1}H NMR spectroscopy by dissolving No. 3 in acetone-d_6 at 23 °C. The equilibration is of first order with rate constants k_1(No. 3 → No. 1) = (4.95 ± 0.02) × 10^{-3} s^{-1}, k_{-1}(No. 1 → No. 3) = (5.05 ± 0.02) × 10^{-3} s^{-1}, $k_{observed}$ = (1.00 ± 0.07) × 10^{-2} s^{-1}, and the equilibrium constant K(No. 1/No. 3) = 0.981 ± 0.003 at 23 °C (in CDCl$_3$?) [21]. However, k_1 = 2.71 × 10^{-5} s^{-1}, k_{-1} = 3.26 × 10^{-5} s^{-1}, $k_{observed}$ = 5.97 × 10^{-5} s^{-1}, and K(No. 1/No. 3) = 1.20 were given in a preliminary report [19].

Thermolysis of a mixture of Nos. 1 and 3 in toluene at 100 °C for 19 h or reaction with aqueous HCl (DCl) in acetone between 45 and 56 °C yielded No. 2 in ca. 70% yield. The formation of No. 2 proceeds by an apparent rotation of the μ-alkylidene group; a mechanism is proposed based on the deuterium labelling study [19, 21].

(μ-H)Os$_3$(μ$_2$,η^1-CHC$_6$H$_4$CH$_3$-4)(CO)$_9$(μ-O)W(η^5-C$_5$H$_5$) (Table 6, No. 5) crystallizes in the triclinic space group P$\bar{1}$–C$_i^1$ (No. 2) with a = 9.763(5), b = 10.759(4), c = 13.336(3) Å, α = 88.00(3)°, β = 108.56(3)°, γ = 100.44(3)°; Z = 2, D$_c$ = 3.034 g/cm^3, R = 0.057. The structure is shown in **Fig. 60**. The C$_6$H$_4$CH$_3$-4 moiety on the μ-alkylidene carbon is oriented syn with respect to the triply edge-bridged Os(1)-W-Os(2) triangle and is disposed towards the bridging μ-oxo ligand. The hydride ligand was not observed crystallographically but is believed to bridge the Os(1)-Os(2) bond due to the enlargement of the bond distance and two of the Os-CO-Os angles. Average bond distances of W-C (of η^5-C$_5$H$_5$), Os-CO and C≡O are 2.35, 1.88 and 1.15 Å, respectively [18].

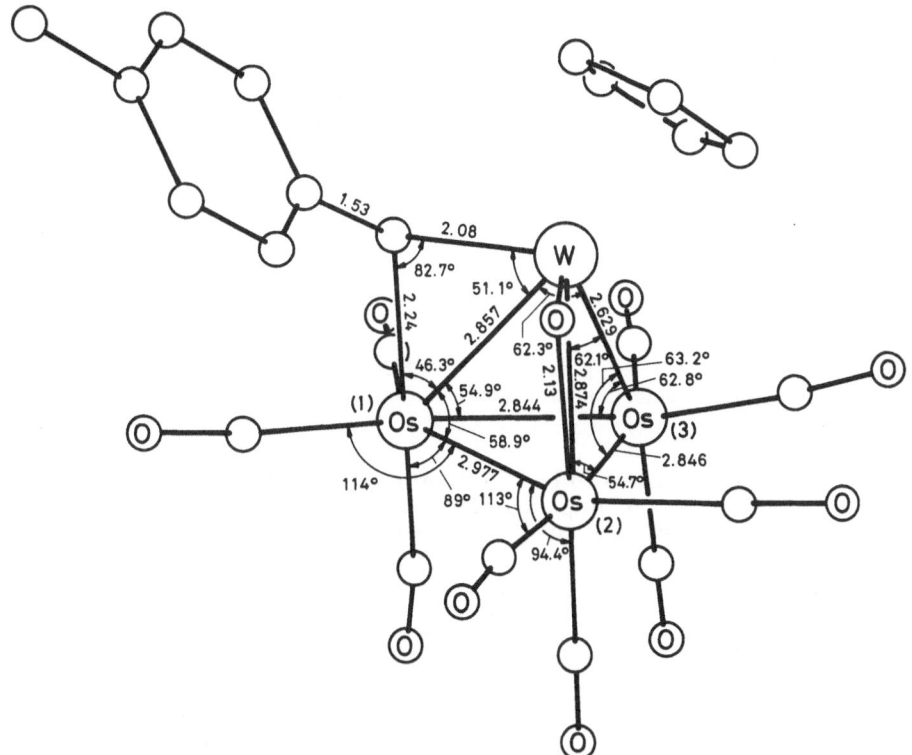

Fig. 60. Molecular structure of (μ-H)Os$_3$(μ$_2$,η^1-CHC$_6$H$_4$CH$_3$-4)(CO)$_9$(μ-O)W(η^5-C$_5$H$_5$) (No. 5) with selected bond distances (in Å) and bond angles [18].

References on p. 208

(μ–H)Os$_3$(μ$_2$,η^1-CHCH$_2$C$_6$H$_4$CH$_3$-4)(CO)$_9$(μ-O)W(η^5-C$_5$H$_5$) (Table **6**, No. **6**) crystallizes in the monoclinic space group Cc–C$_s^4$ (No. 9) with a = 14.1510(27), b = 13.9257(22), c = 13.3179(19) Å, β = 92.023(13)°; Z = 4, D$_c$ = 3.06 g/cm^3, R = 0.035. The solid state structure, shown in **Fig. 61**, corresponds to the anti isomer (Isomer 3 in Formula I, p. 171). The compound as a whole is chiral. However, by virtue of the crystallographic c-glides of the space group, the crystal contains a racemic mixture of the two enantiomers. The considerable differences in metal-metal distances are associated in part with significant variations in formal electron-counts at the individual metal atoms ranging between 17 electrons for W and 19.5 for Os(3). Average bond distances of W–C (of η^5-C$_5$H$_5$), Os–CO and C≡O are 2.376, 1.883 and 1.17 Å, respectively [9].

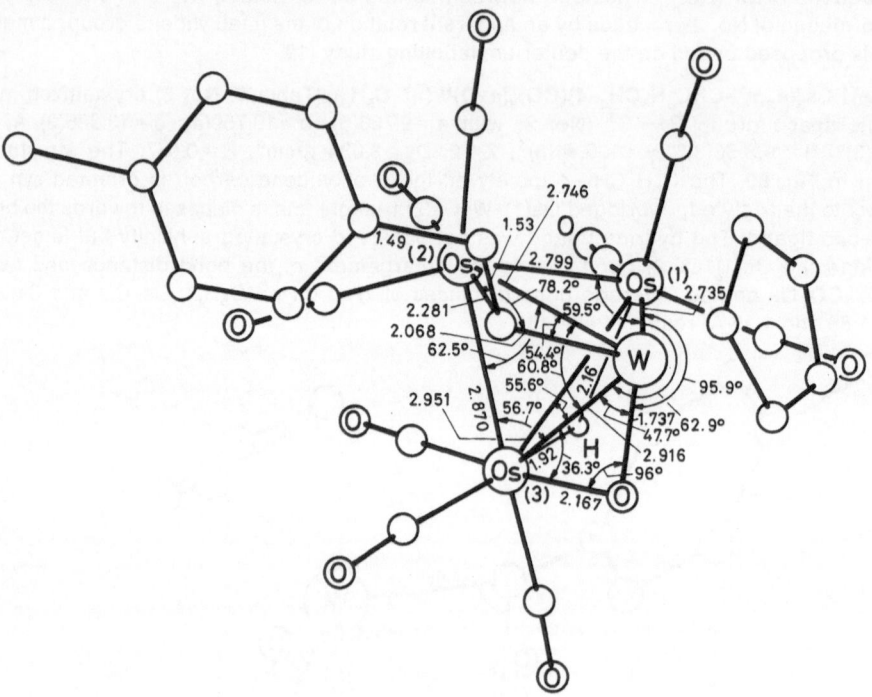

Fig. 61. Molecular structure of (μ-H)Os$_3$(μ$_2$,η^1-CHCH$_2$C$_6$H$_4$CH$_3$-4)(CO)$_9$(μ-O)W(η^5-C$_5$H$_5$) (No. 6) with selected bond distances (in Å) and bond angles [9].

Os$_3$(μ$_2$,η^1-CHCH$_2$C$_6$H$_4$CH$_3$-4)(CO)$_9$(μ-Cl)(μ-O)W(η^5-C$_5$H$_5$) (Table **6**, No. **7**). Deuterated Os$_3$(μ$_2$,η^1-CDCH$_2$C$_6$H$_4$CH$_3$-4)(CO)$_9$(μ-Cl)(μ-O)W(η^5-C$_5$H$_5$) was prepared according to Preparation Method IIIa (p. 174) from No. 17 and BCl$_3$, followed by hydrolysis with D$_2$O [12].

The title compound crystallizes in the triclinic space group P$\bar{1}$–C$_i^1$ (No. 2) with a = 9.2724(32), b = 11.3130(43), c = 14.0040(62) Å, α = 69.790(31)°, β = 77.236(30)°, γ = 83.700(29)°; Z = 2, D$_c$ = 3.07 g/cm^3, R = 0.034. The structure is shown in **Fig. 62**. The compound as a whole is chiral. In addition, μ$_2$,η^1-C behaves as an isolated chiral center. The crystal contains an ordered racemic arrangement of discrete enantiomers. The CH$_2$C$_6$H$_4$CH$_3$-4 substituent of the μ-alkylidene moiety is anti oriented to the triply edge-bridged Os(1)-W-Os(3) face; the Os(1)-Os(3) edge with a nonbonding interatomic distance of 3.747 Å is symmetrically spanned by the μ-Cl. Average bond distances of W–C (of η^5-C$_5$H$_5$), Os–CO and C≡O are 2.378, 1.892 and 1.149 Å, respectively [12].

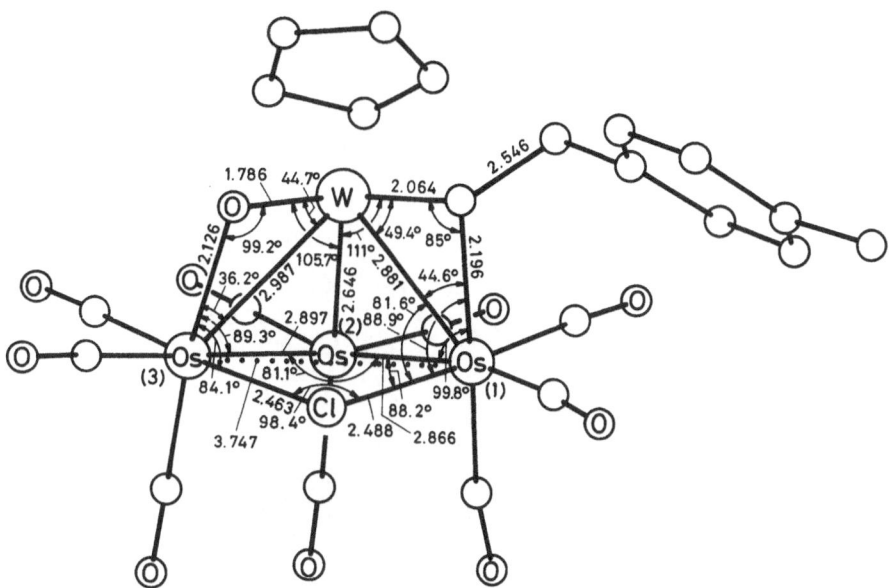

Fig. 62. Molecular structure of $Os_3(\mu_2,\eta^1\text{-}CHCH_2C_6H_4CH_3\text{-}4)(CO)_9(\mu\text{-}Cl)(\mu\text{-}O)W(\eta^5\text{-}C_5H_5)$ (No. 7) with selected interatomic distances (in Å) and bond angles [12].

Pyrolysis in refluxing toluene for 1 h led to the syn isomer No. 8 (Isomer 2 in Formula I); a possible conversion mechanism was discussed. Similarly, deuterated No. 7 led to μ-CD-deuterated No. 8 [12].

One carbonyl group was stereoselectively substituted by treatment with an excess of $P(C_6H_5)_2CH_3$ in refluxing toluene to give No. 9 within 5 min. The specific site of substitution could not be estimated unequivocally from the spectroscopic data, but it is very probable that the carbonyl group at Os(2) (see Formula I, p. 171) which is cis to both the μ-O and the μ-Cl ligand was substituted, being consistent with this cis-labilization of π-donor ligands [12].

$(\mu\text{-}H)Os_3(\mu_2,\eta^1\text{-}C\text{=}CHC_6H_4CH_3\text{-}4)(CO)_9(\mu\text{-}O)W(\eta^5\text{-}C_5H_5)$ (Table **6**, No. **14**) crystallizes in the monoclinic space group $P2_1/c - C_{2h}^5$ (No. 14) with a = 9.522(2), b = 22.376(6), c = 12.761(3) Å, β = 98.276(17)°; Z = 4, D_c = 2.97 g/cm³, R = 0.079. The molecule (see **Fig. 63**) has only C_1 symmetry and is chiral; however, the crystal contains a racemic mixture of the two enantiomers interrelated by operations such as inversion centers and glide planes. The significant differences in the metal–metal bond distances probably result from the appreciable variations in formal electron-counts at the individual metal atoms. Although the hydride is not observed crystallographically, it is believed to bridge the longest edge, the Os(1)–Os(3) bond, due to the bond lengthening effect of μ-H. Average bond distances of W–C (of $\eta^5\text{-}C_5H_5$), Os–CO and C≡O are 2.38, 1.89 and 1.15 Å, respectively [10].

$Os_3(\mu_3,\eta^1\text{-}CCH_3)(CO)_9(\mu\text{-}O)W(\eta^5\text{-}C_5(CH_3)_5)$ (Table **6**, No. **16**) crystallizes in the triclinic space group $P\bar{1} - C_i^1$ (No. 2) with a = 9.524(3), b = 10.110(2), c = 15.550(4) Å, α = 75.09(2)°, β = 77.06(2)°, γ = 63.74(2)°; Z = 2, D_c = 3.057 g/cm³, R, R' = 0.039, 0.043. The structure of the molecule, shown in **Fig. 64**, resembles that of No. 17 and its phosphane substituted derivative No. 18 [23].

References on p. 208

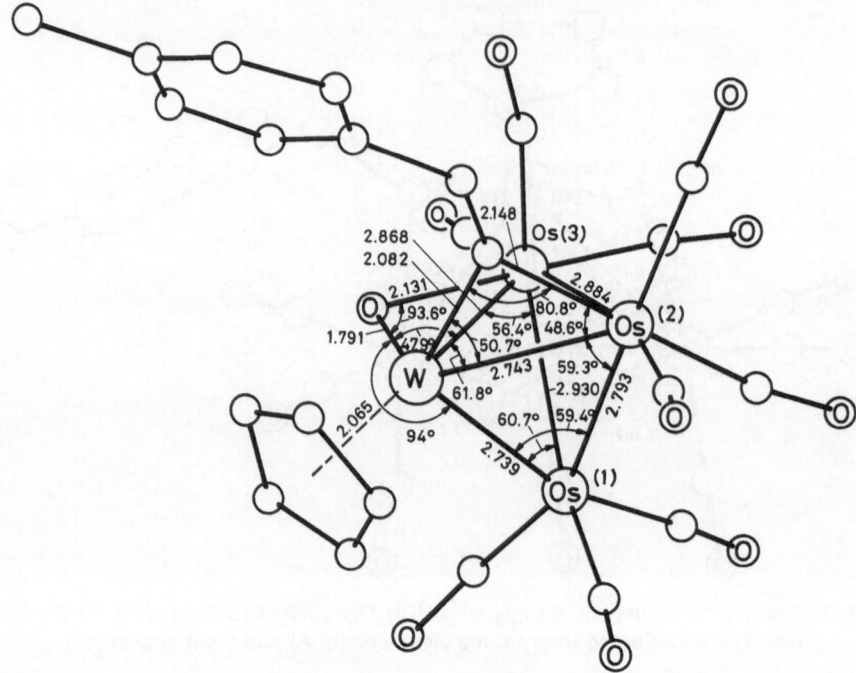

Fig. 63. Molecular structure of $(\mu\text{-H})Os_3(\mu_2,\eta^1\text{-C=CHC}_6H_4CH_3\text{-4})(CO)_9(\mu\text{-O})W(\eta^5\text{-C}_5H_5)$ (No. 14) with selected bond distances (in Å) and bond angles [10].

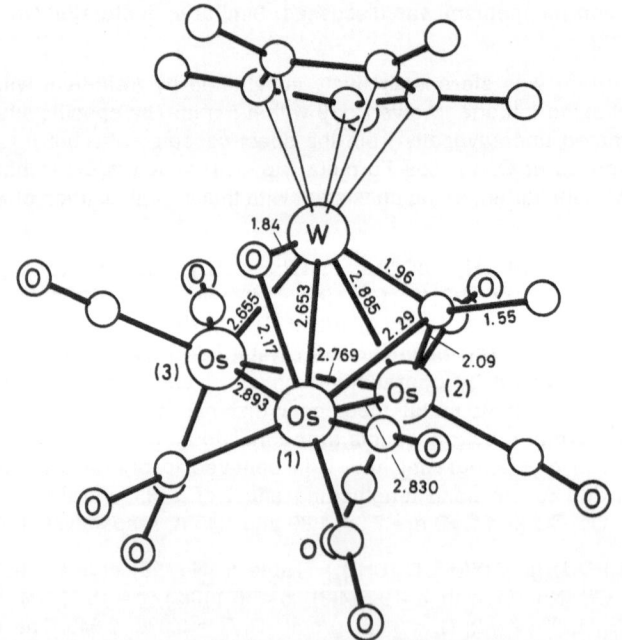

Fig. 64. Molecular structure of $Os_3(\mu_3,\eta^1\text{-CCH}_3)(CO)_9(\mu\text{-O})W(\eta^5\text{-C}_5(CH_3)_5)$ (No. 16) with selected bond distances (in Å) [23]

References on p. 208

Gmelin Handbook
Os-Org. Comp. B6

Os₃(μ₃,η¹-CCH₂C₆H₄CH₃-4)(CO)₉(μ-O)W(η⁵-C₅H₅) (Table **6**, No. **17**) crystallizes in the triclinic space group P$\bar{1}$ – C$_i^1$ (No. 2) with a = 9.7041(25), b = 9.7829(27), c = 14.5992(39) Å, α =
108.106(22)°, β = 93.805(21)°, γ = 94.565(22)°; Z = 2, D$_c$ = 3.06 g/cm³, R = 0.039 [7, 8]. The molecule, shown in **Fig. 65**, is chiral, but the crystal contains an ordered racemic mixture of
the two enantiomers. The significant differences in the metal-metal distances are assigned
to the considerable variations in formal electron-counts at the individual metal atoms. The
interatomic distances between Os(2) and CO′ and between Os(3) and CO″ of ca. 2.89 and
2.93 Å, respectively, indicate some slight semi-bridging character of the two CO ligands.
Average bond distances of W-C (of η⁵-C₅H₅), Os-CO and C≡O are 2.382, 1.909 and 1.139Å
[8].

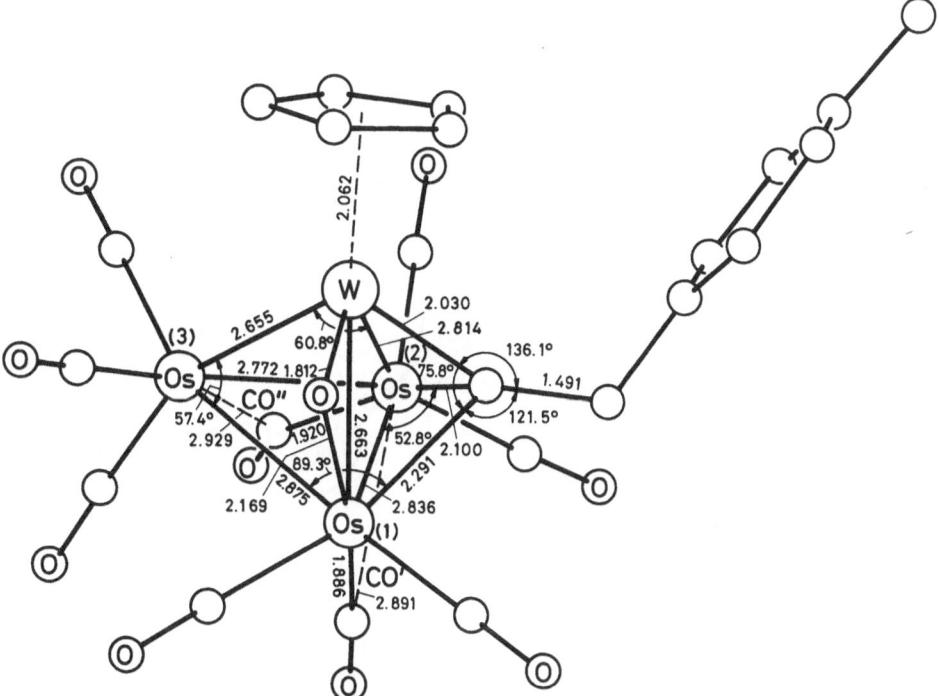

Fig. 65. Molecular structure of Os₃(μ₃,η¹-CCH₂C₆H₄CH₃-4)(CO)₉(μ-O)W(η⁵-C₅H₅) (No. **17**)
with selected bond distances (in Å) and bond angles [7, 8].

Thermolysis in refluxing toluene under an N₂ atmosphere yielded No. 14, whereas hydrogenation by similar conditions led to No. 6; see Preparation Methods IIa and IIIa, p. 174
[7].

The reactions with HCl, HBr or HSC₆H₅ in toluene resulted in No. 7, isomeric mixtures
of Nos. 10 and 11, and of Nos. 12 and 13, respectively; see also Preparation Method IIIa,
p. 174 [12]. Treatment with BCl₃/H₂O gave No. 7 [12].

One carbonyl ligand was substituted by treatment with P(C₆H₅)₂CH₃ or P(OCH₃)₃ in
refluxing toluene to give No. 18 or a mixture of Nos. 19 and 22; see Preparation Method
IIIb, p. 174. The relative amounts of Nos. 19 and 22 are independent of the concentration
of P(OCH₃)₃ utilized with an excess of 14 or 1.5 equivalents, suggesting that the rate laws,
controlling the formation of Nos. 19 and 22 have the same ligand dependence [16].

                          References on p. 208

Carbonylation led to $Os_3(\mu_3,\eta^1-CCH_2C_6H_4CH_3-4)(CO)_{10}(\mu-O)W(\eta^5-C_5H_5)$ (No. 21) [22].

$Os_3(\mu_3,\eta^1-CCH_2C_6H_4CH_3-4)(CO)_8P(C_6H_5)_2CH_3(\mu-O)W(\eta^5-C_5H_5)$ (Table 6, No. **18**) crystallizes in the triclinic space group $P\bar{1} - C_i^1$ (No. 2) with a = 10.111(5), b = 11.732(4), c = 16.836(6)Å, $\alpha = 78.508(25)°$, $\beta = 79.408(31)°$, $\gamma = 66.321(26)°$; Z = 2, $D_c = 2.57$ g/cm^3, R = 0.041. The molecular structure of one of the chiral molecules is shown in **Fig. 66**. The crystal contains an ordered racemic mixture of the two enantiomers. The phosphane ligand is located cis to the bridging oxygen and trans to Os(3); it is displaced out of the W-Os(1)-Os(2) plane in the direction of the μ_3-alkylidyne ligand. Average bond distances of W-C (of $\eta^5-C_5H_5$), Os-CO and C≡O are 2.38, 1.907 and 1.14 Å, respectively [16].

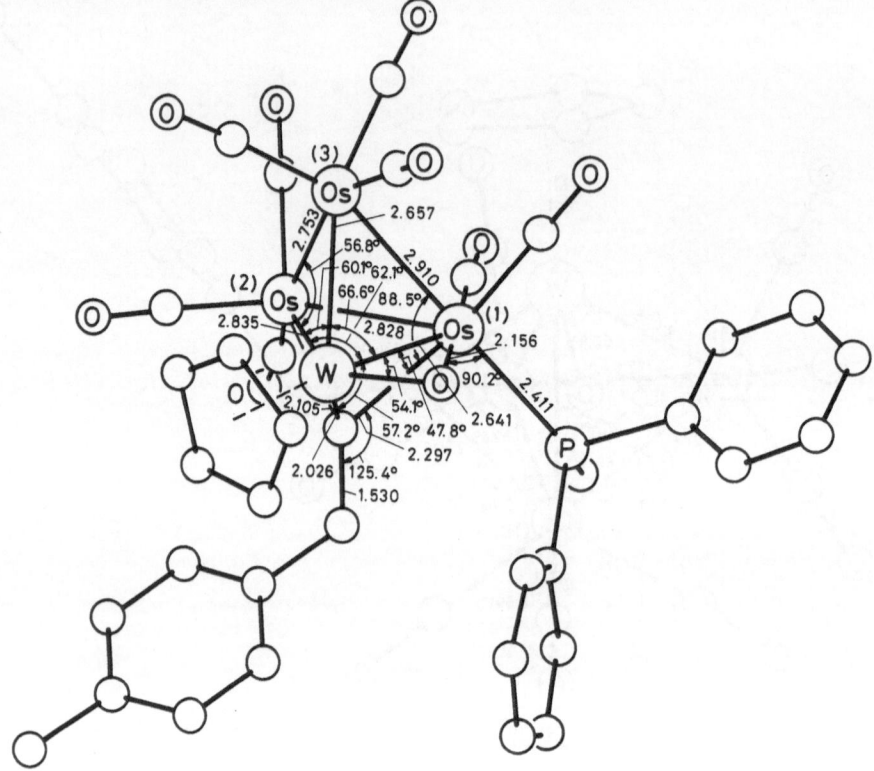

Fig. 66. Molecular structure of $Os_3(\mu_3,\eta^1-CCH_2C_6H_4CH_3-4)(CO)_8P(C_6H_5)_2CH_3(\mu-O)W(\eta^5-C_5H_5)$ (No. 18) with selected bond distances (in Å) and bond angles [16].

$Os_3(\mu_3,\eta^2-OCCH_2C_6H_4CH_3-4)(CO)_{10}W(\eta^5-C_5H_5)(CO)$ (Table 6, No. **24**). The deuterated complex $Os_3(\mu_3,\eta^2-OCCD_2C_6H_4CH_3-4)(CO)_{10}W(\eta^5-C_5H_5)(CO)$ was obtained in 42 to 49% yield from $(\mu-D)_2Os_3(CO)_{10}$ as described for Preparation Method Ia, p. 173. The reaction led to a partially modified product distribution that strongly favors No. 24 at the expense of deuterated No. 35 and No. 36 [11].

The title compound crystallizes from pentane/CCl$_4$ with 0.5 molecule CCl$_4$ in the monoclinic space group $C2/c - C_{2h}^6$ (No. 15) with a = 33.515(5), b = 11.287(2), c = 16.630(3) Å, $\beta = 98.35(1)°$; Z = 8, $D_c = 2.85$ g/cm^3, R = 0.065. The CCl$_4$ molecules of solvate are disordered about the crystallographic twofold axis. Each compound molecule has C_1 symmetry; each metal atom is in a different coordination environment. The molecule contains an almost

planar rhomboidal arrangement of the four metal atoms, see **Fig. 67**, and is consistent
with that expected for a tetranuclear complex with 62 outer valence electrons. The C–O
bond distance in the triply-bridging five-electron donating μ_3,η^2-acyl ligand indicates pre-
dominantly single-bond character suggesting significant activation of this linkage. Interestingly, the bridgehead Os(2)–W bond distance is considerably shorter in comparison to the
Os(1)–W and Os(3)–W bonds which is probably caused by the bridging oxygen atom. Weak
semi-bridging carbonyl interactions are indicated by the relatively close interatomic distances between Os(1) and CO′ as well as CO″ which presumably results from a shift of
electron density from the electron-rich Os(2) and W centers to the electron-poor Os(1).
Average bond distances of W–C (of η^5-C_5H_5), Os–CO and C≡O are 2.35, 1.913 and 1.15Å,
respectively [2, 5].

Hydrogenation in refluxing toluene for ca. 1 h yielded No. 6 in ca. 90% yield; see also
Preparation Method IIIa, p. 174 [7].

Decarbonylation with $(CH_3)_3NO$ in CH_3CN/CH_2Cl_2 at room temperature, followed by refluxing in toluene led to No. 17, while direct thermolysis in refluxing toluene gave a mixture
of Nos. 14 and 17 [7]; see also Preparation Methods IIa and IIb, p. 174. The formation
of No. 17 probably proceeded by the folding of planar No. 24 across the W–Os edge in
conjunction with scission of the η^2-acyl fragment into μ-O=W and μ_3-$CCH_2C_6H_4CH_3$-4,
loss of 2 CO, and formation of an additional Os–Os bond [8].

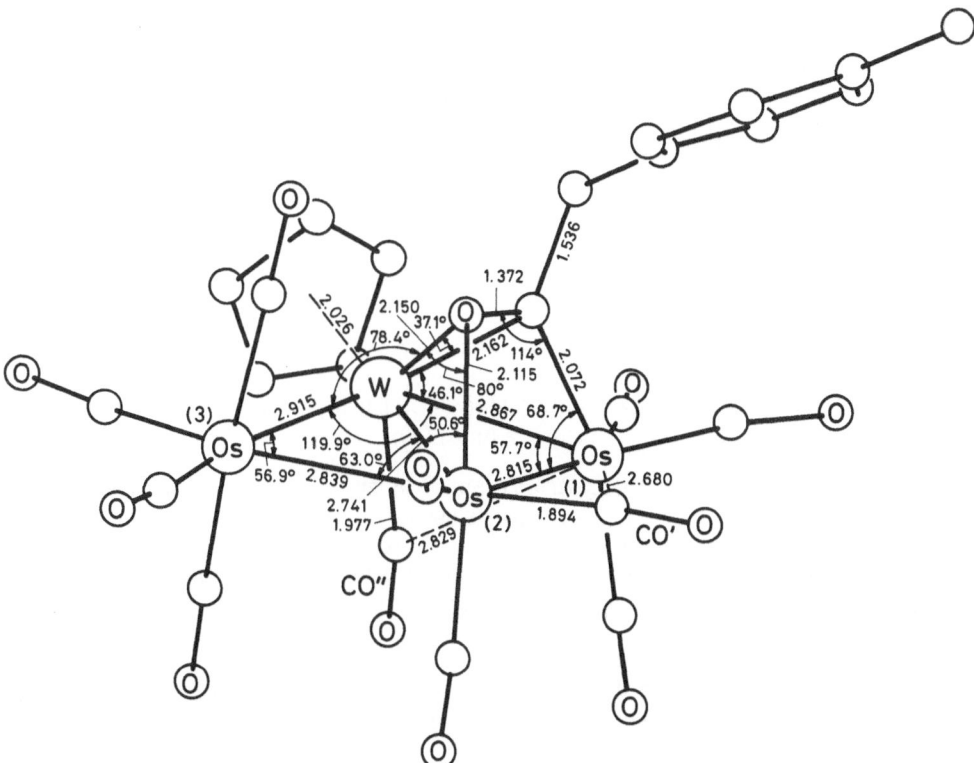

Fig. 67. Molecular structure of the CCl_4 solvate of $Os_3(\mu_3,\eta^2$-$OCCH_2C_6H_4CH_3$-4)-
$(CO)_{10}W(\eta^5$-$C_5H_5)(CO)$ (No. 24) with selected interatomic distances (in Å) and bond angles
[2, 5].

                           References on p. 208

(μ–H)$_2$Os$_3$(μ$_3$,η^2-OCCH$_2$C$_6$H$_4$CH$_3$-4)(CO)$_9$Mo(η^5-C$_5$H$_5$)(CO) (Table **6**, No. **25**) crystallizes in the orthorhombic space group Pna2$_1$ – C$_{2v}^9$ (No. 33) with a = 12.966(9), b = 11.256(3), c = 38.505(10) Å; Z = 8, D$_c$ = 2.71 g/cm^3, R = 0.097. Two crystallographically independent molecules were refined; one of these is shown in **Fig. 68** with the bridging hydrides in their predicted positions which are indicated from the metal–metal bond distances and metal–metal–ligand angles. The molecule contains a planar triangulated rhomboidal arrangement of the four metal atoms [19].

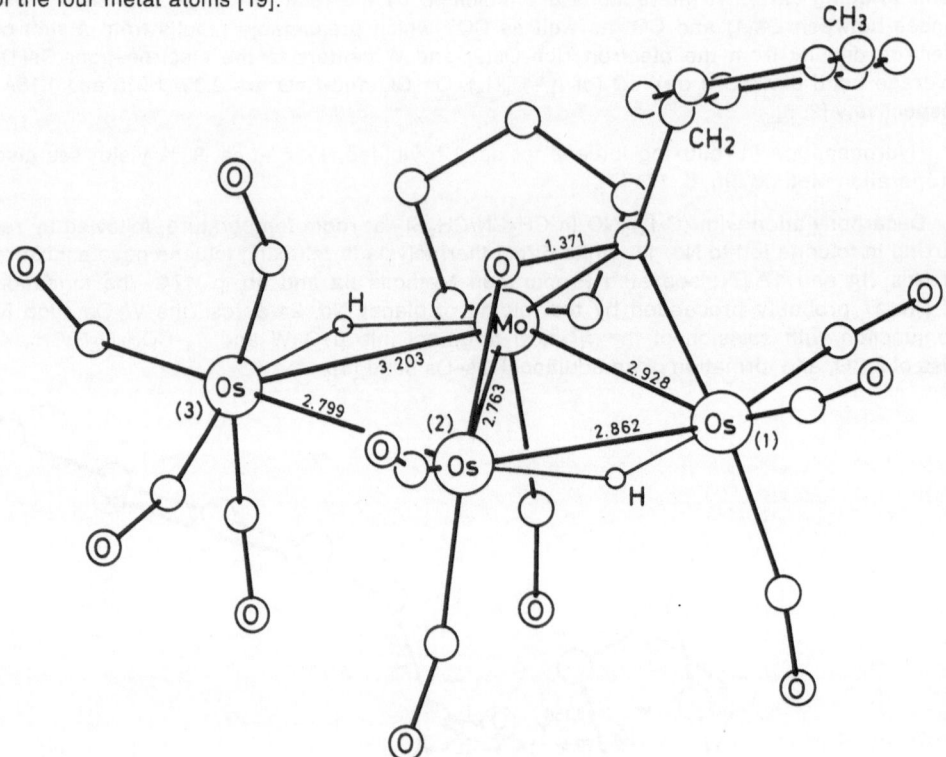

Fig. 68. Molecular structure of one of the two crystallographically independent molecules of (μ–H)$_2$Os$_3$(μ$_3$,η^2-OCCH$_2$C$_6$H$_4$CH$_3$-4)(CO)$_9$Mo(η^5-C$_5$H$_5$)(CO) (No. 25) with selected bond distances (in Å) [19].

Os$_3$(μ$_3$,η^1-CH)(CO)$_9$W(η^5-C$_5$(CH$_3$)$_5$)(CO)$_2$ (Table **6**, No. **26**) crystallizes in the triclinic space group P$\bar{1}$ – C$_i^1$ (No. 2) with a = 11.368(2), b = 16.590.(3), c = 16.614(5) Å, α = 63.06(2)°, β = 83.32(2)°, γ = 71.41(2)°; Z = 4, D$_c$ = 3.039 g/cm^3, R, R′ = 0.053, 0.056. Two molecules, **A** and **B**, have been found in the unit cell; the molecular structure of **A** is shown in **Fig. 69**. The complex adopts a tetrahedral Os$_3$W arrangement with three mutually orthogonal terminal CO ligands at each Os center. One CO group on the W atom exhibits an edge–bridged mode with an Os(3)–(μ–CO) distance of 2.55(3) Å. The shortest non–bonding contact between the μ$_3$,η^1-C of the methylidyne moiety and the CH$_3$ groups of C$_5$(CH$_3$)$_5$ ligand amounts 3.15 to 3.17 Å and are substantially shorter than the sum of the van der Waals radii of 3.65 to 3.7 Å between μ$_3$,η^1-C and a CH$_3$ group [23].

Bond lengths of Molecule **B** (in Å): Os(1)–Os(2) = 2.857(2), Os(1)–Os(3) = 2.824(2), Os(2)–Os(3) = 2.807(2), Os(1)–W = 2.937(2), Os(2)–W = 2.956(2), Os(3)–W = 2.931(2), Os(1)–(μ$_3$,η^1-C) = 2.14(3), Os(2)–(μ$_3$,η^1-C) = 2.08(2), W–(μ$_3$,η^1-C) = 2.13(3) [23].

References on p. 208

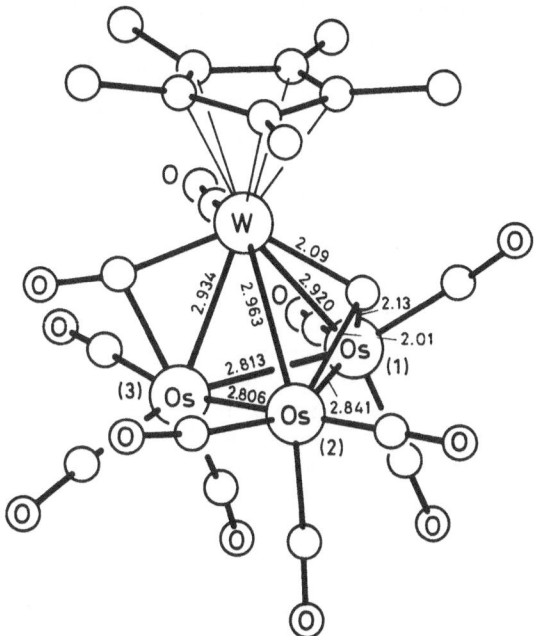

Fig. 69. Molecular structure of Molecule **A** of $Os_3(\mu_3,\eta^1-CH)(CO)_9W(\eta^5-C_5(CH_3)_5)(CO)_2$ (No. 26) with selected bond distances (in Å) [23].

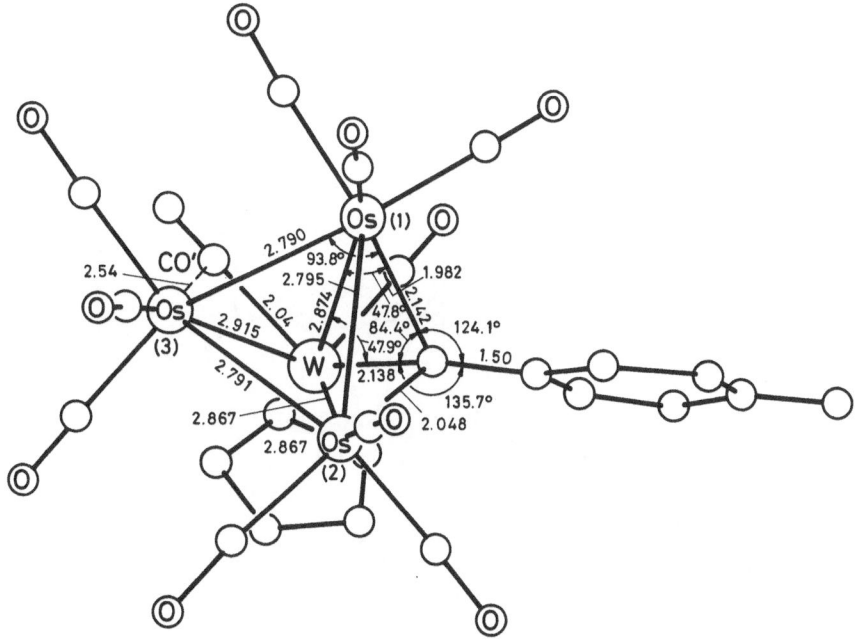

Fig. 70. Molecular structure of $Os_3(\mu_3,\eta^1-CC_6H_4CH_3-4)(CO)_9W(\eta-C_5H_5)(CO)_2$ (No. 29) with selected bond distances (in Å) and bond angles [3]; see also [1].

References on p. 208

$Os_3(\mu_3,\eta^1\text{-}CC_6H_4CH_3\text{-}4)(CO)_9W(\eta^5\text{-}C_5H_5)(CO)_2$ (Table **6**, No. **29**) crystallizes in the mono-clinic space group $P2_1/n - C_{2h}^5$ (No. 14) with a = 14.508(6), b = 13.793(4), c = 13.440(6) Å, β = 101.11(14)°; Z = 4, D_c = 3.1 g/cm³, R = 0.063. The molecule, shown in **Fig. 70**, consists of an approximately tetrahedral arrangement of the four metal atoms. The carbonyl group CO′ bonded·to the tungsten atom interacts with the Os(3) center in a semi-bridging mode. The mean Os–Os separation is ca. 0.09 Å shorter than in $Os_3(CO)_{12}$, supporting the proposal that metal–metal distances in tetranuclear compounds are generally 0.05 to 0.10 Å shorter than in triangular compounds [3]; see also [1].

$Os_3(\mu_3,\eta^1\text{-}CC_6H_4CH_3\text{-}4)_2(CO)_9W(\eta^5\text{-}C_5H_5)H$ (Table **6**, No. **32**). The deuterated complex $Os_3(\mu_3,\eta^1\text{-}CC_6H_4CH_3\text{-}4)_2(CO)_9W(\eta^5\text{-}C_5H_5)D$ was similarly prepared from deuterium-labeled No. 35 as described in Preparation Method IIa, p. 174 [6].

The title compound crystallizes in the monoclinic space group $P2_1/n - C_{2h}^5$ (No. 14) with a = 19.067(3), b = 16.828(3), c = 19.947(3) Å, β = 98.62(1)°; Z = 8, D_c = 2.68 g/cm³, R = 0.054. There are two equivalent molecules, **A** and **B**, in the crystallographic asymmetric unit. **A** and **B** have similar stereochemistry with the exception of the conformation of the CO ligands on the Os(3) center. The terminal hydride ligand could be located crystallographically only for Molecule **B**, which is shown in **Fig. 71**; for Molecule **A**, the hydride is believed to occupy the same position [4, 6]. The tetrametallic compound adopts a tetrahedral geome-try. The two $\mu_3,\eta^1\text{-}CC_6H_4CH_3\text{-}4$ ligands in each molecule are located in almost equivalent environments on opposing sides of a pseudo mirror plane, but the resulting symmetry is disturbed by the twisted arrangement of the CO groups at Os(2) and Os(3). The W–(μ_3–C) distances of ca. 2 Å in **A** and **B** are shorter than in normal W–C single bonds, whereas the Os(1)–(μ_3–C) distances are longer suggesting a shift of charge from the electron rich Os(1) to the electron-poor W center. The short W–Os(1) bond is due to the bond contracting effect of both $\mu_3,\eta^1\text{-}CC_6H_4CH_3\text{-}4$ groups. Average bond distances of W–C (of $\eta^5\text{-}C_5H_5$),

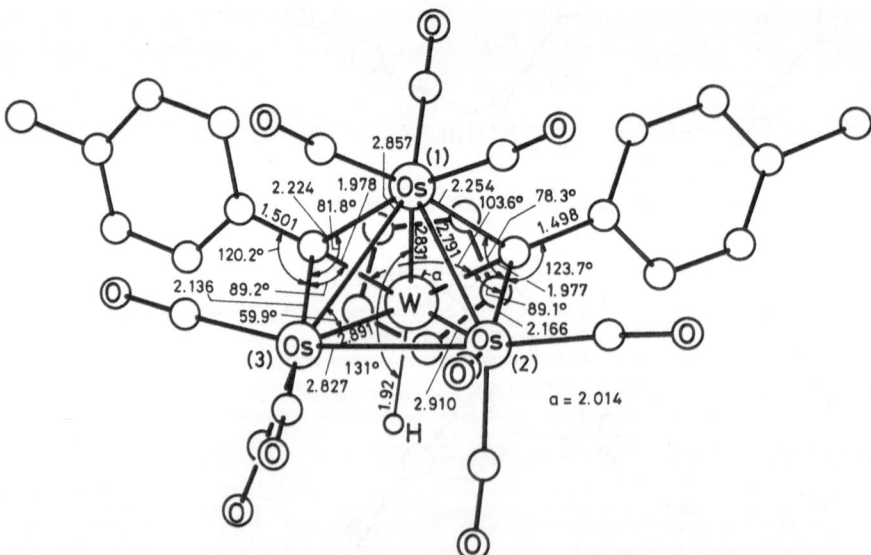

Fig. 71. Molecular structure of Molecule **B** of $Os_3(\mu_3,\eta^1\text{-}CC_6H_4CH_3\text{-}4)_2(CO)_9W(\eta^5\text{-}C_5H_5)H$ (No. 32) with selected bond distances (in Å) and bond angles [6]; see also [4].

References on p. 208

Os–CO and C≡O are 2.342, 1.858 and 1.171 Å, respectively. The W–(η^5-C$_5$H$_5$) mean plane distance is 2.011 Å in Molecule **A** and 2.014 Å in Molecule **B** [6].

Reaction with CCl$_4$ in the presence of t-C$_4$H$_9$OCl at room temperature for 1 h gave No. 33 [4, 6]. Hydrogenation (1 atm H$_2$) in refluxing xylene yielded No. 5 [18].

(μ-H)Os$_3$(μ$_3$,η^1-CC$_6$H$_4$CH$_3$-4)$_2$(μ$_3$-CO)(CO)$_9$W(η^5-C$_5$H$_5$) (Table **6**, No. **35**). (μ-D)Os$_3$-(μ$_3$,η^1-CC$_6$H$_4$CH$_3$-4)$_2$(μ$_3$-CO)(CO)$_9$W(η^5-C$_5$H$_5$) was prepared from (μ-D)$_2$Os$_3$(CO)$_{10}$ as described in Preparation Method Ia, p. 173 [6].

The title compound crystallizes in the monoclinic space group P2$_1$/n – C$_{2h}^5$ (No. 14) with a = 10.610(4), b = 16.000(3), c = 19.394(6) Å, β = 104.33(2)°; Z = 4, D$_c$ = 2.72 g/cm^3, R = 0.071. The molecular structure is shown in **Fig. 72a**. The molecule contains a butterfly arrangement of the four metal atoms. The WOs$_3$ fragment in addition with the alkylidyne carbons defines a capped square pyramid; the base skeleton is defined by Os(1)–Os(2)–Os(3)–C′ (see **Fig. 72b**), the principal apex by W. One of the two μ$_3$-alkylidyne ligands, centered on C″, occupies a "normal" position on the outside of the Os(2)–Os(3)–W triangle, whereas the other one, centered at C′, spans the W bridgehead and two wing-tip Os(1) and Os(3) atoms and is far more distorted than the μ$_3$-C″C$_6$H$_4$CH$_3$-4 moiety. The symmetrical Os–C′ bond lengths are stretched by about 0.07 Å relative to those in Os–C″, whereas the two W–C′ and W–C″ distances are equivalent and at 2.012 Å they are extremely short, presumably indicating some multiple-bond character. The bridging carbonyl ligand spanning the Os(1)–W linkage shows some interaction with the Os(2) center and is referred to as a semi-triply bridging ligand. The bridging hydride is mutually responsible for the elongated Os(2)–Os(3) bond length, whereas the shortness of the Os(1)–W and Os(3)–W distances are probably caused

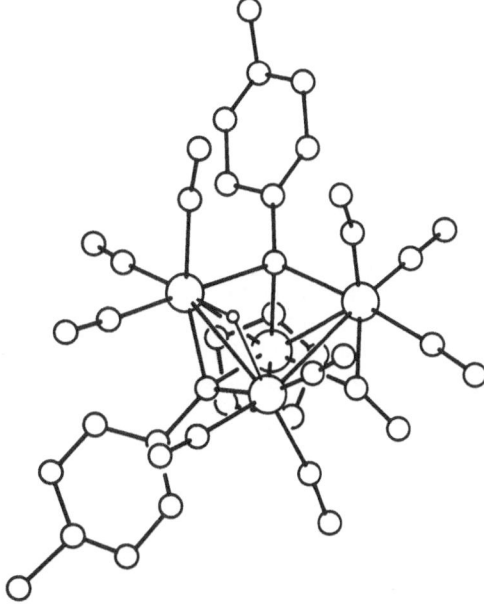

Fig. 72a. Molecular structure of (μ-H)Os$_3$(μ$_3$,η^1-CC$_6$H$_4$CH$_3$-4)$_2$(μ$_3$-CO)(CO)$_9$W(η^5-C$_5$H$_5$) (No. 35) [6].

References on p. 208

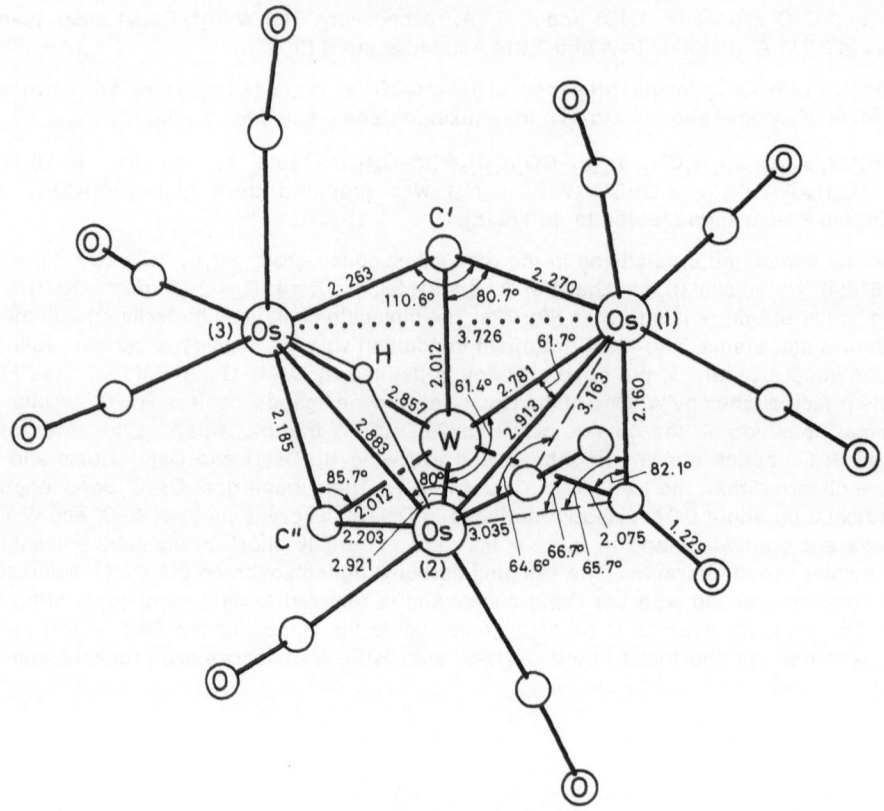

Fig. 72b. $(\mu-H)Os_3(\mu_3,\eta^1-C)_2(\mu_3-CO)(CO)_9W$ skeleton of No. 35 with selected interatomic distances (in Å) and angles [6].

by the alkylidyne ligands and the bridging carbonyl. Average bond distances of W–C (of $\eta^5-C_5H_5$), Os–CO and C≡O are 2.34, 1.937 and 1.129 Å, respectively [6].

$(\mu-H)_2Os_3(\mu_3,\eta^1-CC_6H_4CH_3-4)(CO)_9W(\eta^5-C_5H_5)(CO)_3$ (Table 6, No. 36) crystallizes in the monoclinic space group $P2_1/n-C_{2h}^5$ (No. 14) with a = 12.5313(33), b = 11.0065(20), c = 21.7331(44) Å, β = 105.780(20)°; Z = 4, D_c = 2.90 g/cm³, R = 0.062. The molecular structure is shown in **Fig. 73**. The complex is chiral, and the crystal contains an ordered racemic mixture of the two enantiomers. The bond lengths in the Os_3 triangle are unequal. The two longest distances, Os(2)–Os(3) and Os(1)–Os(2), are believed to be bridged by the μ-hydrides. The difference in the Os–(μ_3,η^1-C) distances indicate that the organic ligand caps the Os_3 core in an asymmetric mode referred to as semi–triply bridging. Average bond distances of W–C (of $\eta^5-C_5H_5$), Os–CO and C≡O are 2.33, 1.885 and 1.17 Å, respectively [11].

The reactions are summarized in Scheme 1 (CO groups represented by dashes) [11]:

In refluxing toluene, hydrogenation led to $(\mu-H)_3Os_3(\mu_3,\eta^1-CC_6H_4CH_3-4)(CO)_9$ and carbonylation to $(\mu-H)Os_3(\mu_3,\eta^1-CC_6H_4CH_3-4)(CO)_{10}$ (both Section 3.1.6.3); both reactions probably proceeded via Intermediate **A** by reductive elimination of $(\eta^5-C_5H_5)W(CO)_3H$ [11].

Treatment with $(\eta^5-C_5H_5)W(\equiv CC_6H_4CH_3-4)(CO)_2$ in CH_2Cl_2 at room temperature gave nearly equivalent amounts of No. 35, $(\mu-H)_3Os_3(\mu_3,\eta^1-CC_6H_4CH_3-4)(CO)_9$ and

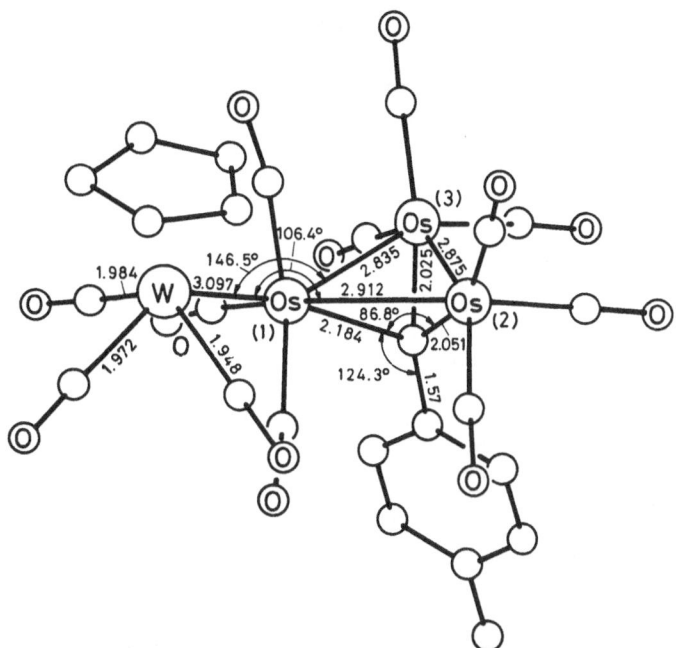

Fig. 73. Molecular structure of $(\mu\text{-H})_2Os_3(\mu_3,\eta^1\text{-}CC_6H_4CH_3\text{-}4)(CO)_9W(\eta^5\text{-}C_5H_5)(CO)_3$ (No. 36) with selected bond distances (in Å) and bond angles [11].

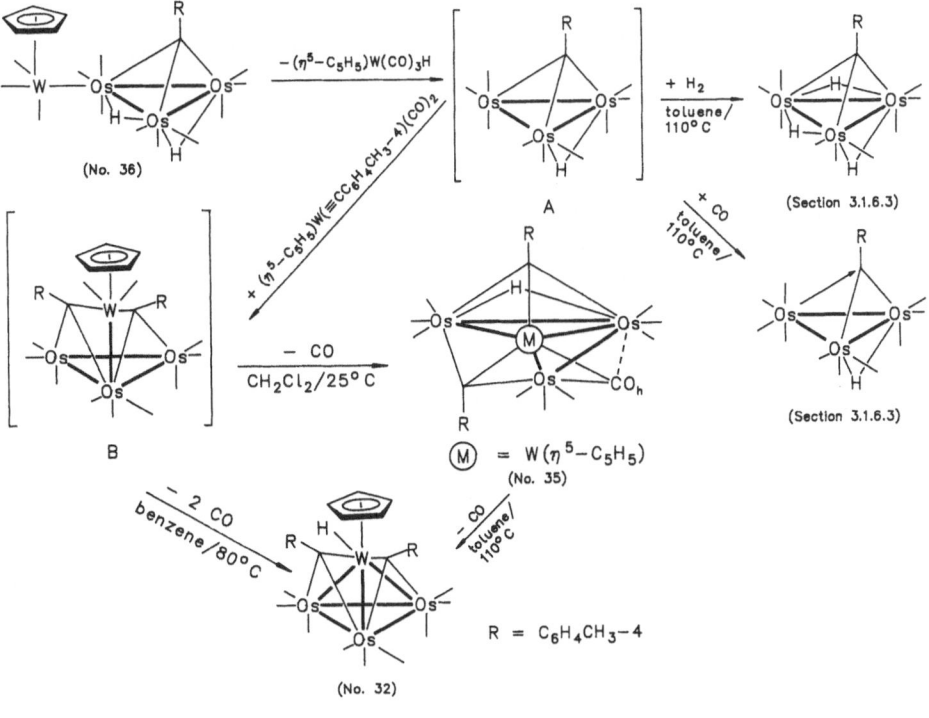

Scheme 1

References on p. 208

$(\mu-H)Os_3(\mu_3,\eta^1-CC_6H_4CH_3-4)(CO)_{10}$ (both Section 3.1.6.3). A similar reaction in refluxing benzene yielded a mixture of No. 32, and the two carbyne complexes mentioned above in a ratio of ca. 4:1:1; see also Preparation Method Ib, p. 174. The formation of the two carbyne complexes as by-products in both reactions is probably consistent with the involvement of unsaturated Intermediate **A**, whereas the formation of Nos. 32 and 35 proceeded apparently via Intermediate **B**, produced by coordination of the W≡C species to **A**, followed by elimination of two or one CO groups, respectively [11].

References:

[1] Busetto, L.; Green, M.; Howard, J. A. K.; Hessner, B.; Jeffery, J. C.; Mills, R. M.; Stone, F. G. A.; Woodward, P. (J. Chem. Soc. Chem. Commun. **1981** 1101/3).

[2] Shapley, J. R.; Park, J. T.; Churchill, M. R.; Bueno, C.; Wasserman, H. J. (J. Am. Chem. Soc. **103** [1981] 7385/7).

[3] Busetto, L.; Green, M.; Hessner, B.; Howard, J. A. K.; Jeffery, J. C.; Stone, F. G. A. (J. Chem. Soc. Dalton Trans. **1983** 519/25).

[4] Park, J. T.; Shapley, J. R.; Churchill, M. R.; Bueno, C. (J. Am. Chem. Soc. **105** [1983] 6182/4).

[5] Park, J. T.; Shapley, J. R.; Churchill, M. R.; Bueno, C. (Inorg. Chem. **22** [1983] 1579/84).

[6] Park, J. T.; Shapley, J. R.; Churchill, M. R.; Bueno, C. (Inorg. Chem. **23** [1984] 4476/86).

[7] Shapley, J. R.; Park, J. T.; Churchill, M. R.; Ziller, J. W.; Beanan, L. R. (J. Am. Chem. Soc. **106** [1984] 1144/5).

[8] Churchill, M. R.; Ziller, J. W.; Beanan, L. R. (J. Organomet. Chem. **287** [1985] 235/46).

[9] Churchill, M. R.; Li, Y.-J. (J. Organomet. Chem. **291** [1985] 61/70).

[10] Churchill, M. R.; Li, Y.-J. (J. Organomet. Chem. **294** [1985] 367/77).

[11] Chi, Y.; Shapley, J. R.; Churchill, M. R.; Li, Y.-J. (Inorg. Chem. **25** [1986] 4165/70).

[12] Chi, Y.; Shapley, J. R.; Ziller, J. W.; Churchill, M. R. (Organometallics **6** [1987] 301/7).

[13] Chi, Y.; Shapley, J. R. (unpublished results, from [12]).

[14] Park, J. T.; Shapley, J. R.; Bueno, C.; Ziller, J. W.; Churchill, M. R. (Organometallics **7** [1988] 2307/16).

[15] Chi, Y.; Lee, G.-H.; Peng, S.-M.; Wu, C.-H. (Organometallics **8** [1989] 1574/6).

[16] Chi, Y.; Shapley, J. R.; Churchill, M. R.; Fettinger, J. C. (J. Organomet. Chem. **372** [1989] 273/85).

[17] Park, J. T.; Shapley, J. R. (Bull. Korean Chem. Soc. **11** [1990] 531/4).

[18] Peng, S.-M.; Lee, G.-H.; Chi, Y. (Polyhedron **9** [1990] 1491/5).

[19] Park, J. T.; Chun, K.-M.; Yun, S.-S.; Kim, S. (Bull. Korean Chem. Soc. **12** [1991] 249/51).

[20] Park, J. T.; Cho, J.-J.; Chun, K.-M.; Yun, S.-S. (J. Organomet. Chem. **433** [1992] 295/303).

[21] Park, J. T.; Chung, M.-K.; Chun, K.-M.; Yun, S.-S.; Kim, S. (Organometallics **11** [1992] 3313/9).

[22] Chi, Y.; Shapley, J. R.; Churchill, M. R. (Organometallics, according to [16], submitted for publication but not published until 1992).

[23] Gong, J.-H.; Chen, C.-C.; Chi, Y.; Wang, S.-L.; Liao, F.-L. (J. Chem. Soc. Dalton Trans. **1993** 1829/34).

3.1.6.4.2 Os₃Pt and Os₃Pt₂ Compounds

The compounds dealt with in this section all consist of an Os_3Pt or an Os_3Pt_2 core; most compounds adopt a closo tetrahedral Os_3Pt arrangement, others have rather open metal frameworks. The structures of nearly all compounds have been confirmed by X-ray analysis.

References on p. 227

Table 7
Os$_3$Pt and Os$_3$Pt$_2$ Compounds.
An asterisk preceding the compound number indicates further information at the end of the table, pp. 216/26.
Explanations, abbreviations, and units on p. X.

No. compound	method of preparation (yield) properties and remarks

*1

from (μ–H)Os$_3$(CO)$_9$(μ–H)Pt(CO)P(C$_6$H$_{11}$–c)$_3$ and 1,1,3,3-tetramethyl-2-thiocarbonylcyclohexane (1:2) in toluene at 90 °C for 12 h under vacuum, followed by chromatography with light petroleum/CH$_2$Cl$_2$ as eluant (45%) [11]
red crystals from light petroleum [11]
^{1}H NMR (CDCl$_3$, −40 °C): 0.82 to 1.78 (m, CH$_2$, CH$_3$ of C$_6$H$_6$(CH$_3$)$_4$, and c-C$_6$H$_{11}$), 2.36 (m, 3 H, c-C$_6$H$_{11}$) [11]
^{31}P NMR (CDCl$_3$, −60 °C): 36.2 (s, P(C$_6$H$_{11}$-c)$_3$; J(P,Pt) = 2770); the sharp singlet remained unchanged between −60 and +25 °C, the reasons therefore were discussed [11]
IR (cyclohexane): 1946, 1961, 1971, 1977, 2023, 2032, 2045, 2069 (all CO) [11]
FAB mass spectrum: [M]$^+$ [11]

*2

from Os$_3$(CO)$_9$(P(CH$_3$)$_2$C$_6$H$_5$)(μ$_3$–S)Pt(P(CH$_3$)$_2$C$_6$H$_5$)$_2$ upon UV irradiation in hexane at 25 °C for 6 h, followed by TLC with CH$_2$Cl$_2$/hexane (ca. 2:1) as eluant (48%), along with 3% Os$_3$(CO)$_8$(P(CH$_3$)$_2$C$_6$H$_5$)(μ$_3$–S)Pt(P(CH$_3$)$_2$C$_6$H$_5$)$_2$ [2]
red crystals from acetone/C$_2$H$_5$OH at −10 °C [2]
^{1}H NMR (CDCl$_3$): −16.84 (m, μ–H), 1.82 (d, CH$_3$; J(H,P) = 10.2), 1.95 (d, CH$_3$; J(H,P) = 9.3), 2.14 (q, CH$_3$; J(H,P) = 10.8), 2.16 (q, CH$_3$; J(H,P) = 10.5), 2.24 (d, 2 CH$_3$; J(H,P) = 9.9), 6.93 to 8.07 (m, C$_6$H$_5$, C$_6$H$_4$) [2]
^{31}P NMR (CDCl$_3$): −55.67 (s, 1 P), −48.30 (t, 1 P; J(P,Pt) = 2694), −25.03 (s, 1 P) [2]
IR (CHCl$_3$): 1894, 1934, 1957, 1992, 2017 (all CO) [2]

References on p. 227

Table 7 (continued)

No. compound	method of preparation (yield) properties and remarks

*3

from $(\mu\text{-}H)_2Os_3(CO)_{10}$ and
  $Pt(CH_3)_2(\eta^4\text{-}C_8H_{12}\text{-}1,5)$ in toluene at reflux
  temperature for 10 min, followed by TLC with
  hexane as eluant (17%, along with 12% of
  No. 4) [6]
from No. 4 and $1,5\text{-}C_8H_{12}$ in CH_2Cl_2 at room
  temperature in the presence of 2 equivalents
  of $(CH_3)_3NO$ dissolved in CH_3OH; work-up by
  TLC (30%) [6]
red-orange crystals from CH_2Cl_2 at $-20\,°C$ [6]
^1H NMR $(CDCl_3)$: -20.6 (m, $\mu\text{-}H$), 2.1 to 2.6 (m,
  CH_2 of C_8H_{12}), 5.5 (m, CH of C_8H_{12}), 6.9 (m,
  CH_2) [6]
IR (CH_2Cl_2): 1957, 1999, 2046, 2060, 2082 (all
  CO) [6]
the electron spectrum for chemical analysis
  showed a molar Os/Pt ratio of 3:0.9 [6]

4

from $(\mu\text{-}H)_2Os_3(CO)_{10}$ and
  $Pt(CH_3)_2(\eta^4\text{-}C_8H_{12}\text{-}1,5)$, see No. 3 (12%) [6]
by carbonylation of No. 3 at room temperature
  under 50 atm of CO, purified by TLC (quantitative yield) [6]
thin, fragile yellow crystals, which decompose
  on dry silica gel within 30 min [6]
^1H NMR $(CDCl_3)$: -21.3 (m, $\mu\text{-}H$), 7.3 (m, CH_2)
  [6]
IR (CH_2Cl_2): 1986, 2015, 2030, 2061, 2082, 2110
  (all CO) [6]
the electron spectrum for chemical analysis
  showed a molar Os/Pt ratio of 3:0.9 [6]
reaction with an excess of $1,5\text{-}C_8H_{12}$ in CH_2Cl_2
  in the presence of $(CH_3)_3NO$ yielded No. 3 [6]

*5

in solution exists an equilibrium with the
  isomeric No. 6, crystals of the mixture can
  be separated by hand [1, 3]
by treatment of $(\mu\text{-}H)_2Os_3(CO)_9Pt(CO)$-
  $P(C_6H_{11}\text{-}c)_3$ with an excess of CH_2N_2 (dissolved in ether) in THF at $0\,°C$ [1] or $-10\,°C$,
  followed by extraction of the residue with
  light petroleum and chromatography on alumina with CH_2Cl_2/light petroleum (1:4) as
  eluant [3] (96% as a mixture with isomeric
  No. 6) [1, 3]
from $(\mu\text{-}H)_2Os_3(\mu_2,\eta^1\text{-}CH_2)(CO)_{10}$ (Section
  3.1.6.2.1) and $(C_2H_4)_2PtP(C_6H_{11}\text{-}c)_3$ in THF or

Table 7 (continued)

No. compound	method of preparation (yield) properties and remarks

toluene at room temperature (as a mixture with isomeric No. 6) [1, 3]

orange crystals [1, 3] from toluene/light petroleum [1]

^{1}H NMR (CDCl$_3$): −20.94 (dt, μ-H; J(H,H) = 2.5, J(H,P) = 5, J(H,Pt) = 18), 1.3 to 2.0 (m, C$_6$H$_{11}$-c) [3], 6.32 (dt, 1H, CH$_2$; J(H,H) = 2.5, 6), 6.40 (ddt, 1 H, CH$_2$; J(H,H) = 2.5, 6, J(H,P) = 2) [1, 3]

^{31}P NMR (CDCl$_3$): 10.8 (s, P(C$_6$H$_{11}$-c)$_3$; J(P,Pt) = 2513) [1, 3]

^{195}Pt NMR (CDCl$_3$): 1045 (d, Pt(CO); J(P,Pt) = 2513) [3]

IR (light petroleum; as a mixture with isomeric No. 6): 1941, 1950, 1957, 1966, 1978, 1996, 2005, 2008, 2025, 2044, 2054, 2081, 2084 (all CO) [3]

FD mass spectrum of the isomeric mixture: [M]$^+$ [3]

*6   (c−C$_6$H$_{11}$)$_3$P   CO

for formation, see No. 5

red crystals [1, 3] from toluene/light petroleum [1]

^{1}H NMR (CDCl$_3$): −21.73 (s, μ-H^1; J(H,Pt) = 20), −13.57 (d, μ-H^2; J(H,P) = 10, J(H,Pt) = 557), 1.3 to 2.0 (m, C$_6$H$_{11}$-c) [3], 6.96 (dd, 1 H, CH$_2$; J(H,H) = 2, 6), 7.46 (d, 1 H, CH$_2$; J(H,H) = 5) [1, 3]

^{31}P NMR (CDCl$_3$): 65.2 (s, P(C$_6$H$_{11}$-c)$_3$; J(P,Pt) = 2661) [1, 3]

^{195}Pt NMR (CDCl$_3$): 172 (d, Pt(CO); J(Pt,P) = 2661) [3]

for IR and mass spectra, see No. 5

*7   (c−C$_6$H$_{11}$)$_3$P   CNC$_6$H$_{11}$−c

by treatment of (μ-H)Os$_3$(CO)$_9$(μ-H)PtP(C$_6$H$_{11}$-c)$_3$CNC$_6$H$_{11}$-c with CH$_2$N$_2$ (dissolved in ether) in toluene at room temperature (83%) [7]

red crystals from ether [7]

^{1}H NMR (CDCl$_3$): −21.82 (s, μ-H^1; J(H^1,Pt) = 22), −15.87 (d, μ-H^2; J(H^2,P) = 7.3, J(H^2,Pt) = 534), 1.12 to 3.96 (m, C$_6$H$_{11}$-c), 6.61 (dd, 1 H of CH$_2$; J(H,H) = 2.3, 6.0), 7.02 (m, 1 H of CH$_2$) [7]; (CD$_2$Cl$_2$): −21.68 (ddd, μ-H^1; J(H^1,P) = 1.9, J(H^1,Pt) = 18), −15.84 (d, μ-H^2; J(H^2,P) = 7.6, J(H^2,Pt) = 533), 6.57 (dd, H^4), 6.95 (ddd, H^3; J(H^3,P) = 1.9,

References on p. 227

Table 7 (continued)

No. compound	method of preparation (yield) properties and remarks

*7 (continued)

$J(H^1,H^4) = 2.4$, $J(H^1,H^3) = 2.7$, $J(H^4,H^3) = 5.9$) [12]

^{13}C NMR (CD_2Cl_2): 58.6 (s, μ_2,η^1-CH_2), 169.9 (d, CO-i; $J(C^i,H^1) = 11.5$, $J(C^i,P) = 1.9$, $J(C^i,Pt) = 4$), 174.4 (s, CO-h; $J(C^h,H^2) = 3.1$, $J(C^h,H^1) = 10.4$, $J(C^h,Pt) = 5$), 175.8 (d, CO-g; $J(C^g,P) = 1.0$, $J(C^g,Pt) = 25$), 176.0 (s, CO-f; $J(C^f,H^2) = 3.7$, $J(C^f,H^1) = 4.3$, $J(C^f,Pt) = 45$), 180.3 (d, CO-e; $J(C^e,H^2) = 8.0$, $J(C^e,H^1) = 2.9$, $J(C^e,P) = 0.9$, $J(C^e,Pt) = 56$), 180.5 (d, CO-d; $J(C^d,P) = 10.5$, $J(C^d,Pt) = 26$), 181.3 (d, CO-c; $J(C^c,H^1) = 2.8$, $J(C^c,P) = 1.0$, $J(C^c,Pt) = 70$), 184.2 (d, CO-b; $J(C^b,H^4) = 2.0$, $J(C^b,P) = 9.0$, $J(C^b,Pt) \approx 4$), 184.2 (s, CO-a; $J(C^a,Pt) = 66$); multiplicities refer to ^1H-decoupled spectra [12]

IR (cyclohexane): 1925, 1950, 1954, 1973, 1989, 2016, 2047, 2068 (all CO), 2181 (CN) [7]

8

by protonation of No. 7 with CF_3CO_2H (based on ^1H NMR spectroscopy but not isolated) [7]

^1H NMR $(CDCl_3?)$: -21.67 (μ-H^1; $J(H,Pt) = 11$), -17.73 (μ-H^2; $J(H,Pt) = 490$) [7]

*9

from $Os_3(\mu_2,\eta^1$-$CH_2)(\mu$-$CO)(CO)_{10}$ (Section 3.1.6.2.1) and $C_2H_4Pt(P(C_6H_5)_3)_2$ (ca. 1:1) in THF at room temperature for 15 min; purified by chromatography with CH_2Cl_2/hexane (3:2) as eluant (68%) [5]

reformed within 2 to 4 h from a THF solution of $Os_3(\mu_2,\eta^1$-$CH_2)(CO)_{11}Pt(CO)P(C_6H_5)_3$ (No. 10) upon removal of the CO atmosphere [5]

yellow platelets from ether; air-stable in the solid state, but air-sensitive in solution decomposing within 15 min [5]

^1H NMR (C_6D_6): 4.50 (d, CH_2; $J(H,P) = 5.5$), 7.00 to 7.53 (m, C_6H_5); the methylene protons couple with only one of the two P atoms of the $P(C_6H_5)_3$ groups, presumably with the group in trans position [5]

^{13}C NMR $(THF$-$d_8)$: 76.5 (dt, CH_2; $J(C,H) = 147$, $J(C,P) = 56$); similar to the CH_2 hydrogens,

References on p. 227

Table 7 (continued)

No. compound	method of preparation (yield) properties and remarks

the methylene carbon couples with only one of the P atoms of the $P(C_6H_5)_3$ groups [5]

^{31}P NMR (C_6D_6): 25.3 (d, $P(C_6H_5)_3$; $J(P,Pt) =$ 4806), 35.6 (d, $P(C_6H_5)_3$; $J(P,P) = 9.8$, $J(P,Pt) =$ 2797) [5]

IR (THF): 1945, 1975, 1989, 2016, 2025, 2053, 2103 (all CO) [5]

FD mass spectrum: $[M]^+$ [5]

10

by treatment of No. 9 with CO (1 atm) in THF at room temperature for 10 min (not isolable) [5]

stable in solution when kept under a CO atmosphere; upon removal of the CO atmosphere or attempted chromatography or crystallization reformed No. 9 within 2 to 4 h [5]

^1H NMR (C_6D_6): 4.30 (m, CH_2), 6.86 to 7.66 (m, C_6H_5) [5]

^{13}C NMR (THF-d_8): 73.1 (t, CH_2; $J(C,H) = 146$); no C,P coupling was observed indicating that probably the $P(C_6H_5)_3$ group in trans position to the methylene ligand was replaced by CO [5]

^{31}P NMR (CD_2Cl_2, -80 °C): -6.6 (s, free $P(C_6H_5)_3$), 20.8 (s, $P(C_6H_5)_3$; $J(P,Pt) = 4088$); upon warming to room temperature the resonances broaden considerably revealing a rapid exchange between free and bonded $P(C_6H_5)_3$ [5]

IR (THF): 1958, 1964, 1982, 1995, 2024, 2043, 2060, 2110 (all CO) [5]

*11

from $(\mu-H)_2Os_3(\mu_3,\eta^1-CCO)(CO)_9$ (Section 3.1.6.3) and $(C_2H_4)_2PtP(C_6H_{11}-c)_3$ (ca. 1:2) in toluene at room temperature for 1 h, followed by chromatography on alumina with light petroleum as eluant, and partial separation of the product mixture by fractional crystallization with the same solvent (ca. 6%, along with ca. 30% of No. 12 as the main-product, based on ^{31}P {^1H} NMR spectrum) [4]

from $(\mu-H)Os_3(\mu_3,\eta^1-CH)(CO)_{10}$ (Section 3.1.6.3) and $(C_2H_4)_2PtP(C_6H_{11}-c)_3$ (ca. 1:1) in toluene at 80 °C for 12 h; work-up as before [4]

bright yellow crystals from light petroleum [4]

References on p. 227

Table 7 (continued)

No. compound	method of preparation (yield) properties and remarks

*11 (continued)

¹H NMR (CDCl₃, −40 °C): −21.52 (s, μ–H, J(H,Pt) = 13), −18.43 (s, μ–H, J(H,Pt) = 35), 1.34 to 1.91 (m, C₆H₁₁–c); the two high–field resonances reveal the presence of two non-equivalent hydride ligands; a broadening of the signals at room temperature is evidence of some fluxional behaviour while the small coupling between μ–H and Pt indicates the absence of a direct H–Pt bond [4]

¹³C {¹H} NMR (CDCl₃, −40 °C): 25.9 to 37.7 (m, C₆H₁₁–c), 162.0, 163.0, 164.9, 168.8, 170.8, 172.0, 174.9, 176.5, 177.7 (all CO), 189.2 (d, PtCO; J(C,Pt) = 1174, J(C,P) = 9), 304.3 (μ₄–C; J(C,Pt) = 780) [4]

³¹P {¹H} NMR (CD₂Cl₂): 34.1 (s, P(C₆H₁₁–c)₃; J(P,Pt) = 3530); the magnitude of the coupling constant indicates a direct P–Pt bond [4]

IR (hexane): 1966, 1986, 2001, 2008, 2018, 2045, 2064, 2089 (all CO) [4]

*12

for formation, see No. 11 [4]

bright yellow crystals from light petroleum [4]

¹H NMR (CDCl₃): −20.67 (s, μ–H; J(H,Pt) = 12), 0.95 to 2.47 (m, C₆H₁₁–c) [4]

¹³C {¹H} NMR (CDCl₃): 26.1 to 38.2 (m, C₆H₁₁–c), 162.1 to 177.3 (all CO), 215.1 (μ–CO; J(C,Pt) = 1153), 322.2 (t, μ₅–C; J(C,Pt) = 318, J(C,P) = 49); the resonance pattern of the μ₅–C ligand indicates that it is bonded to two equivalent ¹⁹⁵Pt³¹P(C₆H₁₁–c)₃ fragments [4]

³¹P {¹H} NMR (CDCl₃): 32.7 (s, P(C₆H₁₁–c)₃; J(P,P) = 10, J(P,Pt) = 4840, 267, J(Pt,Pt) = 2056); the coupling pattern consists of two equivalent ³¹P nuclei [4]

IR (hexane): 1786 (μ–CO), 1960, 1972, 2003, 2008, 2033, 2062, 2085 (all CO) [4]

*13

from (μ–H)Os₃(μ₂,η¹–COCH₃)(CO)₁₀ (Section 3.1.6.2.2) and (C₂H₄)₂PtP(C₆H₁₁–c)₃ (ca. 1:2) in toluene at room temperature for 1.5 h; work–up as at No. 11 (48%) [4]

obtained as a by–product in the preparation of Nos. 11 and 12 from (μ–H)Os₃(μ₃,η¹–CH)(CO)₁₀, indicating that the methylidyne compound was contaminated with (μ–H)Os₃(μ₂,η¹–COCH₃)(CO)₁₀ [4]

References on p. 227

Table 7 (continued)

No. compound	method of preparation (yield) properties and remarks

yellow-orange crystals from CH_2Cl_2/hexane [4]

^1H NMR ($CDCl_3$, $-40\,°C$): -13.24 (dd, μ-H; $^2J(H,P) = 14$, $^3J(H,P) = 4.5$, $^1J(H,Pt) = 656$, $^2J(H,Pt) = 68$), 1.28 to 2.59 (m, C_6H_{11}-c), 3.42 (s, OCH_3); the coupling pattern indicates a μ-H ligand bonded to two nonequivalent pairs of phosphorus and platinum nuclei; the magnitude of the H,^{195}Pt coupling is due to the presence of a direct Pt-H bond [4]

$^{13}C\ \{^1H\}$ NMR ($CDCl_3$): 26.3 to 34.7 (m, C_6H_{11}-c), 175.6, 176.5 (both CO), 182.1 (CO; $J(C,Pt) = 24$), 185.8 (CO; $J(C,Pt) = 37$), 217.3 (μ-CO; $J(C,Pt) = 1062$, 1125), 385.3 (dd, μ_5-C; $J(C,P) = 39$, 43); (CD_2Cl_2, partially decoupled): 74.8 (q, OCH_3; $J(C,H) = 141$), 217.5 (d, μ-CO; $J(C,H) = 15$) [4]

^{31}P ($CDCl_3$): 22.1 (P^b), 42.5 (P^a); AMXY pattern with $^3J(P^a,P^b) = 9.8$, $J(P^aPt^a) = 5483$, $J(P^b,Pt^b) = 3667$, $^2J(P^a,Pt^b) = 341.8$, $^2J(P^b,Pt^a) = 275.9$, $J(Pt^a,Pt^b) = 2093$ (proved by spectral simulation); the coupling pattern is indicative of nonequivalent P ligands each bonded to nonequivalent Pt centers [4]

IR (hexane): 1794 (μ-CO), 1960, 1968, 1978, 1996, 2029, 2044, 2071 (all CO) [4]

*14 $(\mu$-H$)_2Os_3(CNC_4H_9$-t$)(CO)_9Pt(CO)P(C_6H_{11}$-c$)_3$

orange-yellow in solution [7]

^1H NMR ($CDCl_3$, $-50\,°C$): Isomer 1: -17.57 (s, μ-H^1), -9.74 (d, μ-H^2; $J(H,Pt) = 567$, $J(H,P) = 14.6$); Isomer 2: -11.60 (d, μ-H^2; $J(H,Pt) = 503$, $J(H,P) = 10.8$); Isomer 3: -18.98 (s, μ-H^1), -9.65 (d, μ-H^1; $J(H,P) = 12.0$); for assignment, see p. 226 [7]

*15 $(\mu$-H$)_2Os_3(CNC_6H_{11}$-c$)(CO)_9Pt(CO)P(C_6H_{11}$-c$)_3$

orange-yellow in solution [7]

^1H NMR ($CDCl_3$, ca. $-50\,°C$): Isomer 1: -17.63 (d, μ-H^1; $J(H,Pt) = 32$, $J(H,P) = 1.2$), -9.78 (d, μ-H^2; $J(H,Pt) = 565$, $J(H,P) = 14.2$); Isomer 2: -17.56 (d, μ-H^1; $J(H,Pt) = 26$, $J(H,P) = 1.5$), -11.59 (d, μ-H^2; $J(H,Pt) = 503$, $J(H,P) = 11$); Isomer 3: -18.94 (s, μ-H^1; $J(H,Pt) = 16$), -9.70 (d, μ-H^2; $J(H,Pt) = 570$, $J(H,P) = 15$) [7]

^{13}C NMR (CD_2Cl_2/CH_2Cl_2, 1:1, $-55\,°C$): Isomer 1: 110.8 (s, NC), 170.8 (s, CO-h(j);

References on p. 227

Table 7 (continued)

No. compound	method of preparation (yield) properties and remarks
*15 (continued)	J(C,H^1) = 3.4), 172.4 (s, CO–i; J(C,H^1) = 8.8), 172.5 (s, CO–j(h)), 176.7 (d, CO–g; J(C,Pt) = 42, J(C,P) = 3.2, J(C,H^1) = 6.7), 179.6 (d, CO–f; J(C,Pt) = 1784, J(P,C) = 4.6, J(C,H^2) = 28.7), 180.2 (s, CO–e(c); J(C,Pt) = 18, J(C,H^2) = 2.3), 182.3 (d, CO–d; J(C,P) = 1.2, J(C,H^1) = 4.3), 182.9 (s, CO–c(e); J(C,H^2) = 3.9), 184.0, (s, CO–b; J(C,H^2) = 10.4), 184.4 (d, CO–a; J(C,P) = 13.3); for assignment, see Isomer 1, p. 226 [7] ^{31}P(CDCl$_3$, –40 °C): 51.6 (s, PtP; J(P,Pt) = 2520) [7] IR (KBr?): 2175 (CN) [7]

* Further information:

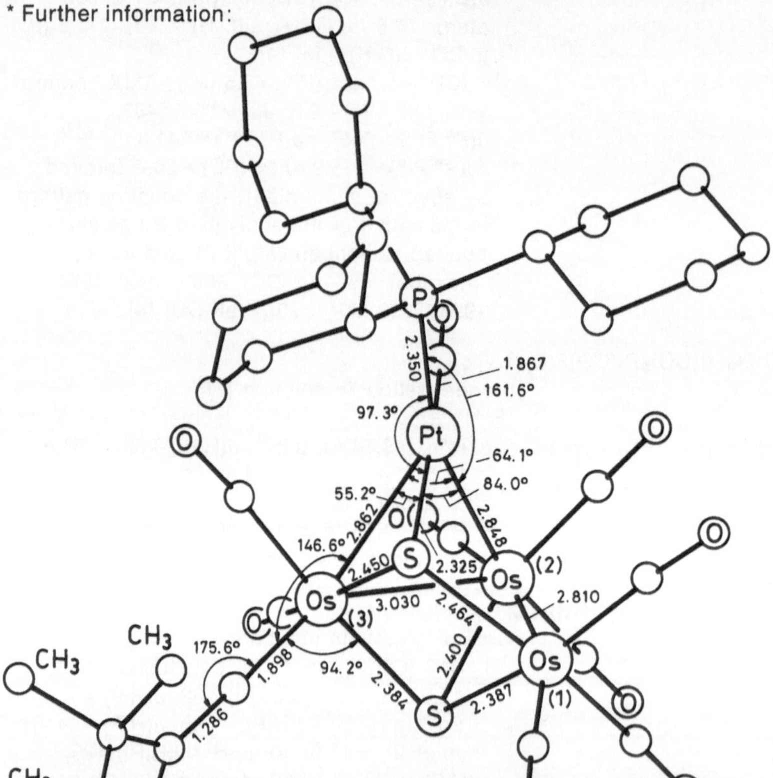

Fig. 74. Molecular structure of Os$_3$(C=C$_6$H$_6$(CH$_3$)$_4$–1,1,3,3)-(CO)$_8$(μ$_3$–S)$_2$Pt(CO)P(C$_6$H$_{11}$–c)$_3$ (No. 1) with selected bond distances (in Å) and bond angles [11].

References on p. 227

Os$_3$(C=C$_6$H$_6$(CH$_3$)$_4$-1,1,3,3)(CO)$_6$(μ_3-S)$_2$Pt(CO)P(C$_6$H$_{11}$-c)$_3$ (Table 7, No. 1) crystallizes in the triclinic space group P$\bar{1}$ – C$_i^1$ (No. 2) with a = 11.935(4), b = 12.755(3), c = 15.228(3) Å, α = 74.84(2)°, β = 80.91(2)°, γ = 84.42(2)°; Z = 2, D$_c$ = 2.28 g/cm^3, R = 0.038. The molecule, shown in **Fig. 74**, is consistent with a 64 outer-valence electron compound having a spiked-triangular metal core of an Os$_2$Pt triangle and an exo-ligated Os atom bonded in a pseudo-axial position on Os(2). The open Os$_3$ and Os$_2$Pt faces are capped by two four-electron donating sulfido groups. The terminal η^1-vinylidene ligand was derived by cleavage of the C=S bond in 1,1,3,3-C$_6$H$_6$(CH$_3$)$_4$C=C=S. The unusual terminal bonding mode of the vinylidene ligand probably results from steric hindrance [11].

(μ-H)Os$_3$(μ_2,η^2-C$_6$H$_4$P(CH$_3$)$_2$)(CO)$_6$(μ_3-S)(P(CH$_3$)$_2$C$_6$H$_5$)Pt(CO)P(CH$_3$)$_2$C$_6$H$_5$ (Table 7, No. 2) crystallizes in the triclinic space group P$\bar{1}$ – C$_i^1$ (No. 2) with a = 10.432(3), b = 12.727(6), c = 14.448(4) Å, α = 98.46(3)°, β = 93.82(2)°, γ = 91.82(3)°; Z = 2, D$_c$ = 2.47 g/cm^3, R = 0.037. The structure is shown in **Fig. 75**. The hydride ligand was not observed crystallographically but it is believed to bridge the Os(1)-Os(3) bond. Average bond distances of Os-CO and C≡O are 1.82 and 1.19 Å, respectively [2].

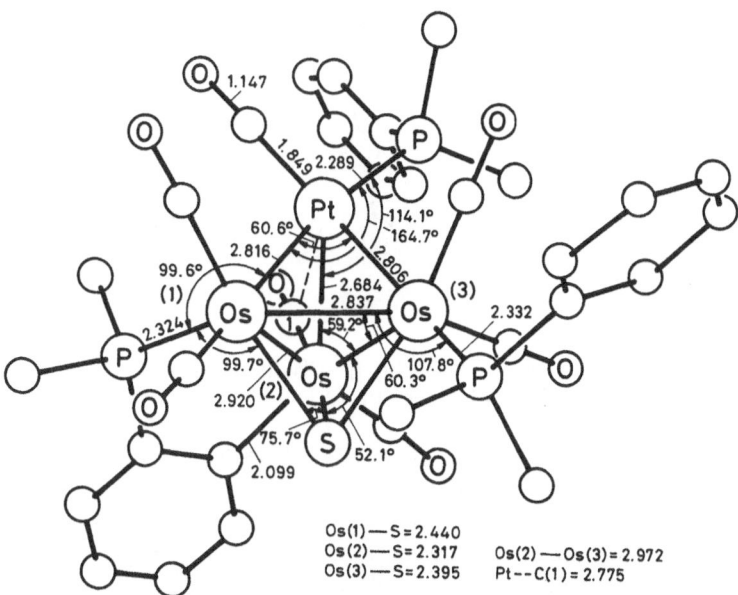

Os(1)—S = 2.440
Os(2)—S = 2.317   Os(2)—Os(3) = 2.972
Os(3)—S = 2.395   Pt--C(1) = 2.775

Fig. 75. Molecular structure of (μ-H)Os$_3$(μ_2,η^2-C$_6$H$_4$P(CH$_3$)$_2$)(CO)$_6$(μ_3-S)(P(CH$_3$)$_2$C$_6$H$_5$)-Pt(CO)P(CH$_3$)$_2$C$_6$H$_5$ (No. 2) with selected interatomic distances (in Å) and bond angles [2].

(μ-H)$_2$Os$_3$(μ_2,η^1-CH$_2$)(CO)$_9$Pt(η^4-C$_8$H$_{12}$-1,5) (Table 7, No. 3) crystallizes in the monoclinic space group P2$_1$/c – C$_{2h}^5$ (No. 14) with a = 15.546(6), b = 10.015(5), c = 15.814(2) Å, β = 114.89(3)°; Z = 4, R = 0.030. The structure is shown in **Fig. 76**. The Os$_3$Pt core is tetrahedral with a pseudo-mirror plane defined by Os(3), Pt, and μ_2,η^1-C. The hydrides were not observed crystallographically but they are believed to bridge the Os(1)-Os(3) and the Os(2)-Os(3) edges due to the elongated bond distances. Average bond distances of Os-CO and C≡O are 1.91 and 1.15 Å, respectively [6].

References on p. 227

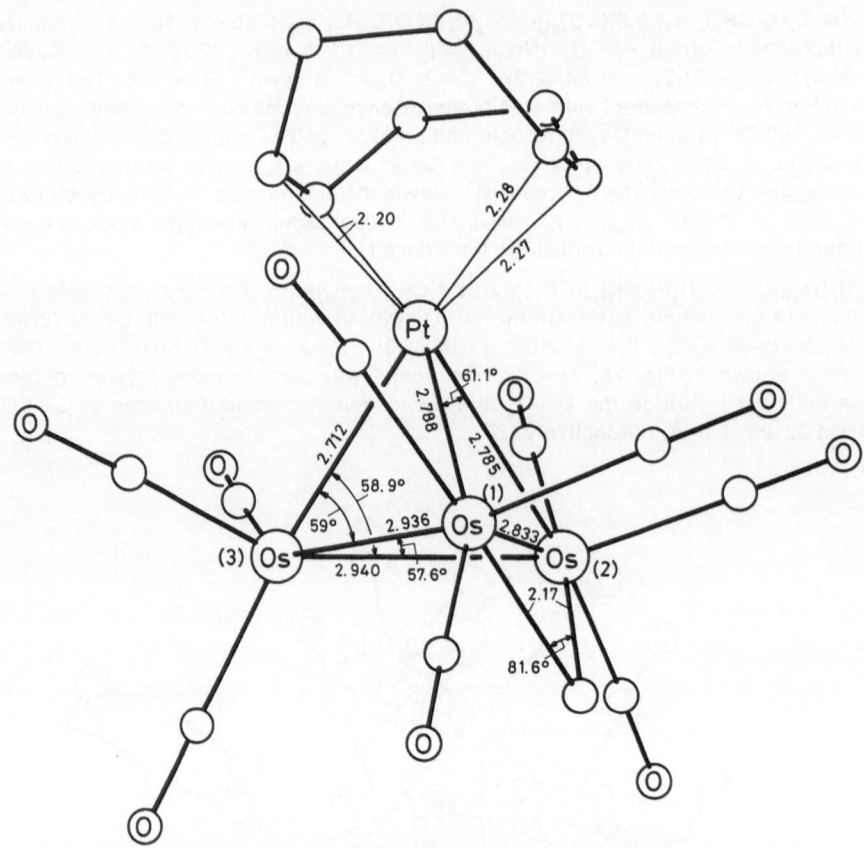

Fig. 76. Molecular structure of $(\mu\text{-}H)_2Os_3(\mu_2,\eta^1\text{-}CH_2)(CO)_9Pt(\eta^4\text{-}C_8H_{12}\text{-}1,5)$ (No. 3) with selected bond distances (in Å) and bond angles [6].

Carbonylation under 50 atm of CO yielded $(\mu\text{-}H)_2Os_3(\mu_2,\eta^1\text{-}CH_2)(CO)_9Pt(CO)_2$ (No. 4) [6].

$(\mu\text{-}H)_2Os_3(\mu_2,\eta^1\text{-}CH_2)(CO)_9Pt(CO)P(C_6H_{11}\text{-}c)_3$ (Table **7**, Nos. **5** and **6**). The reactions of $(\mu\text{-}H)Os_3(CO)_9(\mu\text{-}H)Pt(CO)P(C_6H_{11}\text{-}c)_3$ with CH_2N_2, or of $(\mu\text{-}H)_2Os_3(\mu_2,\eta^1\text{-}CH_2)(CO)_{10}$ (Section 3.1.6.2.1) with $(C_2H_4)_2PtP(C_6H_{11}\text{-}c)_3$ both gave initially No. **5** as the kinetic product, which subsequently converted to the thermodynamically favored isomer No. **6**; however, within 4 to 5 d an equilibrium mixture of ca. 1:4 in favor of No. **6** is reached, based on ^{1}H NMR spectra. The isomerization of No. **5** to No. **6** involves rotation of the $Pt(CO)P(C_6H_{11}\text{-}c)_3$ moiety about an axis perpendicular to the Os_3 plane and hydride transfer from a site bridging an Os–Os bond to one bridging an Os–Pt edge; the ^{1}H NMR gives no evidence of tautomers with bridging CH_3 groups as observed for $(\mu\text{-}H)_2Os_3(\mu_2,\eta^1\text{-}CH_2)$-$(CO)_{10}$ (Section 3.1.6.2.1) [1, 3].

Deuterated species were obtained from $(\mu\text{-}H)Os_3(CO)_9(\mu\text{-}H)Pt(CO)P(C_6H_{11}\text{-}c)_3$ and CD_2N_2, or from $(\mu\text{-}D)Os_3(CO)_9(\mu\text{-}D)Pt(CO)P(C_6H_{11}\text{-}c)_3$ and CH_2N_2. ^{1}H and ^{2}H NMR spectra revealed that in both preparations ca. 30% of the D (from CD_2N_2) or H nuclei (from CH_2N_2), respectively, were transferred from the methylene to the hydride sites in No. **5**. The H/D site exchanges are believed to proceed via intermediates containing $\mu_2,\eta^1\text{-}CH_3$,

μ_2,η^1-CDH$_2$, or μ_2,η^1-CHD$_2$ groups. The H/D distribution remained generally constant for a few days and upon isomerization of No. 5 to No. 6 [1, 3].

No. 5 crystallizes in the triclinic space group $P\bar{1} - C_i^1$ (No. 2) with a = 12.244(5), b = 9.533(4), c = 16.239(6) Å, α = 77.00(3)°, β = 72.79(3)°, γ = 75.84(3)°; Z = 2, D$_c$ = 2.57 g/cm^3, R = 0.037. The structure is shown in **Fig. 77**. The μ-CH$_2$ and μ-H hydrogens were directly located and refined. The orientation of the μ_2,η^1-CH$_2$ group is approximately orthogonal to the metal-metal vector. Average bond distances of Os-CO and C≡O are 1.921 and 1.131Å, respectively [1, 3].

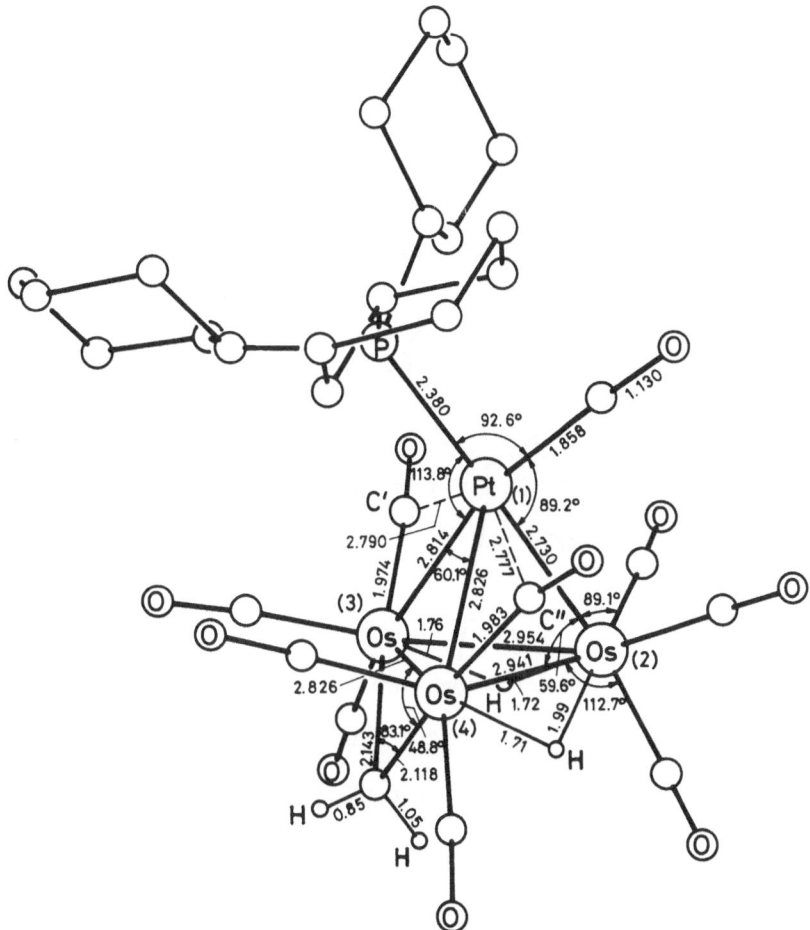

Fig. 77. Molecular structure of (μ-H)$_2$Os$_3$(μ_2,η^1-CH$_2$)(CO)$_9$Pt(CO)P(C$_6$H$_{11}$-c)$_3$ (No. 5) with selected interatomic distances (in Å) and bond angles [1, 3].

No. 6 crystallizes in the monoclinic space group P2$_1$/c – C$_{2h}^5$ (No. 14) with a = 18.708(4), b = 16.664(3), c = 22.875(5) Å, β = 101.82(2)°; Z = 8, D$_c$ = 2.55 g/cm^3, R = 0.047. The structure is shown in **Fig. 78**. The asymmetric unit contains two crystallographically independent molecules, but no significant differences between these molecules were found. The hydride ligands and the μ-CH$_2$ hydrogens were not observed crystallographically, but were assigned on the basis of potential-energy studies and were believed to bridge the Pt(1)-Os(2) and

References on p. 227

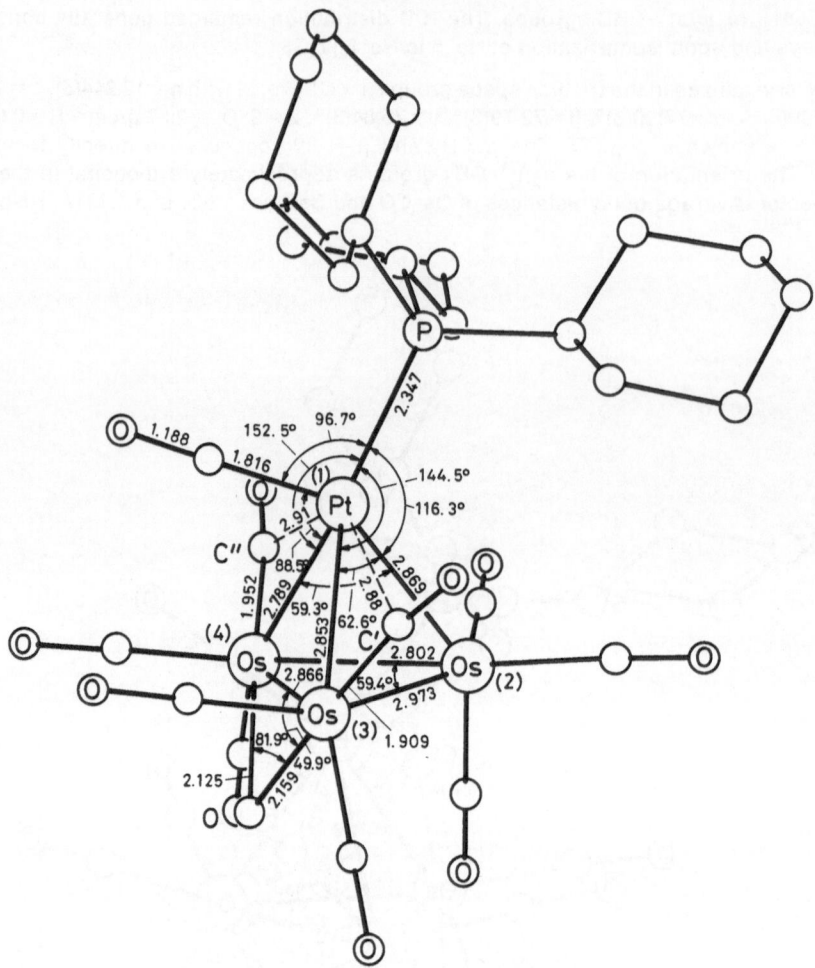

Fig. 78. Molecular structure of one of the two crystallographically independent molecules of $(\mu\text{-}H)_2Os_3(\mu_2,\eta^1\text{-}CH_2)(CO)_9Pt(CO)P(C_6H_{11}\text{-}c)_3$ (No. 6) with selected interatomic distances (in Å) and bond angles [1, 3].

Os(2)–Os(3) edges. The hydride ligand attached to the Pt center lies cis to the bulky phosphane ligand probably for steric reasons. The presence of an Os-(μ-H)-Pt bridge was also confirmed by the ^1H NMR spectrum. Average bond distances of Os–CO and C≡O are 1.901 and 1.155 Å, respectively [1, 3].

The molecular structures of Nos. 5 and 6 (Figs. 77 and 78) involve closed distorted tetrahedral Os_3Pt cores [1, 3] with essentially octahedrally coordinated Os atoms and an 18-electron trigonal-bipyramidally ligated Pt center. The methylene group bridges an Os–Os edge cis to the phosphane group in No. 5 but trans in No. 6. In both isomers the μ-CH_2 ligands occupy sites bridging two Os atoms, but neither is an Os–Pt edge bridged because of kinetic or thermodynamic reasons. All CO groups are terminal and essentially linear; only CO′ and CO″ in both isomers deviate from linearity and interact weakly with the Pt atom [3].

References on p. 227

No. 5 exhibits approximate mirror (C_s) symmetry in the solid state and also in solution, as indicated by the single high-field ^1H resonance for the two μ-H ligands, and the small ^1H, ^{195}Pt coupling is due to the absence of a direct μ-H-Pt bond. The unsymmetrical isomer No. 6 exhibits two high-field resonances corresponding to the two nonequivalent hydride ligands. Similarly, for No. 5 the singlet ^{31}P resonance at 10.8 ppm becomes a triplet in the selectively ^1H-decoupled ^{31}P NMR spectrum, consistent with coupling to two equivalent μ-H ligands, while for No. 6 the singlet in the ^{31}P {^1H} spectrum changed into a doublet due to coupling with only one μ-H-ligand. ^1H NMR measurements up to 100 °C revealed dynamic behaviour with line broadening for No. 6 in solution, whereas No. 5 decomposed under these conditions. Evidently, the μ-H site exchange in No. 6 is a higher-energy process than in the precursor (μ-H)Os$_3$(CO)$_9$(μ-H)Pt(CO)P(C$_6$H$_{11}$-c)$_3$ [1, 3].

Both No. 5 and No. 6 contain closo tetrahedral Os$_3$Pt skeletons but differ in the disposition of the μ-H ligands and the orientation of the Pt(CO)P(C$_6$H$_{11}$-c)$_3$ unit relative to the Os$_3$ triangle. Analogous complexes bearing P(C$_6$H$_5$)$_3$ or CO instead of μ-CH$_2$ adopt an open butterfly arrangement. In an effort to understand this divergent behaviour, extended Hückel molecular orbital calculations on the interactions of the hypothetical model complex [Ru$_3$Pt(CO)$_{10}$(PH$_3$)]$^{2-}$ with CH$_2$ and CO have been carried out. The results indicate a suitable symmetry match for the frontier orbitals of the CH$_2$ fragment in a closo tetrahedral Os$_3$Pt core, and orbital control is the driving force for the formation of Nos. 5 and 6. To the contrary, interactions of the model complex with CO in the same configuration (giving μ-CO) leaves one of the CO π* orbitals essentially unperturbed. This "adduct" would seek further stabilization of the CO π* orbitals by forming an open butterfly arrangement with all CO terminally bound [8 to 10].

(μ-H)Os$_3$(μ$_2$,η^1-CH$_2$)(CO)$_9$(μ-H)PtP(C$_6$H$_{11}$-c)$_3$CNC$_6$H$_{11}$-c (Table 7, No. 7) crystallizes in the triclinic space group P$\bar{1}$ – C$_i^1$ (No. 2) with a = 12.189(3), b = 13.180(4), c = 16.901(6) Å, α = 108.31(4)°, β = 90.62(4)°, γ = 114.20(2)°; Z = 2, D$_c$ = 2.04 g/cm^3, R = 0.059. The compound has a tetrahedral Os$_3$Pt core (see **Fig. 79**). The slightly asymmetric μ$_2$,η^1-CH$_2$ group bridging the Os(2)-Os(3) edge lies near the same face of the tetrahedron as the terminal CNC$_6$H$_{11}$-c ligand. Weak semi-bridging interactions were observed between the Pt center and two carbonyls CO′ and CO″, which are approximately coplanar to the Pt-Os(2)-Os(3) face and trans to the methylene ligand. The hydrides and the methylene hydrogens were not observed crystallographically, and were given in calculated positions assuming pseudotetrahedral geometry for the methylene carbon. Average bond distances of Os-CO and C≡O are 1.89 Å and 1.14 Å, respectively [7].

Magnetization transfer studies revealed a slow exchange of the bridging hydrides at 298 K, and ΔG^{+} = 85.6 ± 0.8 kJ/mol was estimated at 373 K for this process, but no scrambling between the hydrides and the CH$_2$ hydrogens could be observed. No ^{187}Os couplings could be resolved for either hydride. The assignments of the CH$_2$ resonances are based on indirect Nuclear Overhauser Enhancement Effect experiments, and no evidence for mutual exchange between the CH$_2$ hydrogens could be obtained. At higher temperatures the multiplet due to H^3 collapses into a doublet with J(H,H) = 6.0 Hz [12].

Two-dimensional ^{13}C {^1H} (2D EXSY) spectra at 298 and 318 K revealed that CO exchange processes occur. Tripodal rotation processes were observed for the Os(1)(CO)$_3$ group at 298 K, evidenced by cross peaks between the resonances of CO-e, f, and h, and for the Os(3)(CO)$_3$ group at 318 K, probably resulting from an exchange between site a, b and either g or f. In addition, at 318 K a pairwise exchange of the CO ligands a/c, b/d, and i/g occurred indicating that the compound aquired a time-averaged mirror plane and hence racemized. For each of the three CO exchange processes the rate constants k and the

References on p. 227

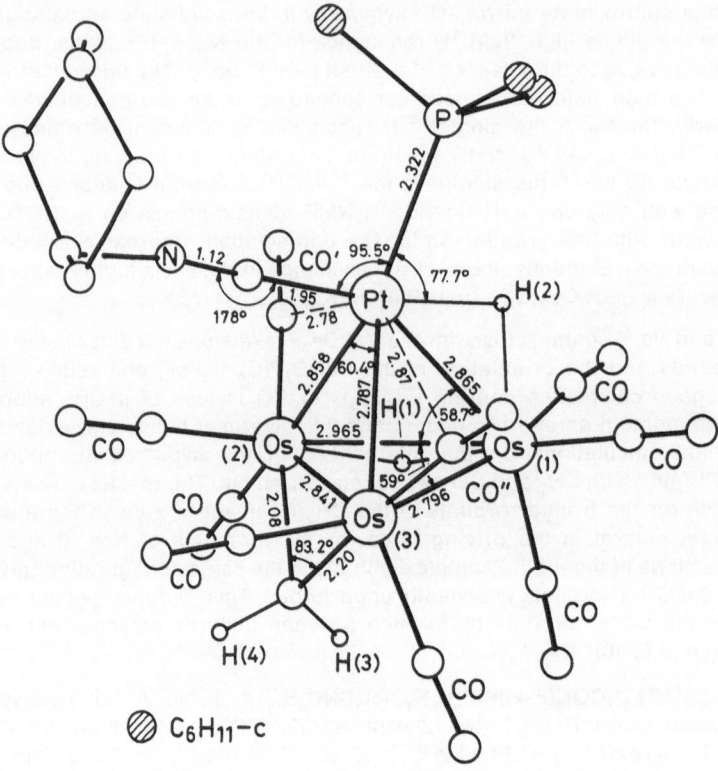

Fig. 79. Molecular structure of $(\mu\text{-H})Os_3(\mu_2,\eta^1\text{-CH}_2)(CO)_9(\mu\text{-H})PtP(C_6H_{11}\text{-c})_3CNC_6H_{11}\text{-c}$ (No. 7) with selected interatomic distances (in Å) and bond angles [7].

free activation energies ΔG^+ at 318 K were estimated from the EXSY spectrum [12]:

CO exchange process	k [s^{-1}] at 318 K			ΔG^+ [kJ/mol]
Os(1)(CO)$_3$	$k_{e,f/f,e} = 0.38(7)$	$k_{e,h/h,e} = 0.43(8)$	$k_{f,h/h,f} = 0.68(9)$	80.5 ± 1.0
Os(3)(CO)$_3$	$k_{a,g/g,a} = 0.02(1)$	$k_{b,g/g,b} = 0.02(1)$	$k_{a,b/b,a}$ not measurable due to overlap	88.4 ± 1.6
pairwise exchange	$k_{a,c/c,a} = 0.07(2)$	$k_{b,d/d,b} = 0.09(2)$	$k_{i,g/g,i} = 0.07(2)$	84.7 ± 1.1

Protonation with CF_3CO_2H gave $[(\mu\text{-H})_2Os_3(\mu\text{-CH}_2)(CO)_9(\mu\text{-H})PtP(C_6H_{11})_3CNC_6H_{11}\text{-c}]^+$ (No. 8) [7].

The compound is unreactive to CO (50 atm) and $P(C_6H_5)_3$ at ambient temperature, while at higher temperature only decomposition was observed [7].

$Os_3(\mu_2,\eta^1\text{-CH}_2)(CO)_{11}Pt(P(C_6H_5)_3)_2$ (Table **7**, No. **9**) crystallizes in the monoclinic space group $C2/c - C_{2h}^6$ (No. 15) with a = 39.15(2), b = 15.202(8), c = 18.542(8) Å, $\beta = 91.15(4)°$; Z = 8, $D_c = 1.941$ g/cm^3, R = 0.083. The compound has a spiked triangular Os_3Pt core (see **Fig. 80**). The coordination geometry about Pt is best described as a distorted square plane [5].

References on p. 227

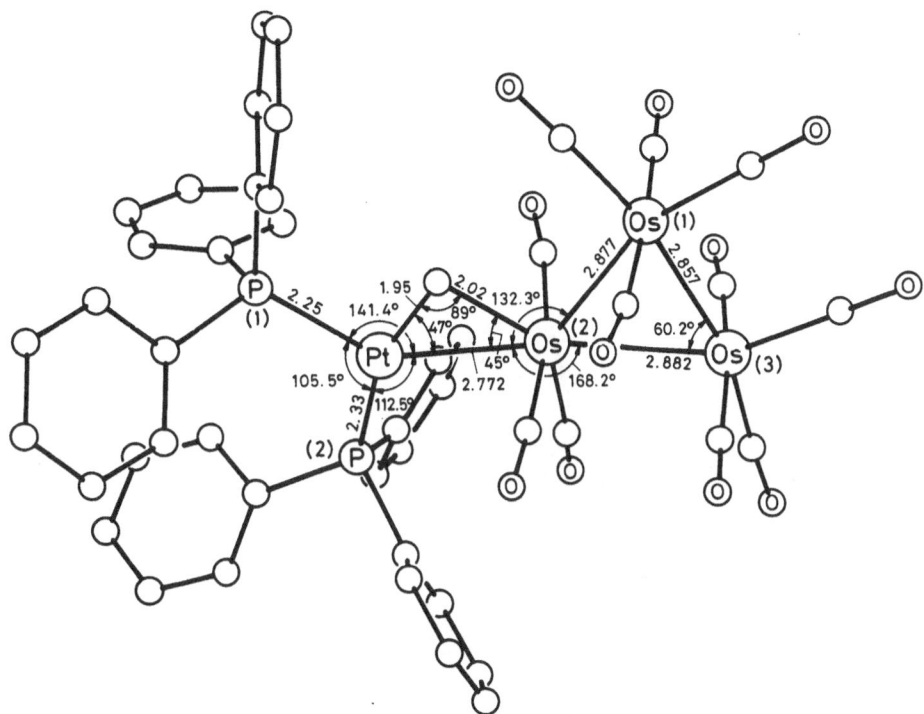

Fig. 80. Molecular structure of Os$_3$(μ$_2$,η^1-CH$_2$)(CO)$_{11}$Pt(P(C$_6$H$_5$)$_3$)$_2$ (No. 9) with selected bond distances (in Å) and bond angles [5].

Carbonylation in THF (1 atm of CO) yielded Os$_3$(μ$_2$,η^1-CH$_2$)(CO)$_{11}$Pt(CO)P(C$_6$H$_5$)$_3$ (No. 10). Treatment with C$_6$H$_5$C≡CC$_6$H$_5$ in THF at 22 °C for 3 d resulted in a red oil that led after chromatographic work-up to C$_6$H$_5$C≡CC$_6$H$_5$ and traces of the starting material as the only bands that eluted; monitoring of the reaction by IR spectroscopy showed that the CO bands of the starting material were replaced by other bands [5].

(μ-H)$_2$Os$_3$(μ$_4$-C)(CO)$_9$Pt(CO)P(C$_6$H$_{11}$-c)$_3$ (Table 7, No. 11) crystallizes in the monoclinic space group P2$_1$/n−C$_{2h}^5$ (No. 14) with a = 17.103(8), b = 17.061(9), c = 12.222(7) Å, β = 95.87(4)°; Z = 4, D$_c$ = 2.52 g/cm^3, R = 0.072. The Os$_3$Pt framework is derived from a butterfly arrangement with the Os(1)−Pt bond broken. The coordination around the Pt is nearly planar being consistent with a 16-electron configuration of the metal center; see **Fig. 81**. The carbido-carbon atom is irregularly bonded to all four metal atoms. The hydride ligands are given in positions determined from potential energy minimization calculations, and are in agreement with the elongated Os(1)−Os(2) and Os(1)−Os(3) bonds. Average bond distances of Os−CO and C≡O are 1.88 Å and 1.16 Å, respectively [4].

(μ-H)$_2$Os$_3$(μ$_5$-C)(CO)$_9$Pt$_2$(μ-CO)(P(C$_6$H$_{11}$-c)$_3$)$_2$ (Table 7, No. 12) crystallizes in the triclinic space group P1̄−C$_i^1$ (No. 2) with a = 12.328(3), b = 12.375(2), c = 18.802(3) Å, α = 101.95(1)°, β = 85.27(2)°, γ = 79.83(2)°; Z = 2, D$_c$ = 2.20 g/cm^3, R = 0.036. The structure is shown in **Fig. 82**. The Os$_3$Pt$_2$ framework can be viewed as derived from a square pyramid with two Pt−Os bonds having been broken. The carbido-carbon atom is approximately equidistant from all five metal atoms. A pseudo mirror plane passes through μ-CO, μ$_5$-C, Os(3), and CO′,

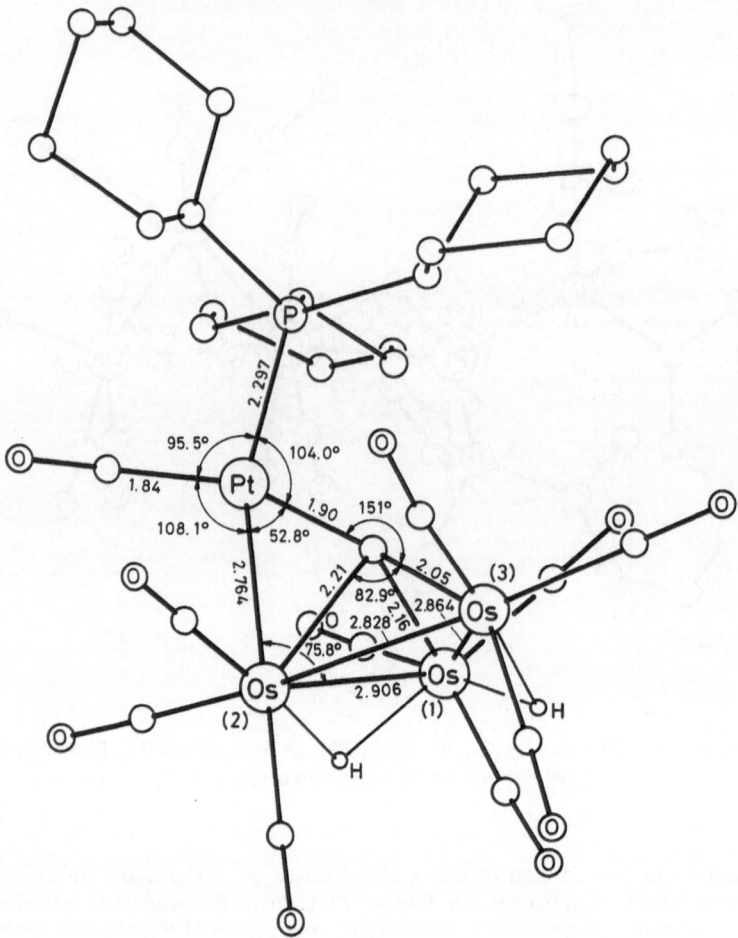

Fig. 81. Molecular structure of $(\mu\text{-H})_2Os_3(\mu_4\text{-C})(CO)_9Pt(CO)P(C_6H_{11}\text{-}c)_3$ (No. 11) with selected bond distances (in Å) and bond angles [4].

thus, the two Pt nuclei including the $P(C_6H_{11}\text{-}c)_3$ ligands, and the two μ-hydrides are equivalent. The angle between the planes defined by the Os_3 triangle and the square $Os(1)\text{-}Os(2)\text{-}Pt(1)\text{-}Pt(2)$ is 76.4°. The hydride ligands were located in positions determined from potential energy minimization calculations; they are in agreement with the elongated $Os(1)\text{-}Os(3)$ and $Os(2)\text{-}Os(3)$ bonds. Average bond distances of Os–CO and C≡O are 1.919 Å and 1.15 Å, respectively [4].

$Os_3(\mu_5\text{-C})(CO)_9(\mu\text{-OCH}_3)(\mu\text{-H})Pt_2(\mu\text{-CO})(P(C_6H_{11}\text{-}c)_3)_2$ (Table **7**, No. **13**) crystallizes in the monoclinic space group $P2_1/n - C_{2h}^5$ (No. 14) with $a = 10.242(5)$, $b = 20.65(1)$, $c = 26.94(2)$ Å, $\beta = 101.55(2)°$; $Z = 4$, $D_c = 2.19$ g/cm^3, $R = 0.054$. The Os_3Pt_2 core exhibits an unusual geometry involving a rather buckled square arrangement of two Pt and two Os atoms, and with $Os(3)$ bonded only to $Os(1)$ in the square. The arrangement of the metal atoms formally arises from the Os_3Pt_2 core of No. 12 by the breaking of the $Os(3)\text{-}Os(1)$ bond, leading to nonequivalent Pt centers; see **Fig. 83**. The carbido carbon is nearly equidistant from

References on p. 227

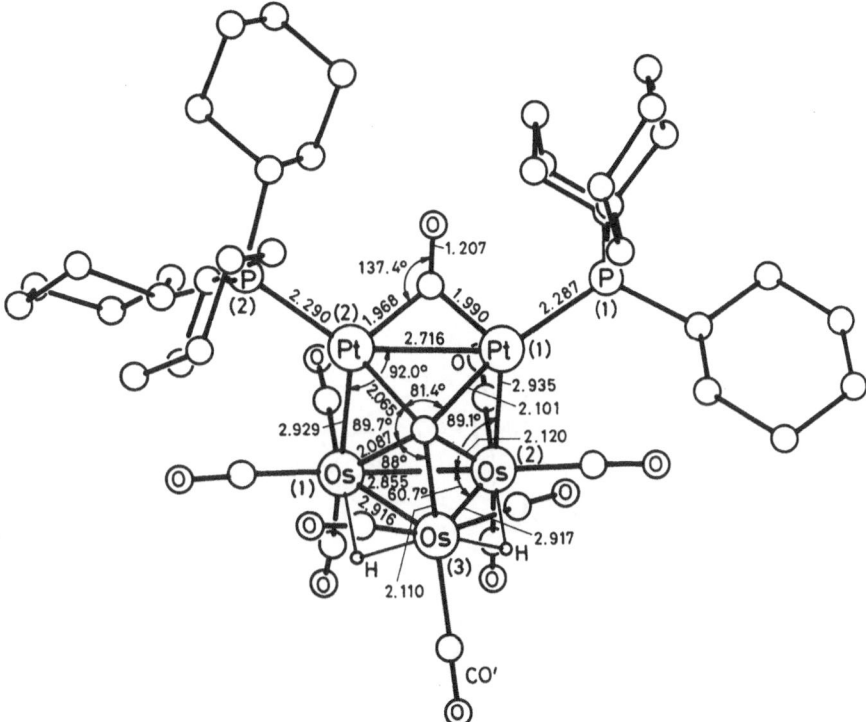

Fig. 82. Molecular structure of $(\mu\text{-}H)_2Os_3(\mu_5\text{-}C)(CO)_9Pt_2(\mu\text{-}CO)(P(C_6H_{11}\text{-}c)_3)_2$ (No. 12) with selected bond distances (in Å) and bond angles [4].

all five metal atoms. The hydride ligand was not observed crystallographically, but potential energy minimization calculations indicate that it bridges the Os(1)–Pt(1) bond. Interestingly, this bond is shorter than the nonbridged Os(2)–Pt(2) bond, in contrast to the elongating effect of bridging hydrides. Average bond distances of Os–CO and C≡O are 1.915 Å and 1.14 Å, respectively [4].

$(\mu\text{-}H)_2Os_3(CNR)(CO)_9Pt(CO)P(C_6H_{11}\text{-}c)_3$ (R = t-C$_4$H$_9$, c-C$_6$H$_{11}$; Table **7**, Nos. **14**, **15**). Both compounds were intermediately formed (based on ^{1}H and ^{13}C NMR data, not isolated) as isomeric mixtures during the preparation of $(\mu\text{-}H)_2Os_3(CO)_9Pt(CNR)P(C_6H_{11}\text{-}c)_3$ from $(\mu\text{-}H)_2Os_3(CO)_{10}PtP(C_6H_{11}\text{-}c)_3$ and CNR in hexane at −78 to 0 °C, then at reflux temperature for 1.5 h. The isomeric composition depended on the initial reaction temperature. At −78 °C, the addition of equivalent amounts of CNR resulted in the formation of mainly Isomer 1, along with traces of Isomer 2, whereas reaction at 0 °C gave Isomer 1 to 3 existing in a thermodynamic equilibrium. The relative proportions of Isomers 1, 2, and 3 varied also depending on the nature of R indicating that Isomers 2 and 3 are formed under kinetic control. At ca. 25 °C, the signals of Isomer 3 disappeared rapidly while the equilibrium between Isomer 1 and 2 is established more slowly. The structures proposed for Isomers 1 to 3 are consistent with variable-temperature ^{1}H and ^{13}C NMR spectroscopic data, suggesting a butterfly Os$_3$Pt core with the CNR ligand coordinated to the Os(CO)$_3$ center that is not involved to the Os(μ-H)Pt bridge [7].

References on p. 227

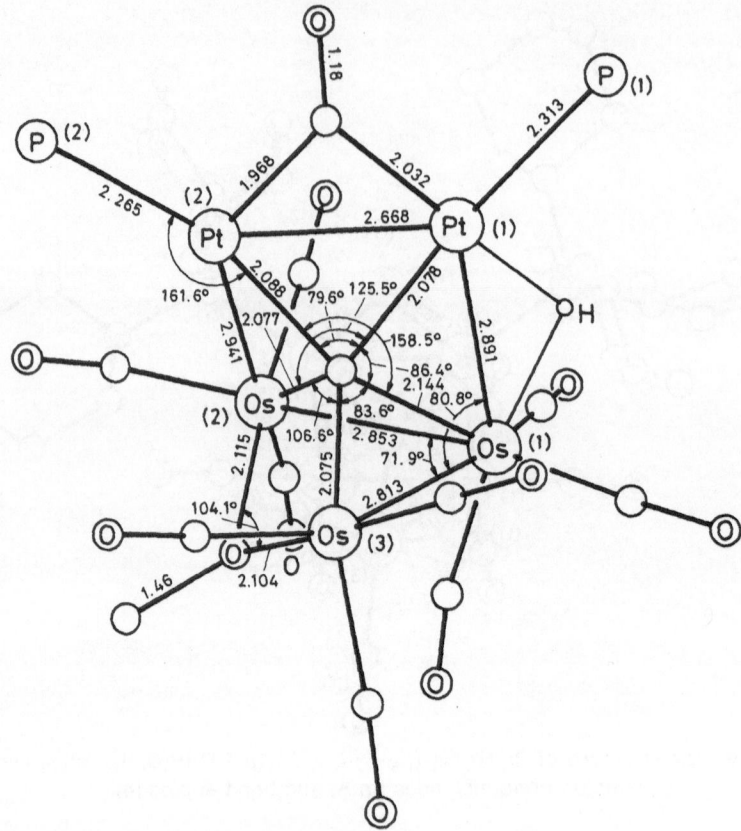

Fig. 83. Molecular structure of $Os_3(\mu_5-C)(CO)_9(\mu-OCH_3)(\mu-H)Pt_2(\mu-CO)(P(C_6H_{11}-c)_3)_2$ (No. 13) with selected bond distances (in Å) and bond angles; cyclohexyl groups are omitted for clarity [4].

Thermolysis of the isomeric mixtures of Nos. 14 or 15 in refluxing hexane for 1 hour gave unsaturated $(\mu-H)_2Os_3(CO)_9Pt(CNR)P(C_6H_{11}-c)_3$ consisting of a tetrahedral Os_3Pt framework with the isocyanide ligand bonded to the Pt center; following treatment with CO at $-78\,°C$ or at ambient temperature reformed initially Isomers 1 of Nos. 14 or 15 as the most stable isomers, but on standing at ambient temperature also Isomers 2 were observed [7].

Isomer 1

Isomer 2

Isomer 3

References on p. 227

References:

[1] Green, M.; Hankey, D. R.; Murray, M.; Orpen, A. G.; Stone, F. G. A. (J. Chem. Soc. Chem. Commun. **1981** 689/91).

[2] Adams, R. D.; Hor, T. S. A. (Inorg. Chem. **23** [1984] 4723/32).

[3] Farrugia, L. J.; Green, M.; Hankey, D. R.; Murray, M.; Orpen, A. G.; Stone, F. G. A. (J. Chem. Soc. Dalton Trans. **1985** 177/90).

[4] Farrugia, L. J.; Miles, A. D.; Stone, F. G. A. (J. Chem. Soc. Dalton Trans. **1985** 2437/47).

[5] Williams, G. D.; Lieszkovszky, M.-C.; Mirkin, C. A.; Geoffroy, G. L.; Rheingold, A. L. (Organometallics **5** [1986] 2228/33).

[6] Norén, B.; Sundberg, P. (J. Chem. Soc. Dalton Trans. **1987** 3103/5).

[7] Ewing, P.; Farrugia, L. J. (Organometallics **7** [1988] 871/8).

[8] Ewing, P.; Farrugia, L. J. (J. Organomet. Chem. **347** [1988] C 31/C 34).

[9] Ewing, P.; Farrugia, L. J. (New J. Chem. **12** [1988] 409/17).

[10] Ewing, P.; Farrugia, L. J. (Organometallics **8** [1989] 1665/73).

[11] Ewing, P.; Farrugia, L. J. (J. Organomet. Chem. **373** [1989] 259/68).

[12] Farrugia, L. J. (Organometallics **8** [1989] 2410/7).

3.1.6.4.3 Compounds with Os Bonded to Sn, Mn, Re, Fe, Ni, or Au Atoms

The compounds dealt with in this section consist of an Os_3 core to which a heterometallic atom such as Sn, Mn, Re, Fe, Ni, or Au is bonded. The heterometallic atoms are coordinated to one, two, or all three Os centers.

Table 8
Compounds with Os Bonded to Sn, Mn, Re, Fe, Ni, or Au Atoms.
An asterisk preceding the compound number indicates further information at the end of the table, pp. 234/46.
Explanations, abbreviations, and units on p. X.

No. compound	method of preparation (yield in %) properties and remarks
Os_3Sn compounds	
*1	from $Os_3(\mu_2,\eta^1\text{-}CH_2)(\mu\text{-}CO)(CO)_{10}$ (Section 3.1.6.2.1) and anhydrous $SnCl_2$ (1:1; dissolved in THF; dropwise addition) in THF at room temperature for 30 to 45 min (91%) [6] yellow crystals from $CDCl_3$ [6] ^1H NMR ($CDCl_3$?): 5.61 (s, CH_2) [6] IR (hexane): 1983, 1993, 2014, 2029, 2052, 2066, 2106, 2141 (all CO) [6] mass spectrum: $[M]^+$ [6]
*2	by thermolysis of $(\mu\text{-}H)Os_3(CO)_{10}(\mu\text{-}H)\text{-}Sn(CH(Si(CH_3)_3)_2)_2$ in refluxing n-heptane; upon cooling to room temperature the complex precipitated (quantitative) [4] crystals, m.p. 177 °C (dec.) [4] ^1H NMR ($CDCl_3$, 30 °C): -19.596, -16.138 (d's, both $\mu\text{-}H$; $^2J(H,H) = 1.21$ Hz), 0.12 (s, 18 H of CH_3), 0.18 (d?, 18 H of CH_3), 1.294, 4.128 (s's, each 1 H of CH) [4]

References on p. 246

Table 8 (continued)

No. compound	method of preparation (yield in %) properties and remarks
*2 (continued)	IR (KBr): 1394 (O=C), 1953, 1970, 1993, 2000, 2010, 2015, 2075, 2100 (all CO) [4]

*3

$((CH_3)_3Si)_2CH$—Sn
$((CH_3)_3Si)_2CH$
OC—Os
Os
Os
CO, CO, CO, O, OCH$_3$, CO_2CH_3

from $(\mu-H)Os_3(CO)_{10}(\mu-H)Sn(CH(Si(CH_3)_3)_2)_2$ and $C_2(CO_2CH_3)_2$ in refluxing n-heptane (low yield) [4]
red crystals from hexane, m.p. 201 to 203 °C (dec.) [4]
^1H NMR (CDCl$_3$, 30 °C): 0.24, 0.301 (s's, each 18 H of CH$_3$), 1.559, 1.563 (s's, each 2 H, CH and CH$_2$), 3.764 (s, 6 H, CH$_3$) [4]
IR (KBr): 1850 (μ-CO), 1967, 1983, 2003, 2047, 2077, 2094 (all CO) [4]

*4 $(CH_3)_3Sn$

$C_6H_5(CH_3)_2P$

from $Os_3(\mu_3,\eta^2-SCH_2)(CO)_8(\mu_3-S)P(CH_3)_2C_6H_5$ ("Organoosmium Compounds" B5, Section 3.1.6.1.5, in preparation) and $(CH_3)_3SnH$ (ca. 1:10) in refluxing heptane for 4 h; work-up by chromatography on alumina with hexane/benzene (3:1) as eluant (less than 3%, along with 36% of $(\mu-H)_2Os_3(\mu_3,\eta^2-SCH_2)-(CO)_7(\mu_3-S)P(CH_3)_2C_6H_5$ [1]
yellow crystals from hexane at −20 °C, m.p. 150 °C (dec.) [1]
^1H NMR (CDCl$_3$): −16.30 (dd, μ-H; J=3.3, J(H,P)=13.3), 0.43 (Sn(CH$_3$)$_3$; $^2J(H,^{117,119}Sn)=50.0$, 47.8), 1.37 (ddd, 1 H, CH$_2$; J=3.3, $^2J(H,H)=11.0$, J(H,P)=1.3), 2.27 (d, PCH$_3$; $^2J(H,P)=10.2$), 2.28 (d, PCH$_3$; $^2J(H,P)=10.1$), 2.42 (d, 1 H, CH$_2$; $^2J(H,H)=11.0$ Hz), 7.58 (m, C$_6$H$_5$) [1]
IR (hexane): 1939, 2000, 2037, 2091 (all CO) [1]

Os$_3$Mn compounds

*5

Mn, CO, C$_6$H$_5$, O, CO, CO, H, CO, CO, OC—Os, Os—CO, CO, OC, CO

from $(\mu-H)_2Os_3(CO)_{10}$ and $(\eta^5-C_5H_5)(CO)_2Mn=C=CHC_6H_5$ (ca. 1:2) in CH$_2$Cl$_2$ at 20 °C for 6 h, followed by chromatography on silica gel with CHCl$_3$/hexane (1:2) [2] or with CHCl$_3$/petroleum ether (1:2 to 11:50) [3] as eluant (60%) [2, 3], along with 5% of orange $(\mu-H)Os_3(\mu_2,\eta^2-CH=CHC_6H_5)(CO)_{10}$, 10% of yellow $(\mu-H)_2Os_3(\mu_3,\eta^2-CH=CC_6H_5)(CO)_9$ (compound III and IV in Scheme 1, p. 239), 4.5% of dark red Os$_3$(CHCC$_6$H$_5$)(CO)$_9$, and traces of Os$_3$(CO)$_{12}$ and $(\eta^5-C_5H_5)Mn(CO)_3$ [3]

References on p. 246

Table 8 (continued)

No. compound	method of preparation (yield in %) properties and remarks
	dark red crystals from 1,4-dioxane at $-15\,°C$, crystallizing with 1.5 molecules dioxane; m.p. 75 to 76 °C [2, 3]

stable in the solid state; readily soluble in organic solvents with decomposition in nonpolar solvents (see Scheme 1, p. 239), but moderately stable in THF and 1,4-dioxane [2, 3]

^1H NMR $(CDCl_3)$: -22.2 (s, μ-H), 3.75 (d, CH; J(H,H) = 14.2), 4.39 (s, C_5H_5), 7.27 (m, C_6H_5), 9.34 (d, OsCH; J(H,H) = 14.2); the coupling constant of 14.2 Hz between the olefin protons is due to the trans configuration [2, 3]

IR (cyclohexane): 1790 (μ-CO), 1933, 1950, 1980, 1997, 2010, 2030, 2047, 2053, 2073, 2127 (all CO) [2, 3]

mass spectrum: no [M$^+$], but $[HOs_3(CH{=}CHC_6H_5)(CO)_n]^+$, n = 10 to 1, $[H_2Os_3(CH{\equiv}CC_6H_5)(CO)_9]^+$, $[(\eta^5\text{-}C_5H_5)Mn(CO)_n]^+$, n = 3 to 0, and fragments of the organic ligand [2, 3]

*6

from $[Os_3(\mu_3,\eta^1\text{-}CCO)(CO)_9][N(P(C_6H_5)_3)_2]_2$ (Section 3.1.6.3) and $[Mn(CO)_3(NCCH_3)_3][PF_6]$ in acetone at 35 °C for 2.5 h, followed by extraction of the resulting oil with ether and precipitation from ether/pentane (82%) [8]

orange air-sensitive microcrystals from CH_2Cl_2/ether/pentane [8]

^{13}C NMR (CD_2Cl_2): 173.7 (CO-f), 179.5 (CO-g), 183.6 (CO-a,b,c), 218.0 (CO-d,e, μ-CO), 331.1 (μ_4-C); $(CD_2Cl_2, -90\,°C)$: 173.8 (CO-f), 179.7 (CO-g), 180.5, 181.0 (each 2 CO, CO-a,b), 191.0 (t, CO-c; J(C,C) = 16), 210.3, 210.8 (each 1 CO, CO-d, μ-CO), 224.8 (CO-e), 330.2 (t, μ_4-C; J(C,C) = 16); upon warming from -90 to $+25\,°C$ the resonances of CO-a,b,c and CO-d,e and μ-CO broaden, collapse, and then coalesce to singlets at 183.6 and 218.0 ppm, respectively; in the low-temperature ^{13}C NMR the large C,C coupling constant observed between μ_4-C and two of the wing-tip carbonyls is indicative of strong bonding interactions between the carbide and Os(1) or Os(3), respectively; spectrum at $-80\,°C$ depicted [8]

References on p. 246

Table 8 (continued)

No. compound	method of preparation (yield in %) properties and remarks

*6 (continued)

IR (ether): 1863 (μ-CO), 1945, 1959, 1985, 2004, 2025, 2071 (all CO) [8]

Os₃Re compounds

7

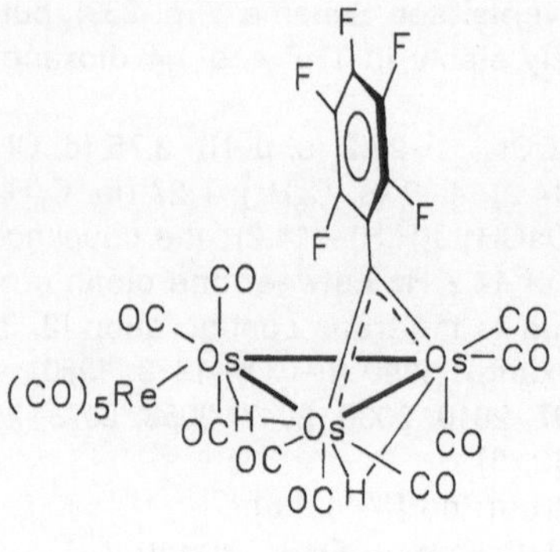

from (μ-H)₃Os₃(μ₃,η¹-CBr)(CO)₉ (Section 3.1.6.3) and Re₂(CO)₁₀ (1:1) in C₆F₆ by UV irradiation under a CO atmosphere for 5 h, followed by TLC with petroleum ether and CH₂Cl₂/petroleum ether as eluants (15%, along with (μ-H)₃Os₃(μ₃,η¹-CC₆F₅)(CO)₉ and (μ-H)₃Os₃(μ₃,η¹-CF)(CO)₉ in 55 and 15% yield, both Section 3.1.6.3, and 3% of {(μ-H)₃Os₃(μ₃,η¹-CC(O)-)(CO)₉}₂) [9]
pale yellow solid [9]
¹H NMR (CDCl₃): −20.70 (s, μ-H), −17.98 (s, μ-H) [9]
IR (cyclohexane): 1951, 1967, 1991, 2012, 2017, 2026, 2043, 2079, 2097, 2123 (all CO) [9]
mass spectrum: [M]⁺ [9]
the proposed structure is based on the known structure of (μ-H)₂Os₃(μ₃,η¹-CC₆H₄CH₃-4)-(CO)₉W(η⁵-C₅H₅)(CO)₃ (Section 3.1.6.4.1) [9]

*8

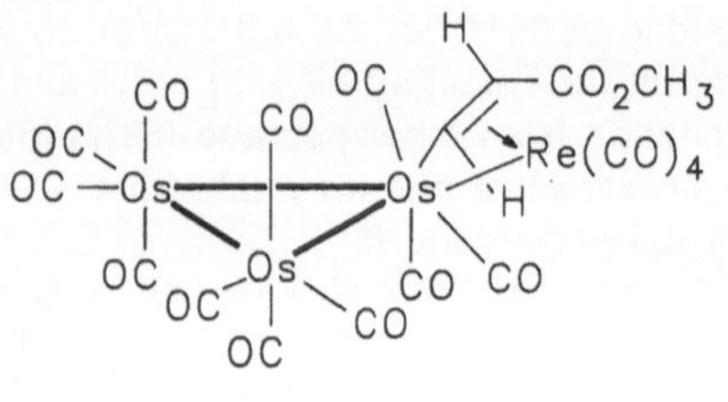

trans
Isomer 1

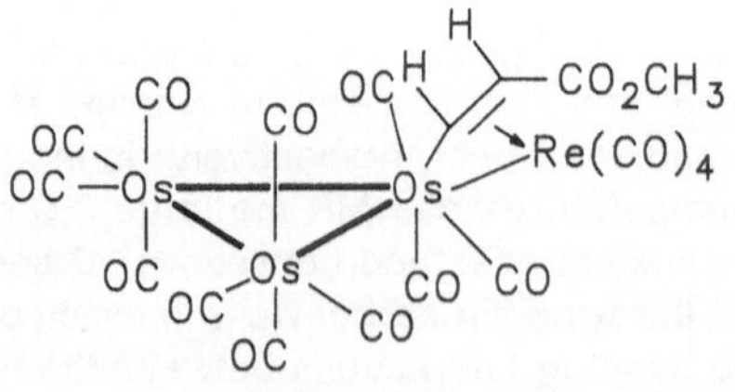

+

cis
Isomer 2

from (μ-H)₂Os₃(CO)₁₀ and (CO)₅ReC≡CCO₂CH₃ (1:1) in benzene at room temperature for 24 h; followed by TLC with benzene as eluant (84:16 ratio of Isomers 1 and 2) [7]
yellow solids [7]
¹H NMR (C₆D₆): Isomer 1: 3.41 (s, CH₃), 4.52 (d, CH; J(CH,OsCH) = 13.3), 9.65 (d, OsCH); Isomer 2: 3.71 (s, CH₃), 5.89 (d, CH; J(CH,OsCH) = 10.5), 8.49 (d, OsCH) [7]

Table 8 (continued)

No.	compound	method of preparation (yield in %) properties and remarks

Os$_3$Fe compounds

*9

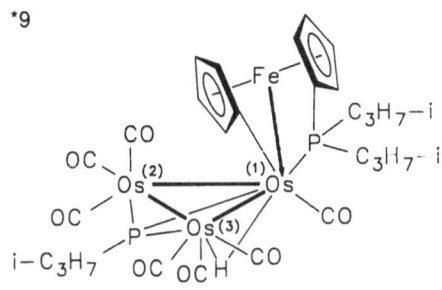

by thermolysis of Os$_3$(CO)$_{10}${Fe(η^5-C$_5$H$_4$P(C$_3$H$_7$-i)$_2$)$_2$} (Formula I, p. 233) in re-fluxing octane for 7.5 h; separation by column chromatography with CH$_2$Cl$_2$/petroleum ether (1:3) as eluant (8 and 15%); by-products: 15 and 3% of (µ H)Os$_3${µ$_2$,η^4-O=CCH$_2$CH(CH$_3$)-{(C$_3$H$_7$-i)P(η^5-C$_5$H$_4$)Fe(η^5-C$_5$H$_4$)P(C$_3$H$_7$-i)$_2$}}-(CO)$_8$ isomers, (see "Organoosmium Compounds" B 5, Section 3.1.6.1.4, in prepa-ration), 10% (µ-H)$_2$Os$_3$-{µ$_3$,η^3-(C$_3$H$_7$-i)$_2$P(η^5-C$_5$H$_4$)Fe-(η^5-C$_5$H$_3$)PC$_3$H$_7$-i}(CO)$_8$, 20% (µ-H)$_2$Os$_3${µ$_3$,η^3-(C$_3$H$_7$-i)$_2$P(η^5-C$_5$H$_4$)Fe-(η^5-C$_5$H$_3$)PC$_3$H$_7$-i}(CO)$_8$, 10% (µ-H)$_2$Os$_3${µ$_3$,η^3-(C$_3$H$_7$-i)$_2$P(η^5-C$_5$H$_3$)Fe-(η^5-C$_5$H$_4$)PC$_3$H$_7$-i}(CO)$_8$ (all "Organoosmium Compounds" B 5, Section 3.1.6.1.6, in prepa-ration), an Os$_3$(µ$_3$,η^3-C$_5$H$_3$FeC$_5$H$_4$) com-pound, and an Os$_6$Fe(C$_5$H$_2$)$_2$ compound [10, 13]
^{1}H NMR (CDCl$_3$): −21.9 (dd, µ-H; ^{1}J=17.5, ^{2}J=7.6), 1.39, 1.47, 1.51, 1.57, 1.84, 1.88 (6 dd's, all CH$_3$), 2.08, 2.56, 3.56 (d of sept's, each CH of C$_3$H$_7$-i), 2.87, 3.22, 3.51, 3.82, 4.86, 4.89, 5.17, 5.21 (br m's, each 1 H of C$_5$H$_4$) [10]
^{31}P {^{1}H} NMR (CDCl$_3$): 16.5 (P(C$_3$H$_7$-i)$_2$), 268.3 (µ$_3$-PC$_3$H$_7$-i) [10]
FAB mass spectrum: [M]$^+$, [M−x CO−y C$_3$H$_7$]$^+$, x=1 to 7, y=1 to 3 [10]

*10

from Os$_3$(CO)$_{12}$ and the ferrocenophane Fe(η^5-C$_5$H$_4$)$_2$PC$_6$H$_5$ (ca. 1:1) in octane at ca. 125 °C for 7 h; purified by column chroma-tography with CH$_2$Cl$_2$/petroleum ether (1:3) as eluant (70%) [14]
red-orange solid; crystallizes as solvate with 0.5 CH$_2$Cl$_2$ [14]
^{1}H NMR (CDCl$_3$): 3.00, 4.62, 4.70, 4.77 (m's, each 2 H of C$_5$H$_4$), 7.15 to 7.55 (m, C$_6$H$_5$) [14]
^{31}P {^{1}H} NMR (CDCl$_3$): 59.9 (br m) [14]
FAB mass spectrum: [M]$^+$ [14]

References on p. 246

Table 8 (continued)

No. compound	method of preparation (yield in %) properties and remarks

*11

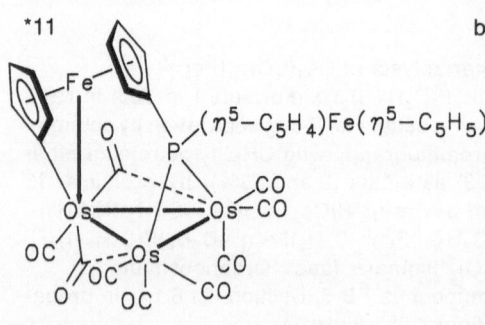

$(\eta^5-C_5H_4)Fe(\eta^5-C_5H_5)$

by thermolysis of $Os_3(CO)_{11}P(C_6H_5)\{(\eta^5-C_5H_4)Fe(\eta^5-C_5H_5)\}_2$ (Formula II) in refluxing octane for 3 h, followed by separation of the product mixture by column chromatography with CH_2Cl_2/petroleum ether (1:4) as eluant (5%) [14]; along with 15% of $(\mu-H_2)Os_3-\{(\mu_3,\eta^3-(\eta^5-C_5H_4)Fe(\eta^5-C_5H_3)P(C_6H_5)-(\eta^5-C_5H_4)Fe(\eta^5-C_5H_5)\}(CO)_8$ [10], 25% of $Os_3\{(\mu_3,\eta^2-(\eta^5-C_5H_3)Fe(\eta^5-C_5H_5)\}-\{\mu_3-P(\eta^5-C_5H_4)Fe(\eta^5-C_5H_5)\}(CO)_9$, and 20% of $Os_3\{(\mu_3,\eta^2-C_6H_4\}\{\mu_3-P(\eta^5-C_5H_4)-Fe(\eta^5-C_5H_5)\}(CO)_9$ [10]

red-brown solid [14]

^1H NMR (CDCl$_3$): 3.12, 3.78 (q's, each 2 H of C_5H_4), 4.07 (s, C_5H_5), 4.22, 4.67, 4.79, 4.91 (m, 3 t's, each 2 H of C_5H_4) [14]

$^{31}P\{^1H\}$ NMR (CDCl$_3$): 39.3 [14]

FAB mass spectrum: $[M]^+$, $[M-x\ CO]^+$, $x=1$ to 9 [14]

Os$_3$Ni$_3$ and Os$_3$Ni$_4$ compounds

12 $[Os_3(\mu_6-C)(\mu-CO)(CO)_9Ni_3(CO)_3][N(P(C_6H_5)_3)_2]_2$

by treatment of a freeze–pump–thaw degassed THF solution of No. 13 with CO (ca. 0.34 atm), in a sealed flask, followed by stirring for 15 min and extraction of the residue with THF (32%) [12]

brown plates by slow diffusion of ether into a CH_2Cl_2 solution [12]

^{13}C NMR (CD$_2$Cl$_2$): 191.8 (9 OsCO), 201.5 (3 NiCO), 226.6 (μ-CO), 366.5 (μ$_6$-C); (CHF$_2$Cl/CD$_2$Cl$_2$ (2:1), −115 °C): 188.2 (3 OsCO), 193.0 (6 OsCO), 200.5 (3 NiCO), 227.6 (μ-CO), 366.0 (μ$_6$-C); the intermetallic site exchange of carbonyls is slow on the NMR time scale at room temperature; above −105 °C the CO resonance at 191.7 ppm splits into two broad resonances at 193 (6 OsCO) and 188 (3 OsCO) ppm; an 86 electron–precise closed octahedral geometry is suggested for the compound in solution retaining C$_{3v}$ symmetry in the absence of inter– and intrametallic exchange of the CO's [12]

IR (THF): 1823 (μ-CO), 1901, 1965, 1981, 2026 (all CO) [12]

References on p. 246

Table 8 (continued)

No.	compound	method of preparation (yield in %) properties and remarks
		FAB mass spectrum: $[M]^+$, $[M^+ - x\ CO]^+$, x=1 to 4 [12]
*13	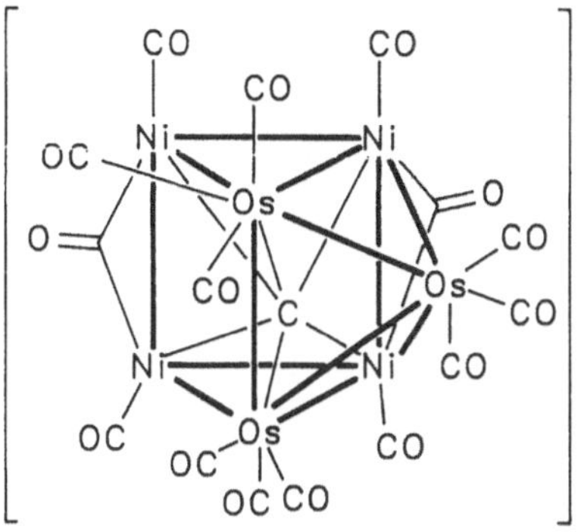	by treatment of a freeze-pump-thaw degassed THF solution of $[Os_3(\mu_3,\eta^1\text{-CCO})(CO)_9]$-$[N(P(C_6H_5)_3)_2]_2$ (Section 3.1.6.3) with an excess of $Ni(1,5\text{-}C_8H_{12}\text{-c})_2$ in refluxing THF for 1.5 h under a CO atmosphere of ca. 0.54 atm, followed by removal of the solvent along with excess $Ni(CO)_4$, and extraction of the residue with THF (86%) [12] black prismatic crystals by diffusion of pentane into the THF solution [12] ^{13}C NMR (CD_2Cl_2, −20 °C): 184.3 (6 OsCO), 194.4 (3 OsCO), 209.1, 209.2 (both 4 NiCO), 210.4, 256.3 (each 1 μ-CO), 333.1 (μ_6-C); at 20 °C the resonances broaden due to carbonyl exchanges; at −90 °C slow intrametal CO exchanges led to a broadening of both OsCO resonances [12] IR (THF): 1811, 1889, 1917, 1953, 1966, 1993 (all CO) [12]
Os_3Au compound		
*14	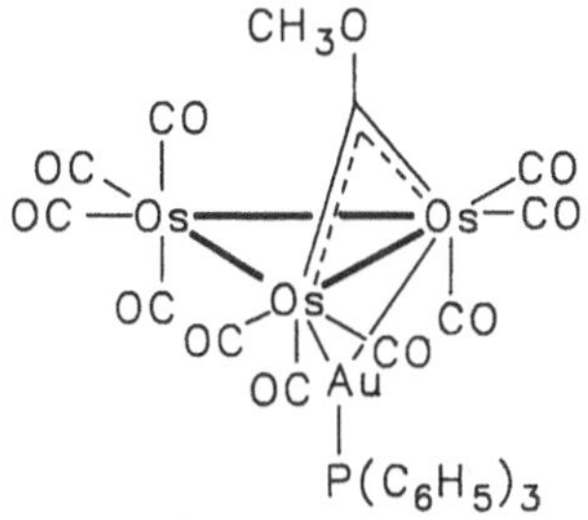	from $(\mu\text{-H})Os_3(\mu_2,\eta^1\text{-COCH}_3)(CO)_{10}$ (Section 3.1.6.2.2) and $AuCH_3P(C_6H_5)_3$ in toluene at 90 °C (no further information) [5] orange-yellow prisms from CH_2Cl_2/hexane [5]

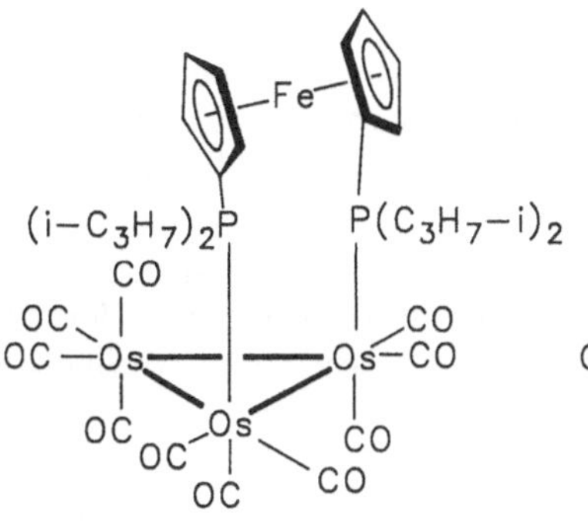

I

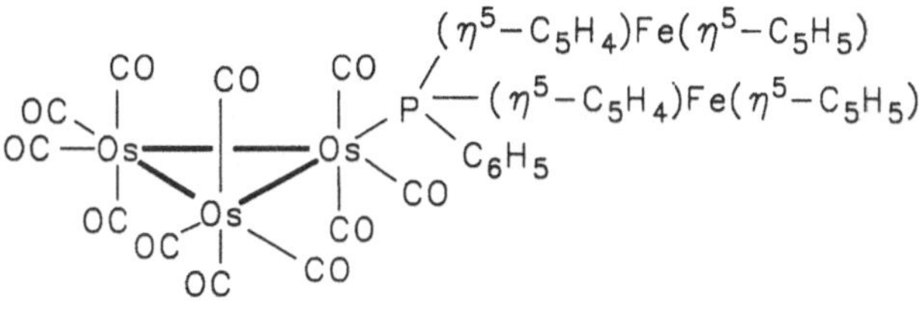

II

References on p. 246

*Further information:

Os$_3$(μ_2,η^1-CH$_2$)(CO)$_{11}$SnCl$_2$ (Table **8**, No. **1**) crystallizes in the monoclinic space group P2$_1$/c − C$_{2h}^5$ (No. 14) with a = 18.655(3), b = 13.668(3), c = 17.042(3) Å, β = 90.926(15)°; Z = 8, D$_c$ = 3.22 g/cm^3, R = 0.0594. The asymmetric unit contains two independent but structurally similar molecules. The compound, shown in **Fig. 84**, has a near-planar butterfly structure having 54 valence electrons within the Os$_3$Sn framework. The coordination about Sn is best described as a distorted trigonal bipyramid; the bond angles within the trigonal plane defined by Sn, both Cl's, and Os(1) average 119.1°. Alternatively, the Sn atom can be considered to be tetrahedrally coordinated by Os(2), both Cl's, and the midpoint of the Os(1)–Os(3) bond, since the bond angles about Sn by these atoms average 108.5° for the two molecules. The actual structure appears to be a compromise between these coordination modes. Selected bond distances and angles for the second molecule are given [6].

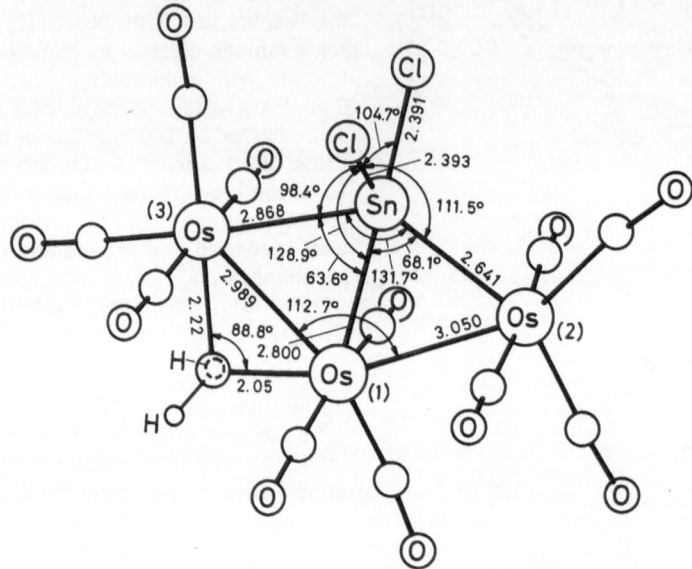

Fig. 84. Molecular structure of one of the two independent molecules of Os$_3$(μ_2,η^1-CH$_2$)(CO)$_{11}$SnCl$_2$ (No. **1**) with selected bond distances (in Å) and bond angles [6].

(μ-H)$_2$Os$_3$(μ_2,η^2-O=CCH(Si(CH$_3$)$_3$)$_2$)(CO)$_9$SnCH(Si(CH$_3$)$_3$)$_2$ (Table **8**, No. **2**). The formation of the title compound from (μ-H)Os$_3$(CO)$_{10}$(μ-H)Sn(CH(Si(CH$_3$)$_3$)$_2$)$_2$ under mild conditions involves an unusual conversion of Sn^{IV} to Sn^{II}; the driving force for the reaction is presumably the reactive Sn-(μ-H)-Os system and the relief of steric strain around the Sn atom [4].

The geometry around the tin is essentially trigonal planar related to the organic ligand and the two Os centers, capped by the oxygen of the bridging acyl moiety. The stereochemistry is in accord with a Sn^{II} center and the tin atom can therefore be viewed as a formal electron pair donor to one of the backbone Os atoms. Thus, the acyl oxygen is coordinated by electron donation into a vacant p orbital of the sp^2 Sn atom. The low IR band at 1394 cm^{-1} reveals an apparent weakening of the acyl C=O bond which is of interest concerning the importance now attached to C=O bond cleavage processes in the early stages of Fischer-Tropsch and related reactions [4].

References on p. 246

The molecular structure is shown in **Fig. 85** but there were no crystallograhic data given. The hydride ligands were directly located in the X-ray analysis [4].

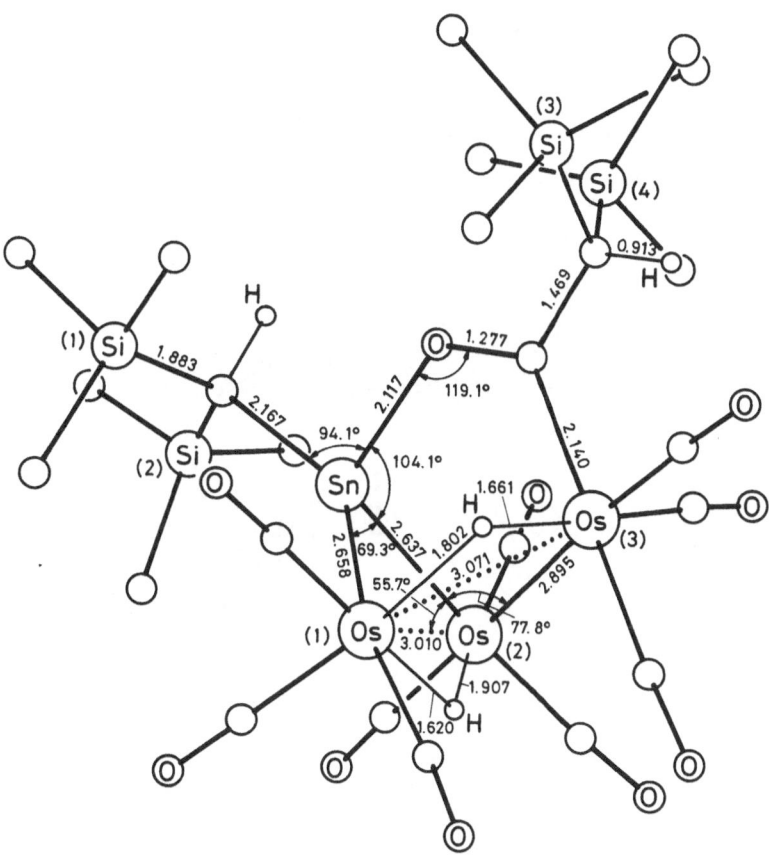

Fig. 85. Molecular structure of $(\mu\text{-H})_2Os_3(\mu_2,\eta^2\text{-O=CCH(Si(CH}_3)_3)_2)(CO)_9SnCH(Si(CH_3)_3)_2$ (No. 2) with selected interatomic distances (in Å) and bond angles [4].

$Os_3(\mu_2,\eta^2\text{-O=C(OCH}_3)CH_2CCO_2CH_3)(CO)_9Sn(CH(Si(CH_3)_3)_2)_2$ (Table 8, No. 3). The molecular structure is shown in **Fig. 86**. The methylene hydrogens were not directly located, but are assigned as one of the signals near 1.56 ppm in the ^{1}H NMR spectrum. Structural features are the semi-bridging CO group and an unusual long Os–Sn bond that probably resulted from a strong trans influence of the bridging sp^3-hybridized μ-C carbon. No crystallographic data were given [4].

References on p. 246

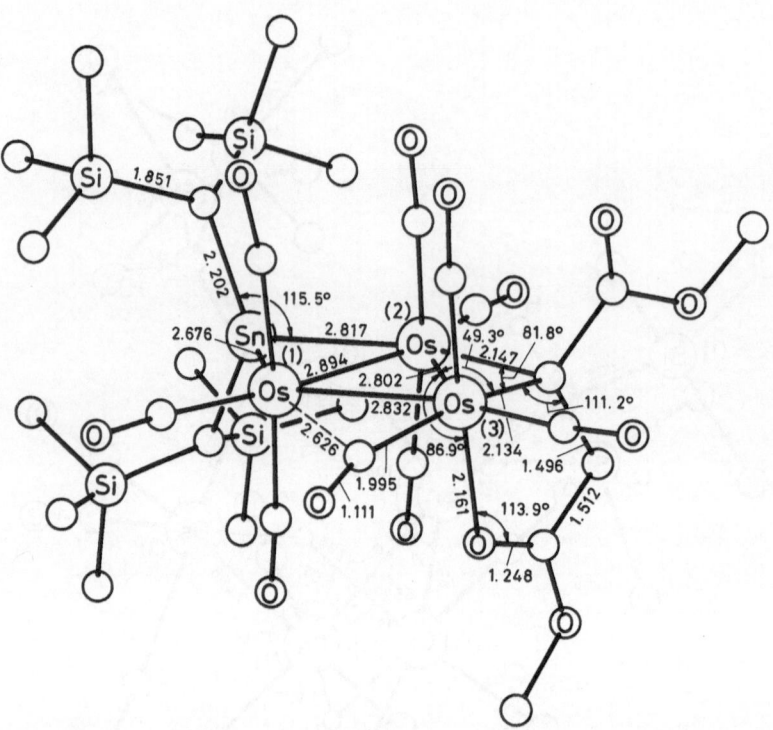

Fig. 86. Molecular structure of $Os_3(\mu_2,\eta^2\text{-}O{=}C(OCH_3)CH_2CCO_2CH_3)(CO)_9Sn(CH(Si(CH_3)_3)_2)_2$ (No. 3) with selected interatomic distances (in Å) and bond angles [4].

$(\mu\text{-}H)Os_3(\mu_3,\eta^2\text{-}SCH_2)(CO)_7(\mu_3\text{-}S)(P(CH_3)_2C_6H_5)Sn(CH_3)_3$ (Table **8**, No. **4**) crystallizes in the monoclinic space group $P2_1/c - C_{2h}^5$ (No. 14) with a = 10.426(2), b = 11.421(2), c = 24.563(4) Å, β = 95.41(2)°; Z = 4, D_c = 2.68 g/cm³, R = 0.034. The compound, shown in **Fig. 87**, consists of an open Os_3Sn framework. The S–C distance is indicative of a single bond. The methylene hydrogens and the μ-H ligand were not observed crystallographically but were confirmed by ¹H NMR spectroscopy; the μ-H ligand is believed to bridge the Os(1)–Os(2) bond. One of the carbonyl ligands at Os(1) exhibits slight semi-bridging interactions to Os(3) [1].

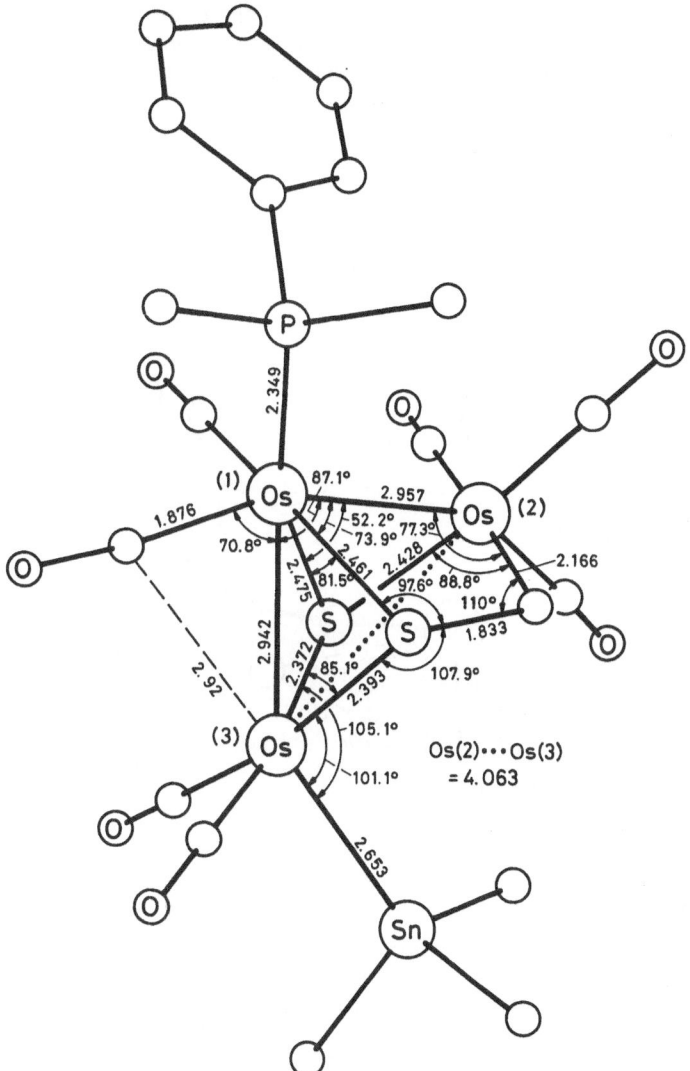

Fig. 87. Molecular structure of $(\mu\text{-H})Os_3(\mu_3,\eta^2\text{-SCH}_2)(CO)_7(\mu_3\text{-S})(P(CH_3)_2C_6H_5)Sn(CH_3)_3$ (No. 4) with selected interatomic distances (in Å) and bond angles [1].

$(\mu\text{-H})Os_3(\mu_2,\eta^2\text{-CH=CHC}_6H_5)(CO)_{10}(\mu\text{-CO})Mn(\eta^5\text{-C}_5H_5)CO$ (Table 8, No. **5**) crystallizes from 1,4-dioxane at $-15\,°C$ with 1.5 molecules 1,4-dioxane in the monoclinic space group $P2_1/c - C_{2h}^5$ (No. 14) with $a = 8.998(1)$, $b = 12.491(4)$, $c = 32.088(8)$ Å, $\beta = 96.55(2)°$; $Z = 4$, $D_c = 2.341$ g/cm^3, $R = 0.046$. The metal framework consists of a Mn-spiked Os_3 triangle; see **Fig. 88**. The hydride ligand was not observed crystallographically but is believed to bridge the Os(1)–Os(3) bond being consistent with the longer bond distance and the greater Os–Os–CO angles. The shortening of the Mn–Os(1) distance is evidently due to the presence of the bridging ligands; the μ-CO is coplanar to the plane of the osmium triangle, whereas the phenylvinyl ligand is perpendicular. The coordination polyhedron of the Os(1) atom can be regarded as a distorted octahedron upon assuming that three atoms (Mn, μ-C′, and C″) together occupy two coordination sites. The phenylvinyl group is σ-bonded to Os(1) and π-bonded to Mn; the elongated C=C double bond is due to the η²-coordination. The solvating 1,4-dioxane molecules, forming a chair conformation, occupy cavities in the crystal packing and have no contact with the compound molecules [2].

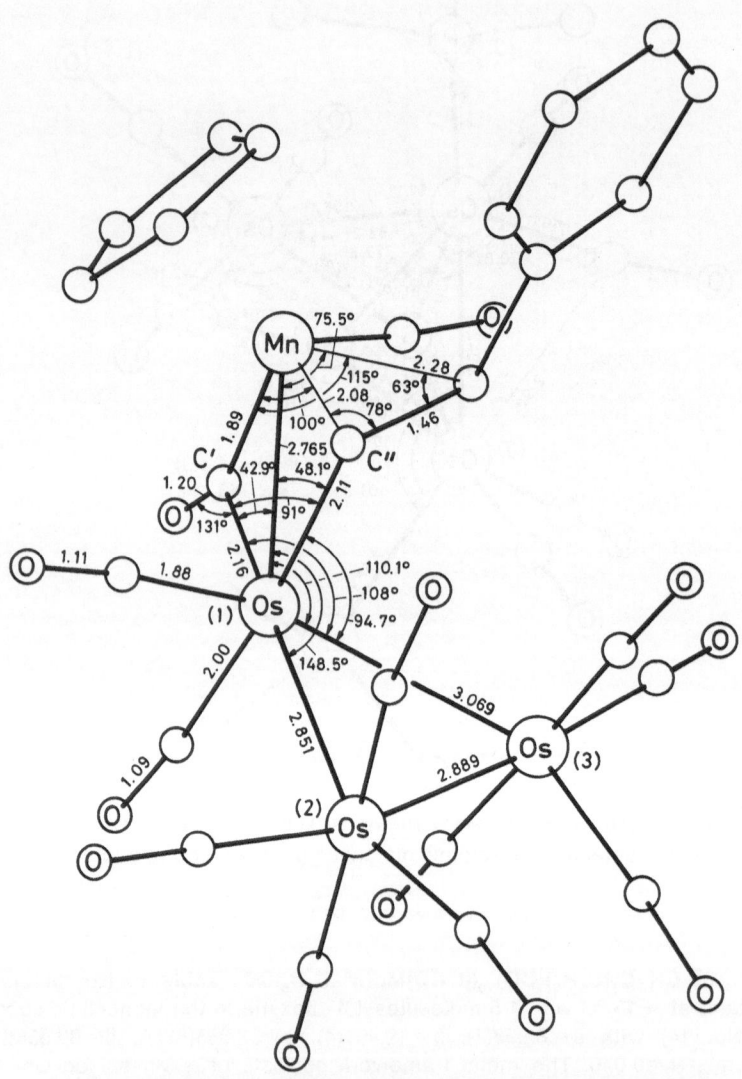

Fig. 88. Molecular structure of the 1,4–dioxane solvate of
$(\mu-H)Os_3(\mu_2,\eta^2-CH=CHC_6H_5)(CO)_{10}(\mu-CO)Mn(\eta^5-C_5H_5)CO$ (No. 5)
with selected bond distances (in Å) and bond angles [2].

Gmelin Handbook
Os-Org. Comp. B6

In aliphatic hydrocarbons No. 5 readily eliminates a Mn-containing fragment with conversion to an Os_3 compound bearing an unsaturated bridging group. The composition of the products depends on the decomposition conditions, see Scheme 1. In pentane No. 5 decomposes at 20 °C within 24 h to compounds III and IV in a mole ratio of ca. 1:2, whereas in octane at 120 °C within 1 h, a mixture of IV and its isomer V (mole ratio ca. 2.3:1) were identified in the reaction mixture. Compounds III and IV were also obtained as by-products in the preparation of No. 5, probably resulting from partial decomposition of No. 5 [3]. The instability of No. 5 is attributed to significant steric strains in the planar system Os(1)–C′–Mn–C″ (see Fig. 88) [2].

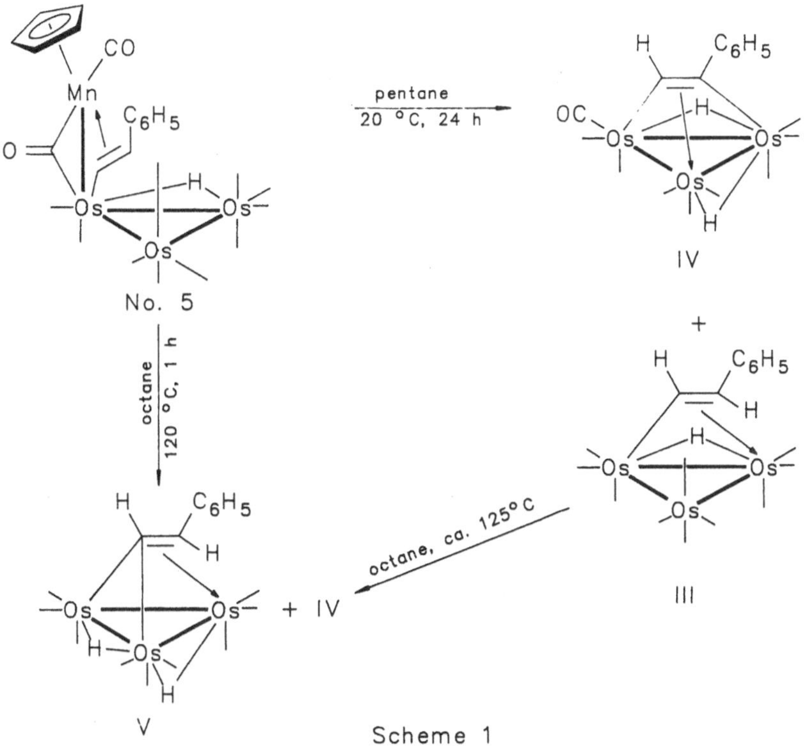

Scheme 1

References on p. 246

$[Os_3(\mu_4-C)(\mu-CO)(CO)_9Mn(CO)_3][N(P(C_6H_5)_3)_2]$ (Table **8**, No. **6**) crystallizes in the triclinic space group $P\bar{1}-C_i^1$ (No. 2) with a = 9.219(2), b = 15.035(2), c = 18.757(5) Å, α = 105.20(2)°, β = 99.18(2)°, γ = 96.96(1)°; Z = 2, D_c = 2.10 g/cm³, R = 0.029. The Os_3Mn core adopts a butterfly configuration with the μ_4-C carbide located over the hinge Os–Mn bond; see **Fig. 89**. The molecule is of C_s symmetry with a mirror plane passing through μ_4-C, Os(2), Mn, and μ–CO. The crystal is disordered resulting in partial occupation of both hinge metal positions with 73.2% osmium and 26.8% manganese at the Os(2) position and vice versa [8].

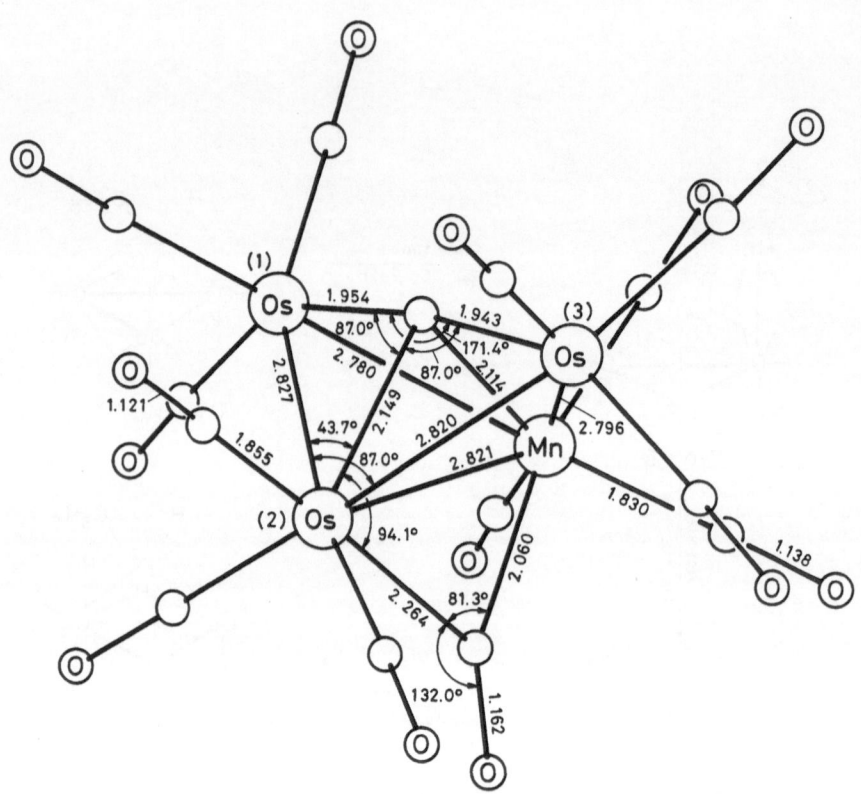

Fig. 89. Molecular structure of the anion of $[Os_3(\mu_4-C)(\mu-CO)(CO)_9Mn(CO)_3][N(P(C_6H_5)_3)_2]$ (No. **6**) with selected bond distances (in Å) and bond angles in its predominating Os/Mn population [8].

$Os_3(\mu_2,\eta^2\text{-CH=CHCO}_2CH_3)(CO)_{11}Re(CO)_4$ (Table **8**, No. **8**). Isomer 1 crystallizes in the monoclinic space group $Cc - C_s^4$ (No. 9) with a = 20.877(6), b = 9.192(2), c = 15.117(4) Å, β = 112.00(3)°; Z = 4, R = 0.0399. The structure of Isomer 1, shown in **Fig. 90**, consists of an almost equilateral Os_3 core with the rhenium linked to one of the Os atoms. The vinyl group is σ-bonded to the osmium and π-bonded to the $Re(CO)_4$ moiety. The Os–Re bond lies close to the Os_3 plane; the deviation is only 0.155(1) Å, whereas in No. 5 the Os–Mn bond is perpendicular to this plane [7].

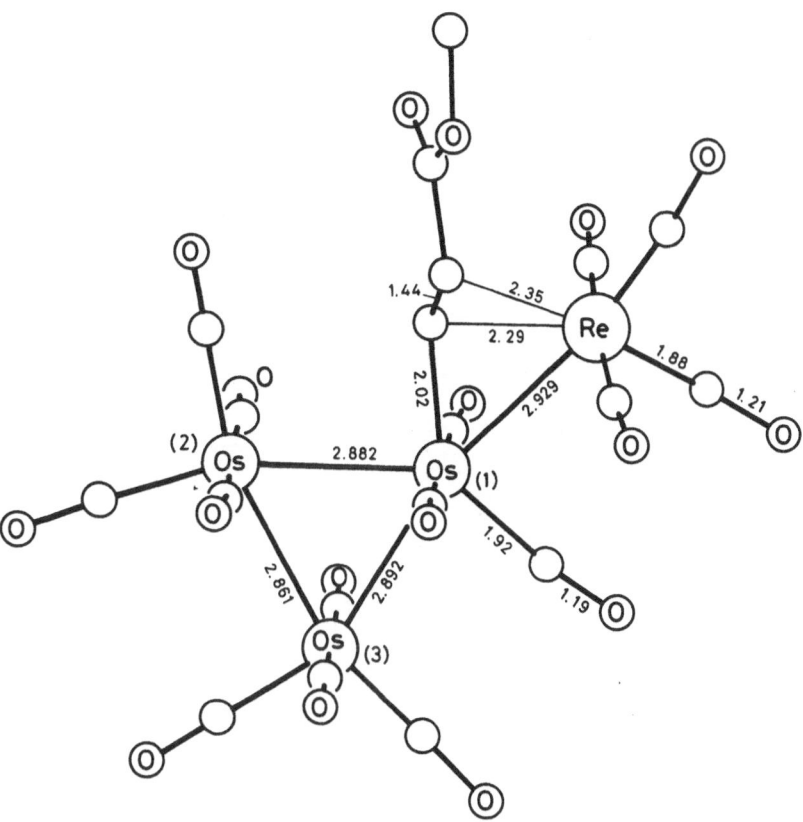

Fig. 90. Molecular structure of Isomer 1 of $Os_3(\mu_2,\eta^2\text{-CH=CHCO}_2CH_3)(CO)_{11}Re(CO)_4$ (No. 8) with selected bond distances (in Å) [7].

References on p. 246

(μ–H)Os₃(CO)₇(μ₃-PC₃H₇-i){C₅H₄FeC₅H₄P(C₃H₇-i)₂} (Table **8**, No. **9**) crystallizes in the monoclinic space group $P2_1/n - C_{2h}^5$ (No. 14) with a = 8.910(3), b = 22.784(2), c = 15.416(2) Å, β = 98.25°; Z = 4, D_c = 2.451 g/cm³, R = 0.032. The structure is shown in **Fig. 91**. The hydride ligand was not observed crystallographically but it is believed to bridge the elongated Os(1)–Os(2) bond. Average bond distances of Os–CO and C≡O are 1.89 and 1.15 Å, respectively [10].

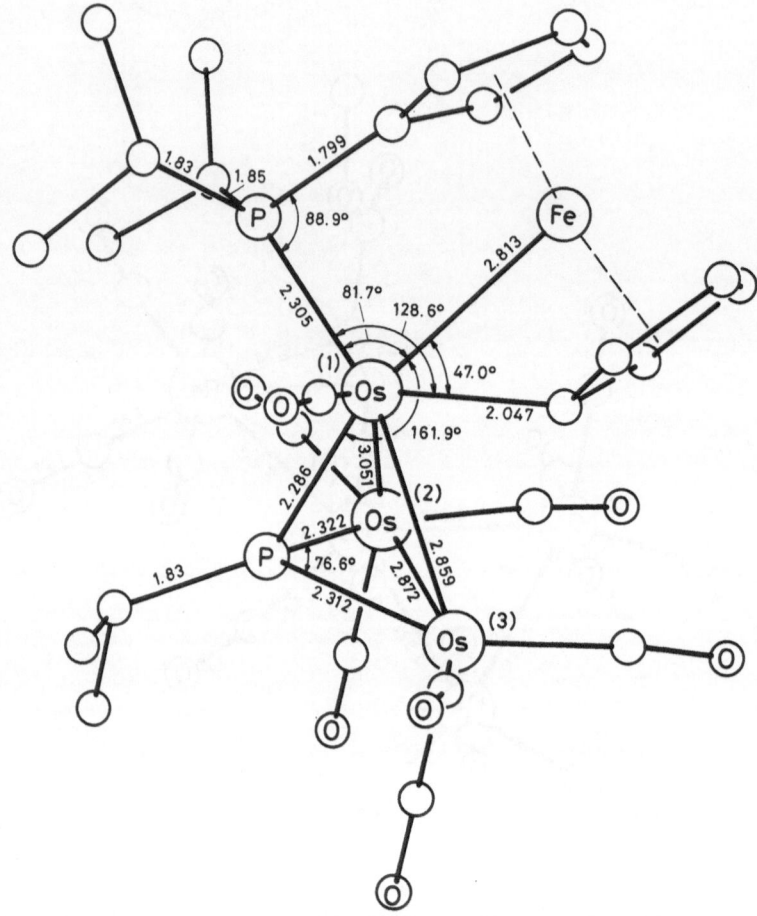

Fig. 91. Molecular structure of (μ–H)Os₃(CO)₇(μ₃-PC₃H₇-i)C₅H₄FeC₅H₄P(C₃H₇-i)₂ (No. 9) with selected interatomic distances and angles [10].

Os₃(CO)₉(μ₃,η³-C₅H₄FeC₅H₄PC₆H₅) (Table **8**, No. **10**) crystallizes with 0.5 molecules of CH₂Cl₂ in the triclinic space group $P\bar{1} - C_i^1$ (No. 2) with a = 12.427(2), b = 13.061(1), c = 9.209(3) Å, α = 96.562(8)°, β = 91.25(1)°, γ = 76.956(9)°; Z = 2, D_c = 2.657 g/cm³, R = 0.031. The structure is shown in **Fig. 92**. The metal framework adopts an Os₃ triangle with two equivalent Os–Os bonds of equal length and one significantly shorter edge which is bridged by the phosphido moiety [14].

References on p. 246

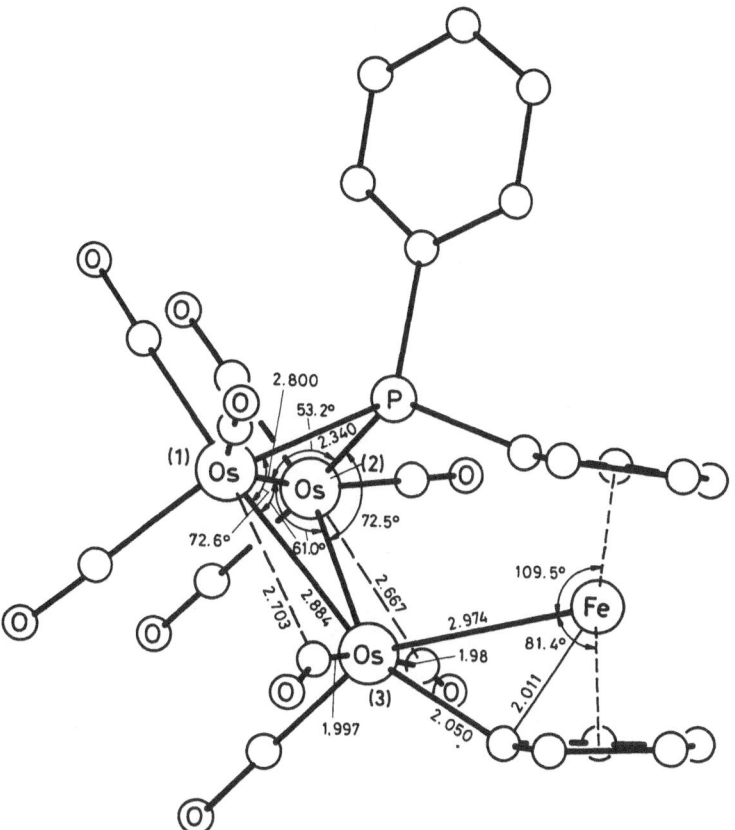

Fig. 92. Molecular structure of $Os_3(CO)_9(\mu_3,\eta^3-C_5H_4FeC_5H_4PC_6H_5)$ (No. 10) with selected interatomic distances (in Å) and bond angles [14].

$Os_3(CO)_9(\mu_3,\eta^3-C_5H_4FeC_5H_4PC_5H_4FeC_5H_5)$ (Table 8, No. **11**) crystallizes in the triclinic space group $P\bar{1}-C_i^1$ (No. 2) with $a = 12.304(1)$, $b = 12.336(2)$, $c = 10.886(2)$ Å, $\alpha = 115.38(2)°$, $\beta = 92.34(1)°$, $\gamma = 79.176(1)°$; $Z = 2$, $D_c = 2.726$ g/cm³, $R = 0.036$. The structure is shown in **Fig. 93**. The Os(2)–Os(3) bond being almost symmetrically bridged by the phosphido moiety, is significantly shorter than the other two bonds of the closed Os_3 triangle. The Os(1) atom is σ-bonded to the C_5H_4 ring and to the Fe center which is situated almost in an axial position with respect to the Os_3 core. The relatively long Fe–Os distance compared to that found in No. 9, indicates weaker interaction. Each Os center is bonded to two equatorial and one axial CO group; two carbonyls are semi-bridging. The Os(1) atom is displaced out of the coordinated C_5H_4 plane by 0.998 Å. The C_5H_4 units of the Os-coordinated ferrocenyl ligand are slightly opened up. As a consequence the C_5H_4–Fe–C_5H_4 angle is 168.4°, with the Fe–C–P distance of 2.105(9) Å being slightly longer and the Fe–C–Os bond of 1.97(1) Å being slightly shorter than the other Fe–C bonds of ca. 2.06 Å [14].

References on p. 246

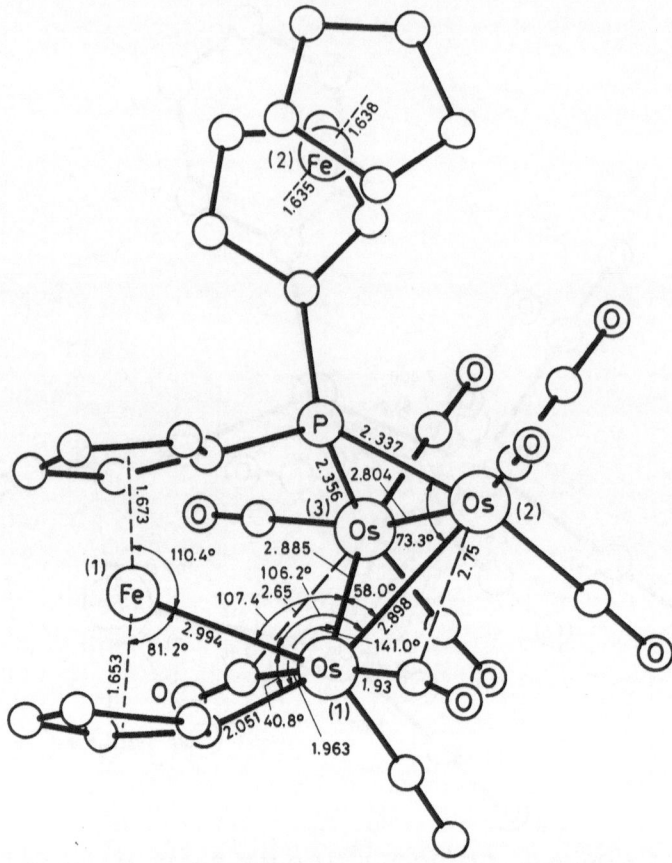

Fig. 93. Molecular structure of $Os_3(CO)_9(\mu_3,\eta^3\text{-}C_5H_4FeC_5H_4PC_5H_4FeC_5H_5)$ (No. 11) with selected interatomic distances (in Å) and bond angles [14].

$[Os_3(\mu_6\text{-}C)(CO)_9Ni_4(\mu\text{-}CO)_2(CO)_4][N(P(C_6H_5)_3)_2]_2$ (Table **8**, No. **13**) crystallizes in the triclinic space group $P\bar{1}-C_i^1$ (No. 2) with a = 13.395(5), b = 13.606(6), c = 25.343(8) Å, α = 88.18(3)°, β = 85.35(3)°, γ = 89.31(3)°; Z = 2, D_c = 1.775 g/cm³. The structure is shown in **Fig. 94**. The carbide ligand is located in the center of a trigonal prismatic Os_2Ni_4 core; the third Os atom caps a square face consisting of two Os and two Ni centers. A plane of symmetry passes through Os(3) perpendicular to the square face prism. The configuration of the metal atoms maximizes the Os–Os and Os–Ni bond strengths, at the expense of weaker Ni–Ni bonding [12].

Gmelin Handbook
Os-Org. Comp. B6

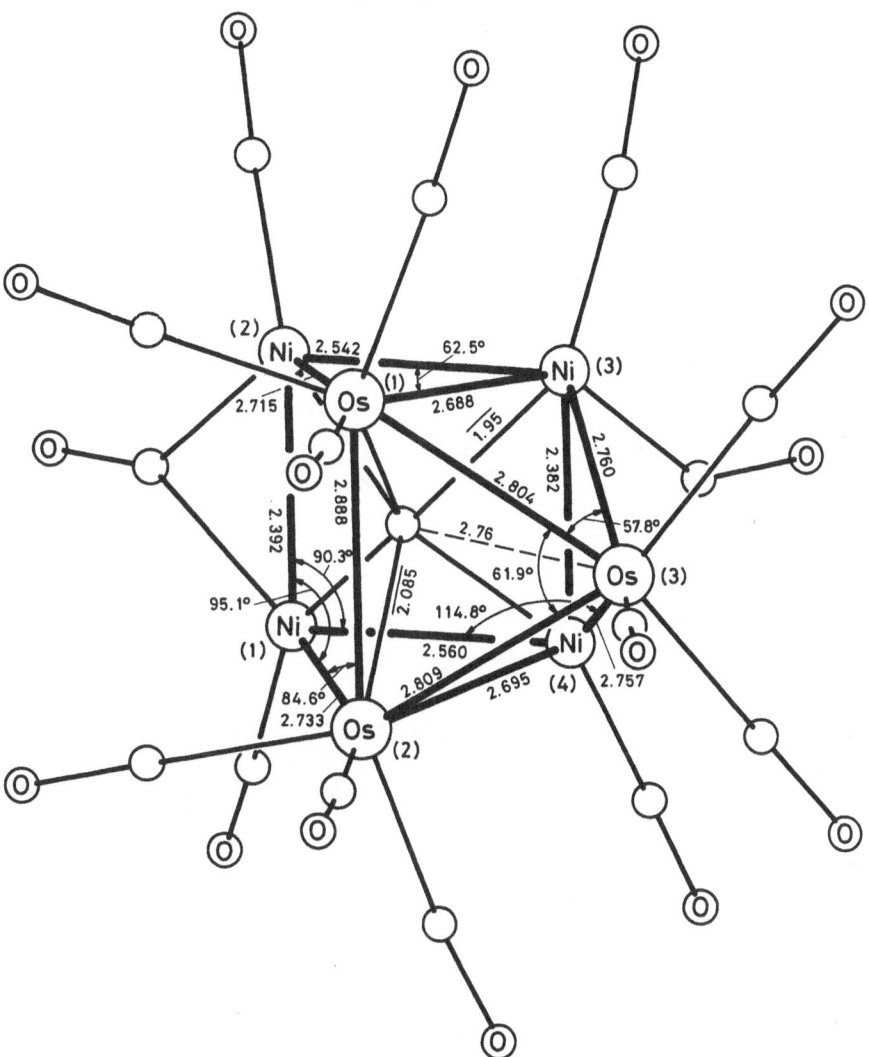

Fig. 94. Molecular structure of the anion of $[Os_3(\mu_6-C)(CO)_9Ni_4(\mu-CO)_2(CO)_4][N(P(C_6H_5)_3)_2]_2$ (No. 13) with selected interatomic distances (in Å) and bond angles [12].

$Os_3(\mu_2,\eta^1-COCH_3)(CO)_{10}AuP(C_6H_5)_3$ (Table **8**, No. **14**) crystallizes in the triclinic space group $P\bar{1}-C_i^1$ (No. 2) with a = 12.940(4), b = 15.680(9), c = 17.462(4) Å, α = 107.70(3)°, β = 95.39(3)°, γ = 90.67(4)°; Z = 4, D_c = 2.67 g/cm³, R = 0.032. Two independent molecules related by a minor twist of the $P(C_6H_5)_3$ group and by a 180° rotation of the μ_2,η^1-COCH_3 bridging unit were found in the asymmetric unit; one of these molecules is shown in **Fig. 95**. A strong trans effect of the alkylidene ligand is indicated by the considerably longer Os–CO vectors trans to this ligand. The CO distance in the alkylidene ligand implies multiple bond character. Weak interactions between Os(3) and the μ–C atom indicate a minimal semi-μ_3 character of the alkylidene ligand [5].

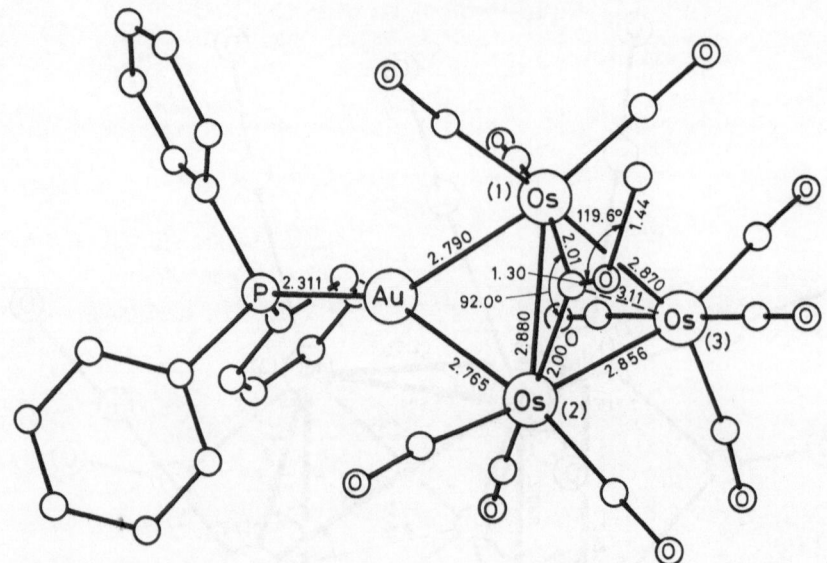

Fig. 95. Molecular structure of $Os_3(\mu_2,\eta^1\text{-}COCH_3)(CO)_{10}AuP(C_6H_5)_3$ (No. 14) with selected interatomic distances (in Å) and bond angles [5].

References:

[1] Adams, R. D.; Katahira, D. A. (Organometallics **1** [1982] 460/6).

[2] Antonova, A. B.; Kovalenko, S. V.; Korniyets, E. D.; Efremova, I. Yu.; Johansson, A. A.; Gubin, S. P. (Izv. Akad. Nauk SSSR Ser. Khim. **1984** 1146/52; Bull. Acad. Sci. USSR Div. Chem. Sci. [Engl. Transl.] **33** [1984] 1052/8).

[3] Antonova, A. B.; Kovalenko, S. V.; Korniyets, E. D.; Johansson, A. A.; Struchkov, Yu. T.; Yanovskii, A. I. (J. Organomet. Chem. **267** [1984] 299/307).

[4] Cardin, C. J.; Cardin, D. J.; Power, J. M.; Hursthouse, M. B. (J. Am. Chem. Soc. **107** [1985] 505/7).

[5] Farrugia, L. J. (Acta Crystallogr. C **42** [1986] 680/2).

[6] Viswanathan, N.; Morrison, E. D.; Geoffroy, G. L.; Geib, S. J.; Rheingold, A. L. (Inorg. Chem. **25** [1986] 3100/2).

[7] Koridze, A. A.; Kizas, O. A.; Struchkov, Yu. T.; Yanovskii, A. I.; Kolobova, N. E. (Metalloorg. Khim. **1** [1988] 831/5; Organomet. Chem. [Engl. Transl.] **1** [1988] 460/3).

[8] Jensen, M. P.; Henderson, W.; Johnston, D. H.; Sabat, M.; Shriver, D. F. (J. Organomet. Chem. **394** [1990] 121/43).

[9] Hadj-Bagheri, N.; Strickland, D. S.; Wilson, S. R.; Shapley, J. R. (J. Organomet. Chem. **410** [1991] 231/9).

[10] Cullen, W. R.; Rettig, S. J.; Zheng, T.-C. (Organometallics **11** [1992] 277/83).

[11] Cullen, R. W.; Rettig, S. J.; Zheng, T.-C. (Organometallics **11** [1992] 928/35).

[12] Karet, G. B.; Espe, R. L.; Stern, C. L.; Shriver, D. F. (Inorg. Chem. **31** [1992] 2658/60).

[13] Cullen, R. W.; Rettig, S. J.; Zheng, T.-C. (Organometallics **12** [1993] 688/96).

[14] Cullen, R. W.; Rettig, S. J.; Zheng, T.-C. (J. Organomet. Chem. **452** [1993] 97/103).

Physical Constants and Conversion Factors

Avogadro constant N_A (or L) = 6.02214×10^{23} mol^{-1}

Faraday constant $\quad\quad$ F = 9.64853×10^4 C/mol

molar gas constant $\quad\quad$ R = 8.31451 J·mol^{-1}·K^{-1}

molar volume (ideal gas) V_m = 2.24141×10^1 L/mol
(273.15 K, 101325 Pa)

Planck constant $\quad$ h = 6.62608×10^{-34} J·s

elementary charge e = 1.60218×10^{-19} C

electron mass $\quad$ m_e = 9.10939×10^{-31} kg

proton mass $\quad$ m_p = 1.67262×10^{-27} kg

1 kg $\;$ = 2.205 pounds

1 m $\;$ = 3.937×10^1 inches = 3.281 feet

1 m^3 = 2.642×10^2 gallons (U.S.)

1 m^3 = 2.200×10^2 gallons (Imperial)

Force	N	dyn	kp
1 N	1	10^5	1.019716×10^{-1}
1 dyn	10^{-5}	1	1.019716×10^{-6}
1 kp	9.80665	9.80665×10^5	1

Pressure	Pa	bar	kp/m^2	at	atm	Torr	lb/in^2
1 Pa = 1 N/m^2	1	10^{-5}	1.019716×10^{-1}	1.019716×10^{-5}	9.86923×10^{-6}	7.50062×10^{-3}	1.450378×10^{-4}
1 bar = 10^6 dyn/cm^2	10^5	1	1.019716×10^4	1.019716	9.86923×10^{-1}	7.50062×10^2	1.450378×10^1
1 kp/m^2 = 1 mm H$_2$O	9.80665	9.80665×10^{-5}	1	10^{-4}	9.67841×10^{-5}	7.35559×10^{-2}	1.422335×10^{-3}
1 at (technical)	9.80665×10^4	9.80665×10^{-1}	10^4	1	9.67841×10^{-1}	7.35559×10^2	1.422335×10^1
1 atm = 760 Torr	1.01325×10^5	1.01325	1.033227×10^4	1.033227	1	7.60×10^2	1.469595×10^1
1 Torr = 1 mmHg	1.333224×10^2	1.333224×10^{-3}	1.359510×10^1	1.359510×10^{-3}	1.315789×10^{-3}	1	1.933678×10^{-2}
1 lb/in^2 = 1 psi	6.89476×10^3	6.89476×10^{-2}	7.03069×10^2	7.03069×10^{-2}	6.80460×10^{-2}	5.17149×10^1	1

Work, Energy, Heat	J	kW·h	kcal	Btu	eV
1 J = 1W·s = 1N·m = 10^7 erg	1	2.778×10^{-7}	2.39006×10^{-4}	9.4781×10^{-4}	6.242×10^{18}
1 kW·h	3.6×10^6	1	8.604×10^2	3.41214×10^3	2.247×10^{25}
1 kcal	4.1840×10^3	1.1622×10^{-3}	1	3.96566	2.6117×10^{22}
1 Btu (British thermal unit)	1.05506×10^3	2.93071×10^{-4}	2.5164×10^{-1}	1	6.5858×10^{21}
1 eV	1.602×10^{-19}	4.450×10^{-26}	3.8289×10^{-23}	1.51840×10^{-22}	1

$1 \text{ cm}^{-1} = 1.239842 \times 10^{-4}$ eV

$1 \text{ hartree} = 27.2114$ eV

$1 \text{ Hz} = 4.135669 \times 10^{-15}$ eV

$1 \text{ eV} \triangleq 23.0578$ kcal/mol

Power	kW	hp	kp·m·s^{-1}	kcal/s
1 kW = 10^3 J/s	1	1.35962	1.01972×10^2	2.39006×10^{-1}
1 hp (horsepower, metric)	7.3550×10^{-1}	1	7.5×10^1	1.7579×10^{-1}
1 kp·m·s^{-1}	9.80665×10^{-3}	1.333×10^{-2}	1	2.34384×10^{-3}
1 kcal/s	4.1840	5.6886	4.26650×10^2	1

References:

Mills, I. (Ed.), International Union of Pure and Applied Chemistry, Quantities, Units and Symbols in Physical Chemistry, Blackwell Scientific Publications, Oxford 1988.

The International System of Units (SI), National Bureau of Standards Spec..Publ. 330 [1972].

Landolt-Börnstein, 6th Ed., Vol. II, Pt. 1, 1971, pp. 1/14.

ISO Standards Handbook 2, Units of Measurement, 2nd Ed., Geneva 1982.

Cohen, E. R., Taylor, B. N., Codata Bulletin No. 63, Pergamon, Oxford 1986.